METHODS IN MOLECULAR BIOLOGY

Series Editor
John M. Walker
School of Life and Medical Sciences
University of Hertfordshire
Hatfield, Hertfordshire, AL10 9AB, UK

For further volumes:
http://www.springer.com/series/7651

High-Throughput Glycomics and Glycoproteomics

Methods and Protocols

Edited by

Gordan Lauc

Faculty of Pharmacy and Biochemistry, University of Zagreb, Zagreb, Croatia

Manfred Wuhrer

Center for Proteomics and Metabolomics, Leiden University Medical Center, Leiden, The Netherlands

Humana Press

Editors
Gordan Lauc
Faculty of Pharmacy and Biochemistry
University of Zagreb
Zagreb, Croatia

Manfred Wuhrer
Center for Proteomics and Metabolomics
Leiden University Medical Center
Leiden, The Netherlands

ISSN 1064-3745 ISSN 1940-6029 (electronic)
Methods in Molecular Biology
ISBN 978-1-4939-8217-2 ISBN 978-1-4939-6493-2 (eBook)
DOI 10.1007/978-1-4939-6493-2

Cover Illustration: Original drawing by Ana Badrić, graphic designer, Zagreb, Croatia

Printed on acid-free paper

This Humana Press imprint is published by Springer Nature
The registered company is Springer Science+Business Media LLC
The registered company address is: 233 Spring Street, New York, NY 10013, U.S.A.

Preface

Carbohydrates—or glycans—are a highly abundant class of biomolecules and can be found in various shapes and sizes in all kingdoms of life. Glycans often exist in the form of oligo- or polysaccharides as well as glycoconjugates, i.e., with a covalent linkage of the glycan portion to proteins or lipids. Glycans significantly contribute to the structure and function of the majority of proteins. Interindividual variability in both glycome composition and glycome changes after environmental challenge is large, and these differences strongly affect many physiological functions. Due to chemical and biological complexity of glycans, the knowledge about glycans is lagging significantly behind the knowledge about genes and proteins. However, recent comprehensive report endorsed by the US National Academies concluded that "glycans are directly involved in the pathophysiology of every major disease" and the interest in glycomics studies is growing exponentially.

The analysis of glycans is an often challenging endeavor, due to their heterogeneity, their non-template-based synthesis, and the abundance of functionally distinct isomers which are challenging for separation and characterisation using current analytical techniques. Technological challenges in glycan analysis are generally seen as the major obstacle for revealing insights into functional relevance of glycans. In the last few years several robust methods that enable high-throughput analysis of glycans have been developed and glycomics is now ready to join genomics and other omics in genetic, epidemiological, and clinical studies.

This volume of Methods in Molecular Biology entitled *High-Throughput Glycomics and Glycoproteomics* brings detailed protocols for the analysis of glycosylation at the level of free glycans and glycopeptides, an essential reference for researchers planning to enter this rapidly evolving field. The importance of glycans in eukaryotic life, with a focus on mammals and particularly humans, is highlighted in Chapter 1 of this issue. The following chapters largely have a focus on recent technologies that allow the characterization of larger sets of samples, and major attention has been paid to method repeatability and robustness as well as higher throughput with respect to sample preparation, measurement, and data analysis. The chapters span a wide range of techniques, including the analysis of fluorescently labeled glycans with HPLC, LC-MS analysis of glycopeptides both for glycosylation profiling and for in-depth tandem mass spectrometric analysis of protein glycosylation, and the analysis of glycans with fluorescent labeling and capillary electrophoresis in conjunction with laser-induced fluorescence detection. The chapters also pay attention to specific sample types including brain tissues, N-glycans from in vitro cell cultures, milk oligosaccharides, invertebrate and protist N-glycans as well as plant glycans.

We thank all the contributors, who are leading researchers in analytical glycomics and developers as well as expert users of the presented methods, for sharing their protocols in this volume. We thank Dr. John Walker, the Editor-in-Chief of the Methods in Molecular Biology series, for his guidance and continuous support.

Zagreb, Croatia — *Gordan Lauc*
Leiden, The Netherlands — *Manfred Wuhrer*

Contents

Contributors

JODIE L. ABRAHAMS • *Biomolecular Frontiers Research Centre, Macquarie University, North Ryde, NSW, Australia*

BARBARA ADAMCZYK • *Department of Medical Biochemistry and Cell Biology, Institute of Biomedicine, Sahlgrenska Academy, University of Gothenburg, Gothenburg, Sweden; NIBRT GlycoScience Group, NIBRT–The National Institute for Bioprocessing Research and Training, Blackrock, Co., Dublin, Ireland*

FELIX AGAKOV • *Pharmatics Limited, Edinburgh, UK*

ANNA AGAKOVA • *Pharmatics Limited, Edinburgh, UK*

DAVIDE ALOCCI • *Proteome Informatics Group, SIB Swiss Institute of Bioinformatics, Geneva, Switzerland*

ANA I. BELO • *Department of Molecular Cell Biology and Immunology, VU University Medical Centre, Amsterdam, The Netherlands*

HARRY CAMPBELL • *Center for Global Health Research, Usher Institute of Population Health Sciences and Informatics, University of Edinburgh, Edinburgh, UK*

MATTHEW P. CAMPBELL • *Biomolecular Frontiers Research Centre, Macquarie University, North Ryde, NSW, Australia*

BARBARA ECKMAIR • *Department für Chemie, Universität für Bodenkultur, Vienna, Austria*

DAVID FALCK • *Center for Proteomics and Metabolomics, Leiden University Medical Center, Leiden, The Netherlands*

JONATAN ULRIK FANGEL • *Section for Plant Glycobiology, Department for Plant and Environmental Sciences, University of Copenhagen, Frederiksberg C, Denmark*

IVAN GUDELJ • *Genos Glycoscience Research Laboratory, Zagreb, Croatia*

ANDRAS GUTTMAN • *Horvath Csaba Laboratory of Bioseparation Sciences, University of Debrecen, Debrecen, Hungary; MTA-PE Translational Glycomics Group, University of Pannonia, Veszprem, Hungary; SCIEX, Brea, CA, USA*

OLGA GORNIK • *Faculty of Pharmacy and Biochemistry, University of Zagreb, Zagreb, Croatia*

NOORTJE DE HAAN • *Center for Proteomics and Metabolomics, Leiden University Medical Center, Leiden, The Netherlands*

HANNES HINNEBURG • *Department of Biomolecular Systems, Max Planck Institute of Colloids and Interfaces, Potsdam, Germany; Department of Biology, Chemistry, Pharmacy, Institute of Chemistry and Biochemistry, Freie Universität Berlin, Berlin, Germany*

STEPHANIE HOLST • *Center for Proteomics and Metabolomics, Leiden University Medical Center, Leiden, The Netherlands*

ALBA HYKOLLARI • *Department für Chemie, Universität für Bodenkultur, Vienna, Austria*

BAS C. JANSEN • *Center for Proteomics and Metabolomics, Leiden University Medical Center, Leiden, The Netherlands*

CHUNSHENG JIN • *Department of Medical Biochemistry and Cell Biology, Institute of Biomedicine, Sahlgrenska Academy, University of Gothenburg, Gothenburg, Sweden*

NICLAS G. KARLSSON • *Department of Medical Biochemistry and Cell Biology, Institute of Biomedicine, Sahlgrenska Academy, University of Gothenburg, Gothenburg, Sweden*
THOMAS KLARIĆ • *Genos Glycoscience Research Laboratory, Zagreb, Croatia*
LUCIJA KLARIĆ • *Genos Glycoscience Research Laboratory, Zagreb, Croatia*
DANIEL KOLARICH • *Department of Biomolecular Systems, Max Planck Institute of Colloids and Interfaces, Potsdam, Germany*
PETRA KORAĆ • *Division of Molecular Biology, Department of Biology, Faculty of Science, University of Zagreb, Zagreb, Croatia*
STJEPAN KREŠIMIR KRAČUN • *Section for Plant Glycobiology, Department for Plant and Environmental Sciences, University of Copenhagen, Frederiksberg C, Denmark*
JASMINKA KRIŠTIĆ • *Genos Glycoscience Research Laboratory, Zagreb, Croatia*
GORDAN LAUC • *Faculty of Pharmacy and Biochemistry, University of Zagreb, Zagreb, Croatia; Genos Glycoscience Research Laboratory, Zagreb, Croatia*
CARLITO B. LEBRILLA • *Department of Chemistry, University of California, Davis, CA, USA*
FREDERIQUE LISACEK • *Proteome Informatics Group, SIB Swiss Institute of Bioinformatics, Geneva, Switzerland*
MIGUEL A. ROJAS MACIAS • *Department of Medical Biochemistry and Cell Biology, Institute of Biomedicine, Sahlgrenska Academy, University of Gothenburg, Gothenburg, Sweden*
JULIEN MARIETHOZ • *Proteome Informatics Group, SIB Swiss Institute of Bioinformatics, Geneva, Switzerland*
YEHIA MECHREF • *Department of Chemistry and Biochemistry, Texas Tech University, Lubbock, TX, USA*
WILMA E. MESKER • *Department of Surgery, Leiden University Medical Center, Leiden, The Netherlands*
ELAINE MULLEN • *The MITRE Corporation, McLean, VA, USA*
RÓISÍN O'FLAHERTY • *NIBRT GlycoScience Group, NIBRT–The National Institute for Bioprocessing Research and Training, Blackrock, Co., Dublin, Ireland*
NICOLLE H. PACKER • *Biomolecular Frontiers Research Centre, Macquarie University, North Ryde, NSW, Australia*
KATHARINA PASCHINGER • *Department für Chemie, Universität für Bodenkultur, Vienna, Austria*
TAMARA PAVIĆ • *Faculty of Pharmacy and Biochemistry, University of Zagreb, Zagreb, Croatia*
HENRIETTE LODBERG PEDERSEN • *Section for Plant Glycobiology, Department for Plant and Environmental Sciences, University of Copenhagen, Frederiksberg C, Denmark*
MAJA PUČIĆ-BAKOVIĆ • *Genos Glycoscience Research Laboratory, Zagreb, Croatia*
KARLI R. REIDING • *Center for Proteomics and Metabolomics, Leiden University Medical Center, Leiden, The Netherlands*
YOANN ROMBOUTS • *Center for Proteomics and Metabolomics, Leiden University Medical Center, Leiden, The Netherlands; University Lille, CNRS, UMR 8576, UGSF, Unite de Glycobiologie Structurale et Fonctionnelle, Lille, France*
PAULINE M. RUDD • *NIBRT GlycoScience Group, NIBRT–The National Institute for Bioprocessing Research and Training, Blackrock, Co., Dublin, Ireland*
L. RENEE RUHAAK • *Department of Translational Molecular Pathology, UT MD Anderson Cancer Center, Houston, TX, USA; Department of Chemistry, UC Davis, Davis, CA, USA*
MAJA GRO RYDAHL • *Section for Plant Glycobiology, Department for Plant and Environmental Sciences, University of Copenhagen, Frederiksberg C, Denmark*

FALKO SCHIRMEISTER • *Department of Biomolecular Systems, Max Planck Institute of Colloids and Interfaces, Potsdam, Germany; Department of Biology, Chemistry, Pharmacy, Institute of Chemistry and Biochemistry, Freie Universität Berlin, Berlin, Germany*
JONAS STÅHLE • *Department of Organic Chemistry, Arrhenius Laboratory, Stockholm University, Stockholm, Sweden*
KATHRIN STAVENHAGEN • *Division of BioAnalytical Chemistry, VU University Amsterdam, Amsterdam, The Netherlands; Center for Proteomics and Metabolomics, Leiden University Medical Center, Leiden, The Netherlands*
HENNING STÖCKMANN • *NIBRT GlycoScience Group, NIBRT–The National Institute for Bioprocessing Research and Training, Blackrock, Co., Dublin, Ireland; AbbVie Inc., North Chicago, IL, USA*
MARTON SZIGETI • *Horvath Csaba Laboratory of Bioseparation Sciences, University of Debrecen, Debrecen, Hungary*
ROB A. TOLLENAAR • *Department of Surgery, Leiden University Medical Center, Leiden, The Netherlands*
IVO UGRINA • *Genos Glycoscience Research Laboratory, Zagreb, Croatia*
IRMA VAN DIE • *Department of Molecular Cell Biology and Immunology, VU University Medical Center, Amsterdam, The Netherlands*
GABI W. VAN PELT • *Department of Surgery, Leiden University Medical Center, Leiden, The Netherlands*
SILVIA VIDAL-MELGOSA • *Section for Plant Glycobiology, Department for Plant and Environmental Sciences, University of Copenhagen, Frederiksberg C, Denmark*
FRANO VUČKOVIĆ • *Genos Glycoscience Research Laboratory, Zagreb, Croatia*
GÖRAN WIDMALM • *Department of Organic Chemistry, Arrhenius Laboratory, Stockholm University, Stockholm, Sweden*
WILLIAM GEORGE TYCHO WILLATS • *Section for Plant Glycobiology, Department for Plant and Environmental Sciences, University of Copenhagen, Frederiksberg C, Denmark*
IAIN B.H. WILSON • *Department für Chemie, Universität für Bodenkultur, Vienna, Austria*
KERRY M. WOODING • *Department of Chemistry and Biochemistry, Texas Tech University, Lubbock, TX, USA*
LAUREN D. WU • *Department of Chemistry, University of California, Davis, CA, USA*
MANFRED WUHRER • *Division of BioAnalytical Chemistry, VU University Amsterdam, Amsterdam, The Netherlands; Center for Proteomics and Metabolomics, Leiden University Medical Center, Leiden, The Netherlands*
SHIYUE ZHOU • *Department of Chemistry and Biochemistry, Texas Tech University, Lubbock, TX, USA*

Chapter 1

Ubiquitous Importance of Protein Glycosylation

Jasminka Krištić and Gordan Lauc

Abstract

More than half of all proteins are glycosylated. The attached glycans provide proteins with important structural and functional properties and glycan parts of glycoproteins have essential roles in many key biological processes. This chapter describes the effect of glycosylation on the structure and function of proteins, with emphasis on regulation of protein half-life and modulation of protein function by alternative glycosylation. In addition, this chapter highlights the importance of glycan–lectin interactions, the ability of glycans to block phosphorylation of proteins, and the importance of glycans in disease.

Key words Glycans, Protein structure, Protein function, Lectins, Protein half-life, Alternative glycosylation, IgG glycosylation, *O*-GlcNAc, Glycosylation in disease

1 Glycans Are Important Structural and Functional Components of the Majority of Proteins

Glycosylation of proteins is the enzymatic process in which complex sugar molecules (glycans) are attached to proteins. Protein glycosylation is a very common co-translational and posttranslational modification of proteins. Moreover, it has been estimated that more than half of all proteins are glycosylated [1], with glycosylation being especially widespread among cell surface and extracellular proteins [2]. There are several different ways in which glycans can be added to glycoproteins (Fig. 1). The most abundant way of protein glycosylation is *N*-glycosylation which is characterized by the attachment of the glycan to the nitrogen of asparagine residue within the Asn-X-Ser/Thr motif of a protein. *O*-glycosylation is another type of protein glycosylation and is also quite common protein modification with estimated abundance of 10 % of all glycoproteins [1]. In *O*-glycosylated proteins glycans are attached to the oxygen of serine or threonine, but, contrary to *N*-glycosylation, *O*-glycosylation is not sequence specific. Similar to *O*-linked glycoproteins, carbohydrate part of another group of glycoproteins called proteoglycans is also bound

Gordan Lauc and Manfred Wuhrer (eds.), *High-Throughput Glycomics and Glycoproteomics: Methods and Protocols*, Methods in Molecular Biology, vol. 1503, DOI 10.1007/978-1-4939-6493-2_1, © Springer Science+Business Media New York 2017

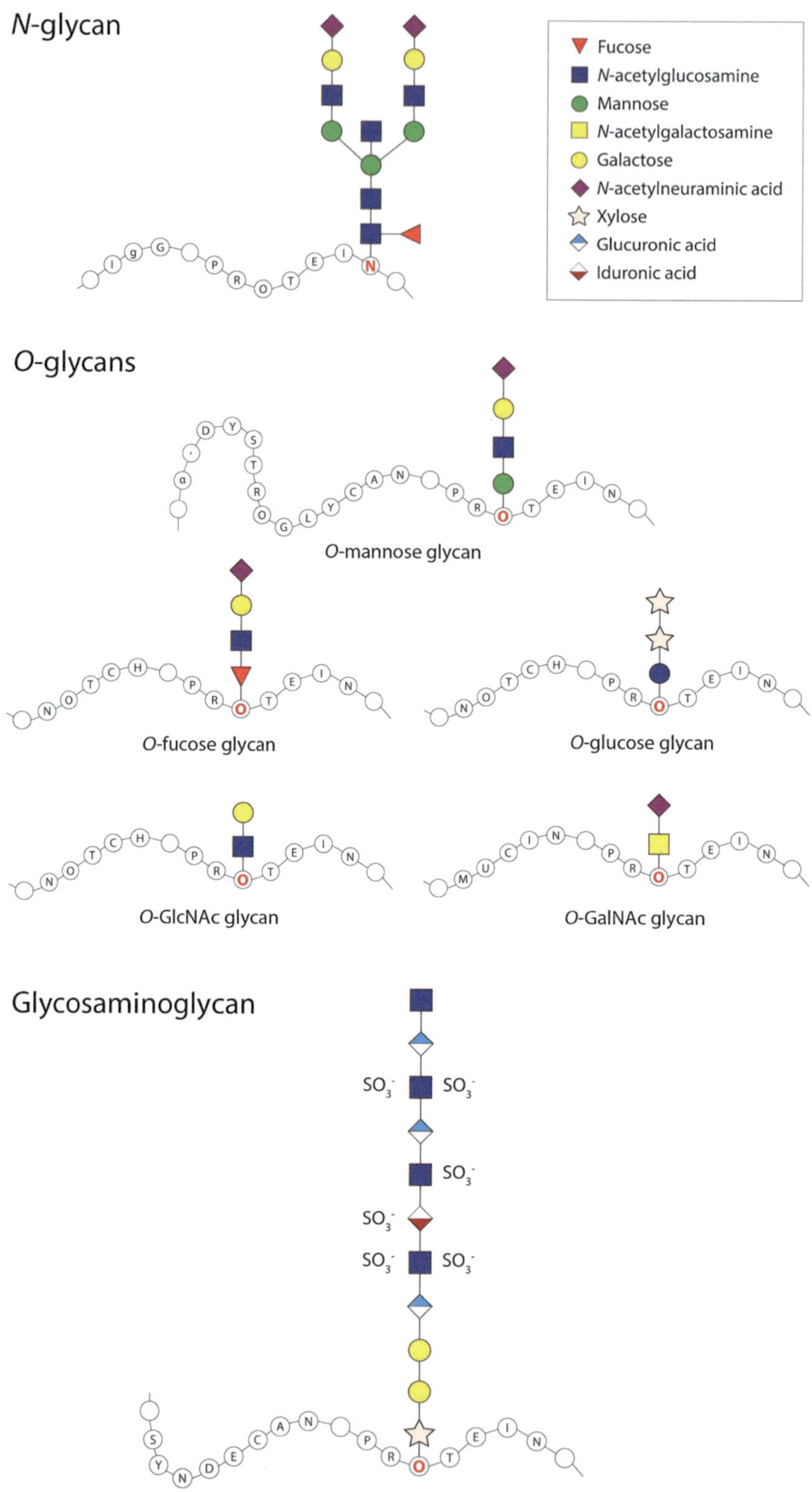

Fig. 1 Structures of some *N*-glycans, *O*-glycans, and glycosaminoglycans attached to different proteins. *N*-glycan attached to the immunoglobulin G (IgG) protein; *O*-mannose glycan attached to the α-dystroglycan protein; *O*-fucose, *O*-glucose, and *O*-GlcNAc glycans attached to the Notch protein; *O*-GalNAc glycan attached to the mucin protein and heparan sulfate glycosaminoglycan attached to the syndecan protein

to serine or threonine residues in a protein; however, proteoglycans consist of long, linear carbohydrate structures (glycosaminoglycans) that are often highly sulfated [3].

Glycosylation significantly contributes to structural heterogeneity of proteins. Several thousand different glycan determinants can be made by varying number, order, and type of sugar monomers, as well as inter-monomer binding (a numeric configuration, position, and branching) [4]. Additionally, proteins can be glycosylated at multiple sites and a single protein can contain both *N*- and *O*-glycans which further increases the number of ways that proteins can be modified by glycans [1]. The high prevalence of glycosylated proteins and huge complexity of the glycosylation pathway indicate the significant importance of glycosylation for modified proteins. Glycosylation greatly affects the physical properties of proteins (solubility, conformation, folding, stability, etc.) as well as biological functions of proteins [5, 6]. Therefore, since final molecular structure of glycoproteins that perform function is a combination of both polypeptide and glycan parts, glycosylated proteins should be studied in the context a combined peptide and glycan structure. In contrary to the polypeptide part of a glycoprotein, which is defined by the corresponding gene, glycan part of a glycoprotein is not directly encoded in the genome. Instead, glycans are synthesized in the complex biosynthetic pathway in which a number of different enzymes and their substrates, transcriptional factors, and other proteins together with environmental factors determine the final glycan structure [7]. Numerous important roles of glycan parts of glycoproteins are described in the literature. Glycans are involved in protein–protein interactions, protein folding and protein sorting. Furthermore, they regulate protein half-life, determine biological activity of a protein and participate in many other molecular processes [6]. The following sections describe and provide examples of some key roles played by glycans.

2 Most Cell Surface Interactions Are Mediated by Glycans

The surface of all cells in nature is heavily coated with glycans. This glycan coat (called glycocalyx) is composed of glycan part of glycoproteins, glycolipids, and proteoglycans attached to the cell membrane [8]. Glycocalyx is much thicker than the membrane itself and thus dominates the cell surface [9]. Therefore, virtually all initial interactions of a cell with a protein, another cell, extracellular matrix or with a microorganism are mediated by glycans. Cell surface glycans often serve as ligands for glycan-binding proteins [10]. For instance, in cell–cell interactions specific surface glycan of one cell is recognized by corresponding glycan-binding protein present on the surface of another cell. One of the best-known groups of glycan-binding proteins are lectins. Lectins represent diverse group of proteins that possess carbohydrate-recognition domain by which they selectively recognize and bind specific glycan structure [11, 12]. Such glycan–lectin interactions are crucial for many biological processes, including cell adhesion and migration, signal transduction, pathogen attachment, self/nonself recognition, and many others [6, 13, 14].

Probably the best studied glycan–lectin interactions are those between selectin family of lectins and their glycan ligands. L-selectin is the glycan-binding protein that is expressed on lymphocytes. L-selectin recognizes and binds sulfated *N*- and *O*-glycoprotein ligands which are expressed on endothelial cells of high endothelial venules of secondary lymphoid organs [15, 16]. This interaction between L-sectin and sulfated glycans enables lymphocytes to migrate from blood to lymph nodes. Moreover, selectin–glycan interactions are also important during inflammation when they govern movement of lymphocytes from circulation into inflamed tissues [15].

Fertilization is just another example where contact between two cells is initiated via cell surface glycans. Sperm–egg interaction that leads to successful fertilization is mediated by the binding of lectin-like sperm proteins with specific glycans that are present on egg's extracellular glycocalyx coat called the *zona pellucida* [17, 18].

Nearly all pathogens start their invasion by attachment to host cell surface glycans. The influenza virus has a lectin called hemagglutinin on its envelope. Hemagglutinin recognizes and binds sialic acid residue of host cell surface glycoproteins [19]. Other examples of pathogens that initiate infection by binding to host glycans include noroviruses [20], rotaviruses [21], herpesviruses [22], and bacterium *Helicobacter pylori* [23].

3 Glycans Control Half-Life of Proteins

As was mentioned before, a single protein can be glycosylated at multiple glycosylation sites, which can differ in glycans attached to them. In addition, the same glycosylation site of a single glycoprotein can bear different glycans. Therefore, the same glycoprotein can exist in many glycosylation variants that differ only with respect to glycan part of a glycoprotein. It is known that differences in terminal sugar residues and degree of branching of attached glycans affects half-life of some membrane and circulating glycoproteins [24, 25]. For example, membrane retention of cytokine and growth factor receptors depends on the structure of *N*-glycans attached to those receptors [24]. Cytokine and growth factor receptors are membrane glycoproteins, which upon ligand binding, can active various signaling pathways involved in proliferation, differentiation, migration, survival, apoptosis, and other cellular processes [26]. Therefore, differential glycosylation of cell surface receptors regulates membrane dynamics of those receptors, which may consequently influence cell fate. Half-life of GABBA (inhibitory neurotransmitter receptor) and GLUT2 (Glucose transporter 2) membrane receptors is also regulated by glycosylation [27, 28].

Glycans also control clearance of glycoproteins from the circulation. Glycoprotein hormones follicle-stimulating hormone and luteinizing hormone have *N*-glycans that terminate with sialic acid and/or sulfonated *N*-acetylgalactosamine. Glycoprotein variants of these hormones which are more sialylated and less sulfated have longer circulation half-life which is linked to rapid clearance of sulfated hormone molecules mediated by receptors in a liver [29, 30].

The ability of glycans to alter half-life of proteins to which they are attached has found practical application in the production of therapeutics. One such example is production of recombinant erythropoietin, which is used to treat anemia. Circulatory half-life of recombinant erythropoietin can be improved by increasing the number of terminal sialic acid residues or by adding extra *N*-glycosylation sites to recombinant erythropoietin [31–34].

4 Glycosylation Modulates Biological Function of Proteins

Alternative glycosylation, addition of different glycans to the same attachment site on the protein, is observed in many glycoproteins. Depending on attached glycans protein can exhibit different biological functions. There are numerous examples where differential glycosylation of a protein modulates its function. Effects of differential glycosylation of Notch and immunoglobulin G (IgG) proteins on their biological function have been studied in great detail.

The Notch protein is essential component of Notch signaling pathway. Notch signaling pathway controls differentiation of many cell types during embryonic and adult development [35]. Moreover, this pathway is aberrantly activated in many human cancers [36]. The main components of Notch signaling pathways are Notch protein, which acts as transmembrane receptor, and its ligands Delta and Jagged. Notch signaling pathway is activated when Notch ligand on one cell binds to Notch receptor on a neighboring cell. Upon ligand binding, intracellular domain of Notch receptor is released and transported to the nucleus where it activates Notch target genes [35]. The extracellular domain of Notch receptor, which is involved in ligand binding, is glycosylated with *O*-fucose, *O*-glucose, and *N*-glycans. Of particular interest are *O*-fucose glycans, which can be elongated by the addition of an *N*-acetylglucosamine (GlcNAc) by the enzyme called Fringe [37].

Such elongation of *O*-fucose glycans modulates the ability of Notch ligands to activate the Notch signaling pathway. Notch receptor that contains elongated O-fucose glycans is activated upon binding of Delta ligand, while binding of Jagged ligand to Notch receptor with elongated *O*-fucose glycans does not activate Notch signaling [38–40]. On the other hand, in the absence of elongation of *O*-glycans by Fringe enzyme, Notch signaling pathway can be activated by both Delta and Jagged ligands [39]. Therefore, depending on structure of *O*-fucose glycans attached to Notch protein Notch signaling pathway can be activated or inhibited and such modulation of Notch signaling by glycosylation enables fine-tuning of the cell differentiation process [41].

Glycoprotein immunoglobulin G (IgG) is another example of a protein whose function is regulated by glycosylation. IgG is the most abundant class of antibody in the blood which play important role in protecting the body from invading pathogens. IgG molecule consist of two fragments—Fab and Fc fragment. Fab fragment (antigen-binding fragment) is responsible for recognition of different pathogens like bacteria or viruses. Fc fragment provides effector functions which means that this fragment determines how immune system will respond or react to presence of antigen. These effector functions are enabled through interaction of Fc fragment with different Fc receptors. Fc fragment of IgG molecule contains two conserved *N*-glycosylation sites and glycosylation of Fc fragment affects binding affinity to Fc receptors. Differences in glycosylation of Fc fragment alter binding affinity to different Fc receptors and thus modulate IgG effector functions (Fig. 2) [42]. For example, Fc glycosylation acts as a switch between pro- or anti-inflammatory effector function of IgG molecule. IgG molecules that contain terminal sialic acid residues through interaction with specific Fc receptor activate anti-inflammatory effector pathway while IgG molecules that lack sialic acid and galactose residues act in a pro-inflammatory way [43]. Core fucose of *N*-glycans attached to the Fc fragment of IgG molecule has role in modulation of antibody-dependent cellular cytotoxity (ADCC). IgG glycans that lack core fucose are more efficient in activating ADCC than fucosylated variants of the IgG molecule [44–46].

5 Glycosylation as a Regulatory Modification

A type of protein *O*-glycosylation, *O*-GlcNAc, has some unique properties not found in any other type of protein glycosylation [47, 48]. Among them, the ability of *O*-GlcNAc to block phosphorylation of proteins indicates regulatory role of *O*-GlcNAc [47–50]. Phosphorylation is an abundant posttranslational modification which regulates activity of many proteins (i.e., activates or

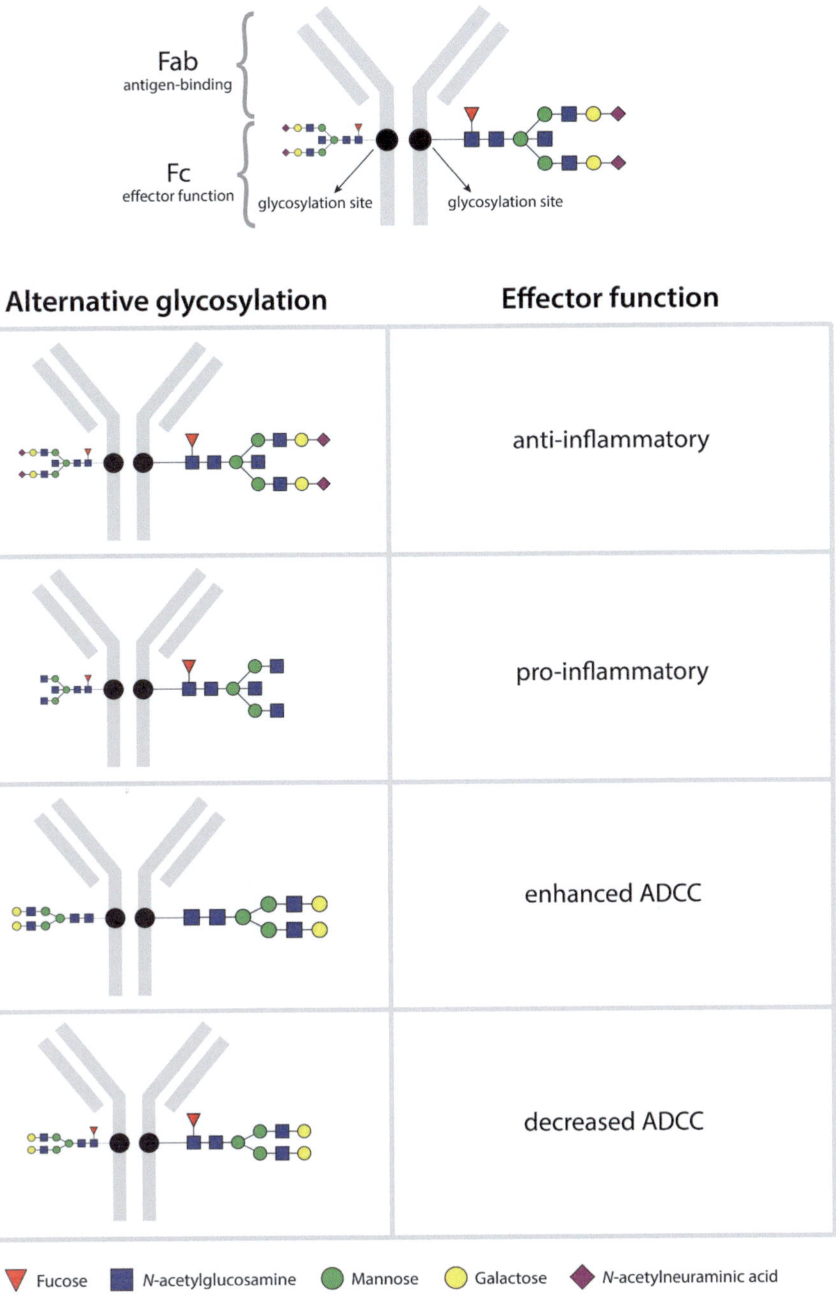

Fig. 2 Modulation of immunoglobulin G (IgG) effector function by alternative glycosylation. ADCC, antibody-dependent cellular cytotoxity

inactivates the target proteins), especially proteins involved in signal transduction [51]. Both phosphorylation and *O*-GlcNAc modification occurs on serine or threonine residue of a protein. Therefore, *O*-GlcNAc can prevent phosphorylation of specific protein by occupying phosphorylation site or site adjacent to phosphorylation site which can sterically block phosphorylation of a protein [52]. For example, RNA polymerase is modified by *O*-GlcNAc at the preinitiation step of transcription but transcription elongation can proceed only after *O*-GlcNAc residues are removed and RNA polymerase becomes phosphorylated [53, 54]. Many nuclear and cytoplasmic proteins involved in various cellular processes are modified by *O*-GlcNAc [55, 56]. *O*-GlcNAc modification regulates function of many of those proteins either by blocking phosphorylation or directly [48]. The regulation of protein function by *O*-GlcNAc modification is important for numerous cellular processes, including transcription, translation, protein degradation, signal transduction, cell cycle progression, and others [53, 57]. Moreover, the absence of enzyme that adds *O*-GlcNAc to proteins is lethal at the single cell level [58]. Furthermore, since *O*-GlcNAc is reversible modification which means that a single protein can have *O*-GlcNAc added and remove multiple times and since *O*-GlcNAc cycling often occurs very fast, *O*-GlcNAc modification enables cells to quickly adapt to changes in the nutrient status or stress [47, 53, 57].

6 Glycans Are Implicated in Various Diseases

As exemplified above, glycan parts of glycoproteins are involved in virtually all biological events, from fertilization, through cell proliferation, cell signaling, cell differentiation, and development, to immunity, host–pathogen interactions, hormonal activities, and others. Complete loss of glycosylation is lethal [58, 59] and defects in glycan synthesis cause a number of serious multisystemic diseases known as congenital disorders of glycosylation [60]. Moreover, changes in glycosylation have been observed in various diseases [61], including cancer [62], diabetes [63], autoimmune and inflammatory diseases [64–66], neurodegenerative diseases [67], and infectious diseases [68]. Considering recent development of methods for high-throughput analysis of glycan which enable systematic analysis of large patient cohorts [69, 70], new insight into disease associated glycan changes and their role in disease pathophysiology can be expected. Many glycans from total plasma glycoproteins or from individual glycoproteins were identified as potential disease biomarkers, especially cancer biomarkers, and can be used for diagnostic and therapeutic purpose [71–74]. Furthermore, population studies have shown that interindividual differences in glycosylation are large [75, 76] and these differences associate with disease susceptibility, disease course, and response to therapy [77–80]. Therefore, glycans have the potential to help stratify patients according to disease predisposition, prognosis, and response to treatment.

References

1. Apweiler R, Hermjakob H, Sharon N (1999) On the frequency of protein glycosylation, as deduced from analysis of the SWISS-PROT database. Biochim Biophys Acta 1473:4-8
2. Roth J (2002) Protein N-Glycosylation along the secretory pathway: relationship to organelle topography and function, protein quality control, and cell interactions. Chem Rev 102:285–304
3. Esko JD, Selleck SB (2002) Order out of chaos: assembly of ligand binding sites in heparan sulfate. Annu Rev Biochem 71:435–471
4. Cummings RD (2009) The repertoire of glycan determinants in the human glycome. Mol Biosyst 5:1087–1104
5. Varki A (1993) Biological roles of oligosaccharides: all of the theories are correct. Glycobiology 3:97-130.
6. Ohtsubo K, Marth JD (2006) Glycosylation in cellular mechanisms of health and disease. Cell 126:855–867
7. Krištić J, Zoldoš V, Lauc G (2014) Complex genetics of protein N-glycosylation. In: Endo T, Seeberger PH, Hart GW, Wong CH, Taniguchi N (eds) Glycoscience: biology and medicine. Springer, Japan
8. Varki A (2011) Evolutionary forces shaping the Golgi glycosylation machinery: why cell surface glycans are universal to living cells. Cold Spring Harb Perspect Biol 3:a005462
9. Vogel J, Sperandio M, Pries AR, Linderkamp O, Gaehtgens P, Kuschinsky W (2000) Influence of the endothelial glycocalyx on cerebral blood flow in mice. J Cereb Blood Flow Metab 20:1571–1578
10. Cohen M (2015) Notable aspects of glycan-protein interactions. Biomolecules 5:2056–2072
11. Varki A, Etzler ME, Cummings RD et al (2009) Discovery and classification of glycan-binding proteins. In: Varki A, Cummings RD, Esko JD et al (eds). Essentials of glycobiology, 2nd edn. Cold Spring Harbor, New York. http://www.ncbi.nlm.nih.gov/books/NBK1923/
12. Cummings RD, Esko JD (2009) Principles of glycan recognition. In: Varki A, Cummings RD, Esko JD et al (eds). Essentials of glycobiology, 2nd edn. Cold Spring Harbor, New York. http://www.ncbi.nlm.nih.gov/books/NBK1950/
13. Nagae M, Yamaguchi Y (2015) Sugar recognition and protein-protein interaction of mammalian lectins conferring diverse functions. Curr Opin Struct Biol 34:108–115
14. Bochner BS, Zimmermann N (2015) Role of siglecs and related glycan-binding proteins in immune responses and immunoregulation. J Allergy Clin Immunol 135:598–608
15. Sperandio M, Gleissner CA, Ley K (2009) Glycosylation in immune cell trafficking. Immunology 230:97–113
16. Hirakawa J, Tsuboi K, Sato K, Kobayashi M, Watanabe S, Takakura A, Imai Y, Ito Y, Fukuda M, Kawashima H (2010) Novel anti-carbohydrate antibodies reveal the cooperative function of sulfated N- and O-glycans in lymphocyte homing. J Biol Chem 285:40864–40878
17. Diekman AB (2003) Glycoconjugates in sperm function and gamete interactions: how much sugar does it take to sweet-talk the egg? Cell Mol Life Sci 60:298-308
18. Pang PC, Chiu PC, Lee CL, Chang LY, Panico M, Morris HR, Haslam SM, Khoo KH, Clark GF, Yeung WS, Dell A (2011) Human sperm binding is mediated by the Sialyl-Lewis(x) oligosaccharide on the zona pellucida. Science 333:1761–1764
19. Suzuki Y (2005) Sialobiology of influenza: molecular mechanism of host range variation of influenza viruses. Biol Pharm Bull 28: 399–408
20. Hu l, Crawford SE, Czako R, Cortes-Penfield NW, Smith DF, Le Pendu J, Estes MK, Prasad BV (2012) Cell attachment protein VP8* of a human rotavirus specifically interacts with A-type histo-blood group antigen. Nature 485:256-259
21. Glass RI, Parashar UD, Estes MK (2009) Norovirus gastroenteritis. N Engl J Med 361:1776–1785
22. Shukla D, Spear PG (2001) Herpesviruses and heparan sulfate: an intimate relationship in aid of viral entry. J Clin Invest 108:503–510
23. Ilver D, Arnqvist A, Ogren J, Frick IM, Kersulyte D, Incecik ET, Berg DE, Covacci A, Engstrand L, Borén T (1998) Helicobacter pylori adhesin binding fucosylated histo-blood group antigens revealed by retagging. Science 279:373–377
24. Partridge EA, Le Roy C, Di Guglielmo GM, Pawling J, Cheung P, Granovsky M, Nabi IR, Wrana JL, Dennis JW (2004) Regulation of cytokine receptors by Golgi N-glycan processing and endocytosis. Science 306:120–124
25. Wide L, Eriksson K, Sluss PM, Hall JE (2009) Serum half-life of pituitary gonadotropins is decreased by sulfonation and increased by sialylation in women. J Clin Endocrinol Metab 94:958–964
26. Becker V, Timmer J, Klingmüller U (2012) Receptor dynamics in signaling. Adv Exp Med Biol 736:313–323

27. Lo WY, Lagrange AH, Hernandez CC, Harrison R, Dell A, Haslam SM, Sheehan JH, Macdonald RL (2010) Glycosylation of {beta}2 subunits regulates GABAA receptor biogenesis and channel gating. J Biol Chem 285:31348–31361
28. Ohtsubo K, Takamatsu S, Minowa MT, Yoshida A, Takeuchi M, Marth JD (2005) Dietary and genetic control of glucose transporter 2 glycosylation promotes insulin secretion in suppressing diabetes. Cell 123:1307–1321
29. Wide L, Naessén T, Sundström-Poromaa I, Eriksson K (2007) Sulfonation and sialylation of gonadotropins in women during the menstrual cycle, after menopause, and with polycystic ovarian syndrome and in men. J Clin Endocrinol Metab 92:4410–4417
30. Mi Y, Lin A, Fiete D, Steirer L, Baenziger JU (2014) Modulation of mannose and asialoglycoprotein receptor expression determines glycoprotein hormone half-life at critical points in the reproductive cycle. J Biol Chem 289:12157–12167
31. Egrie JC, Browne JK (2001) Development and characterization of novel erythropoiesis stimulating protein (NESP). Br J Cancer 84:3–10
32. Byrne B, Donohoe GG, O'Kennedy R (2007) Sialic acids: carbohydrate moieties that influence the biological and physical properties of biopharmaceutical proteins and living cells. Drug Discov Today 12:319–326
33. Jeong YT, Choi O, Lim HR, Son YD, Kim HJ, Kim JH (2008) Enhanced sialylation of recombinant erythropoietin in CHO cells by human glycosyltransferase expression. J Microbiol Biotechnol 18:1945–1952
34. Su D, Zhao H, Xia H (2010) Glycosylation-modified erythropoietin with improved half-life and biological activity. Int J Hematol 91:238–244
35. Artavanis-Tsakonas S, Rand MD, Lake RJ (1999) Notch signaling: cell fate control and signal integration in development. Science 284:770–776
36. Bolós V1, Grego-Bessa J, de la Pompa JL (2007) Notch signaling in development and cancer. Endocr Rev 28:339-363
37. Stanley P (2007) Regulation of Notch signaling by glycosylation. Curr Opin Struct Biol 17:530–535
38. Hicks C, Johnston SH, diSibio G, Collazo A, Vogt TF, Weinmaster G (2000) Fringe differentially modulates Jagged1 and Delta1 signalling through Notch1 and Notch2. Nat Cell Biol 2:515–520
39. Yang LT, Nichols JT, Yao C, Manilay JO, Robey EA, Weinmaster G (2005) Fringe glycosyltransferases differentially modulate Notch1 proteolysis induced by Delta1 and Jagged1. Mol Biol Cell 16:927–942
40. Taylor P, Takeuchi H, Sheppard D, Chillakuri C, Lea SM, Haltiwanger RS, Handford PA (2014) Fringe-mediated extension of O-linked fucose in the ligand-binding region of Notch1 increases binding to mammalian Notch ligands. Proc Natl Acad Sci U S A 111:7290–7295
41. Boareto M, Jolly MK, Lu M, Onuchic JN, Clementi C, Ben-Jacob E (2015) Jagged–Delta asymmetry in Notch signaling can give rise to a Sender/Receiver hybrid phenotype. Proc Natl Acad Sci U S A 112:E402–E409
42. Schwab I and Nimmerjahn F (2013) Intravenous immunoglobulin therapy: how does IgG modulate the immune system? Nat Rev Immunol 13:176-189
43. Kaneko Y, Nimmerjahn F, Ravetch JV (2006) Anti-inflammatory activity of immunoglobulin G resulting from Fc sialylation, vol 313. Science, pp. 670–673
44. Niwa R, Hatanaka S, Shoji-Hosaka E, Sakurada M, Kobayashi Y, Uehara A, Yokoi H, Nakamura K, Shitara K (2004) Enhancement of the antibody-dependent cellular cytotoxicity of low-fucose IgG1 Is independent of Fcgamma RIIIa functional polymorphism. Clin Cancer Res 10:6248–6255
45. Iida S, Misaka H, Inoue M, Shibata M, Nakano R, Yamane-Ohnuki N, Wakitani M, Yano K, Shitara K, Satoh M (2006) Nonfucosylated therapeutic IgG1 antibody can evade the inhibitory effect of serum immunoglobulin G on antibody-dependent cellular cytotoxicity through its high binding to FcgammaRIIIa. Clin Cancer Res 12:2879–2887
46. Masuda K, Kubota T, Kaneko E, Iida S, Wakitani M, Kobayashi-Natsume Y, Kubota A, Shitara K, Nakamura K (2007) Enhanced binding affinity for FcgammaRIIIa of fucose-negative antibody is sufficient to induce maximal antibody-dependent cellular cytotoxicity. Mol Immunol 44:3122–3131
47. Hart GW, Akimoto Y. The O-GlcNAc modification (2009) In: Varki A, Cummings RD, Esko JD et al (eds). Essentials of glycobiology, 2nd edn. Cold Spring Harbor, New York. http://www.ncbi.nlm.nih.gov/books/NBK1954/
48. Hayes BK, Hart GW (1998) Protein O-GlcNAcylation: potential mechanisms for the regulation of protein function. In: Axford JS (ed) Glycoimmunology, Advances in experimental medicine and biology, vol vol 435, 2 edn. Springer, New York, pp. 85–94
49. Slawson C, Hart GW (2003) Dynamic interplay between O-GlcNAc and O-phosphate: the

sweet side of protein regulation. Curr Opin Struct Biol 13:631–636

50. Zeidan Q, Hart GW (2010) The intersections between O-GlcNAcylation and phosphorylation: implications for multiple signaling pathways. J Cell Sci 123:13–22

51. Cohen P (2000) The regulation of protein function by multisite phosphorylation—a 25 year update. Trends Biochem Sci 25:596–601

52. Whelan SA, Hart GW (2003) Proteomic approaches to analyze the dynamic relationships between nucleocytoplasmic protein glycosylation and phosphorylation. Circ Res 93:1047–1058

53. Hardivillé S, Hart GW (2014) Nutrient regulation of signaling, transcription, and cell physiology by O-GlcNAcylation. Cell Metab 20:208–213

54. Lewis BA (2013) O-GlcNAcylation at promoters, nutrient sensors, and transcriptional regulation. Biochim Biophys Acta 1829:1202-1206

55. Comer FI, Hart GW (2000) O-Glycosylation of nuclear and cytosolic proteins: DYNAMIC INTERPLAY BETWEEN O-GlcNAc ANDO-PHOSPHATE. J Biol Chem 275:29179–29182

56. Hanover JA (2001) Glycan-dependent signaling: O-linked N-acetylglucosamine. FASEB J 15:1865–1876

57. Bond MR, Hanover JA (2015) A little sugar goes a long way: the cell biology of O-GlcNAc. J Cell Biol 208:869–880

58. Shafi R, Iyer SP, Ellies LG, O'Donnell N, Marek KW, Chui D, Hart GW, Marth JD (2000) The O-GlcNAc transferase gene resides on the X chromosome and is essential for embryonic stem cell viability and mouse ontogeny. Proc Natl Acad Sci U S A 97:5735–5739

59. Marek KW, Vijay IK, Marth JD (1999) A recessive deletion in the GlcNAc-1-phosphotransferase gene results in peri-implantation embryonic lethality. Glycobiology 9:1263–1271

60. Freeze HH (2001) Update and perspectives on congenital disorders of glycosylation. Glycobiology 11:129R-143R

61. Lauc G, Pezer M, Rudan I, Campbell H (2015) Mechanisms of disease: the human N-glycome. Biochim Biophys Acta. doi:10.1016/j.bbagen.2015.10.016

62. Stowell SR, Ju T, Cummings RD (2015) Protein glycosylation in cancer. Annu Rev Pathol 10:473–510

63. Thanabalasingham G, Huffman JE, Kattla JJ, Novokmet M, Rudan I, Gloyn AL, Hayward C, Adamczyk B, Reynolds RM, Muzinic A, et al. (2013) Mutations in HNF1A result in marked alterations of plasma glycan profile. Diabetes 62:1329–1337

64. Vučković F, Krištić J, Gudelj I, Teruel M, Keser T, Pezer M, Pučić-Baković M, Štambuk J, Trbojević-Akmačić I, Barrios C et al (2015) Association of systemic lupus erythematosus with decreased immunosuppressive potential of the IgG glycome. Arthritis Rheumatol 67:2978-2989

65. Trbojević Akmačić I, Ventham NT, Theodoratou E, Vučković F, Kennedy NA, Krištić J, Nimmo ER, Kalla R, Drummond H, Štambuk J, et al. (2015) Inflammatory bowel disease associates with proinflammatory potential of the immunoglobulin G glycome. Inflamm Bowel Dis 21:1237–1247

66. Parekh RB, Dwek RA, Sutton BJ, Fernandes DL, Leung A, D. Stanworth D, Rademacher TW, Mizuochi T, Taniguchi T, Matsuta K (1985) Association of rheumatoid arthritis and primary osteoarthritis with changes in the glycosylation pattern of total serum IgG. Nature 316:452-457

67. Hwang H, Zhang J, Chung KA, Leverenz JB, Zabetian CP, Peskind ER, Jankovic J, Su Z, Hancock AM, Pan C, Montine TJ, Pan S, Nutt J, Albin R, Gearing M, Beyer RP, Shi M, Zhang J (2010) Glycoproteomics in neurodegenerative diseases. Mass Spectrom Rev 29:79-125

68. Kreisman LS, Cobb BA (2012) Infection, inflammation and host carbohydrates: a Glyco-Evasion hypothesis. Glycobiology 22:1019–1030

69. Huffman JE, Pučić-Baković M, Klarić L, Hennig R, Selman MH, Vučković F, Novokmet M, Krištić J, Borowiak M, Muth T, et al. (2014) Comparative performance of four methods for high-throughput glycosylation analysis of immunoglobulin G in genetic and epidemiological research. Mol Cell Proteomics 13:1598–1610

70. Zoldos V, Horvat T, Lauc G (2013) Glycomics meets genomics, epigenomics and other high throughput omics for system biology studies. Curr Opin Chem Biol 17:34-40

71. Taniguchi N, Kizuka Y (2015) Glycans and cancer: role of N-glycans in cancer biomarker, progression and metastasis, and therapeutics. Adv Cancer Res 126:11–51

72. Drake PM, Cho W, Li B, Prakobphol A, Johansen E, Anderson NL, Regnier FE, Gibson BW, Fisher SJ (2010) Sweetening the pot: adding glycosylation to the biomarker discovery equation. Clin Chem 56:223–236

73. Fuster MM, Esko JD (2005) The sweet and sour of cancer: glycans as novel therapeutic targets. Nat Rev Cancer 5:526–542

74. Dube DH, Bertozzi CR (2005) Glycans in cancer and inflammation–potential for therapeutics and diagnostics. Nat Rev Drug Discov 4:477–488

75. Pucic M, Knezevic A, Vidic J, Adamczyk B, Novokmet M, Polasek O, Gornik O, Supraha-Goreta S, Wormald MR, Redzic I, et al. (2011) High throughput isolation and glycosylation analysis of IgG-variability and heritability of the IgG glycome in three isolated human populations. Mol Cell Proteomics 10:M111 010090
76. Knežević A, Polašek O, Gornik O, Rudan I, Campbell H, Hayward C, Wright A, Kolčić I, O'Donoghue N, Bones J, et al. (2009) Variability, heritability and environmental determinants of human plasma N-glycome. J Proteome Res 8:694–701
77. Ogata S, Shimizu C, Franco A, Touma R, Kanegaye JT, Choudhury BP, Naidu NN, Kanda Y, Hoang LT, Hibberd ML, Tremoulet AH, Varki A, Burns JC (2013) Treatment response in Kawasaki disease is associated with sialylation levels of endogenous but not therapeutic intravenous immunoglobulin G. PLoS One 8:e81448
78. Rombouts Y, Ewing E, van de Stadt LA, Selman MH, Trouw LA, Deelder AM, Huizinga TW, Wuhrer M, van Schaardenburg D, Toes RE, Scherer HU (2013) Anti-citrullinated protein antibodies acquire a pro-inflammatory Fc glycosylation phenotype prior to the onset of rheumatoid arthritis. Ann Rheum Dis. doi:10.1136/annrheumdis-2013-203565
79. Ackerman ME, Crispin M, Yu X, Baruah K, Boesch AW, Harvey DJ, Dugast AS, Heizen EL, Ercan A, Choi I, Streeck H, Nigrovic PA, Bailey-Kellogg C, Scanlan C, Alter G (2013) Natural variation in Fc glycosylation of HIV-specific antibodies impacts antiviral activity. J Clin Invest 123:2183–2192
80. Ercan A, Cui J, Chatterton DE, Deane KD, Hazen MM, Brintnell W, O'Donnell CI, Derber LA, Weinblatt ME, Shadick NA, Bell DA, Cairns E, Solomon DH, Holers VM, Rudd PM, Lee DM (2010) Aberrant IgG galactosylation precedes disease onset, correlates with disease activity, and is prevalent in autoantibodies in rheumatoid arthritis. Arthritis Rheum 62:2239–2248

Chapter 2

Laboratory Experimental Design for a Glycomic Study

Ivo Ugrina, Harry Campbell, and Frano Vučković

Abstract

Proper attention to study design before, careful conduct of procedures during, and appropriate inference from results after scientific experiments are important in all scientific studies in order to ensure valid and sometimes definitive conclusions can be made. The design of experiments, also called experimental design, addresses the challenge of structuring and conducting experiments to answer the questions of interest as clearly and efficiently as possible.

Key words Randomization, Blocking, Replication, Experimental design

1 Introduction

The general principles of study design and analysis of 'omics studies (including glycomic studies) in epidemiology research is covered in a number of recent reviews [1]. In addition, these principles have been formulated as reporting guidelines to ensure that key aspects of the study which aid interpretation and review are reported. These also ensure that key data are presented in a standard format in order to promote data synthesis in systematic reviews. Examples of these reporting guidelines include:

1. STROBE ME—STrengthening of Reporting of Observational studies in Epidemiology: Molecular Epidemiology studies [2]
2. STARD—STAndards for Reporting Diagnostic studies [3]
3. REMARK—Guidelines for REporting tumor MARKers [4] and
4. GRIPS—Guidelines for the reporting of Genetic Risk Prediction Studies. [5]

However, much less has been published on the detailed design of laboratory procedures to ensure valid and reliable 'omic data are generated for analysis. This review focuses on this important aspect of glycomic study design.

Gordan Lauc and Manfred Wuhrer (eds.), *High-Throughput Glycomics and Glycoproteomics: Methods and Protocols*, Methods in Molecular Biology, vol. 1503, DOI 10.1007/978-1-4939-6493-2_2, © Springer Science+Business Media New York 2017

The main aim of the high-throughput analysis is to analyse very large numbers of samples in a cost-efficient manner and in a relatively short time. Nevertheless, the very large number of samples often necessitates that an experiment lasts several weeks. This can lead to changes over time in the setup of a lab analysis (e.g., change of column in a UPLC machine) of glycans. These changes may distort later results leading to some variables falsely appearing to be correlated, i.e., leading to bias and/or confounding. These are not necessarily an artifact of changes within the laboratory, since samples usually come in batches and these problems may exist even before they enter the lab (e.g., bias introduced due to cases being in one batch, controls in another). Variables affecting results and possibly leading to bias and/or confounding are often called *nuisance factors*. An effective approach to reduce or even eliminate the effects of nuisance factors can be achieved with a proper application of the theory of experimental design.

The "design of experiments" was first described by Ronald A. Fischer in 1920 [6] to improve agricultural work and results. Although it was originally developed for agriculture, the main ideas and methods have since been applied in numerous fields and are therefore called the fundamental principles of the experimental design. The three most important principles for experimental design, relevant to high-throughput glycomics experiments, are:

1. Randomization
2. Blocking
3. Replication

Randomization is a method that guards against unknown nuisance factors affecting the results of the experiment. An example of a bias that can be introduced is a change in an instrument used for chemical analysis (e.g., change of a column in a UPLC machine). If all control samples from a case–control study are analysed first and then all the case samples are analysed subsequently, the observed difference between analytical results could be due to the instrument change. In the worst case scenario, the change would occur during the time between analyses of control and case samples. However, if samples are run in a random case–control order (e.g., case, control, control, case, case, …) then any change in the instrument during the experiment should equally affect both the cases and controls and not lead to bias.

With known nuisance factors the blocking method can be applied to increase the precision of results and aid in future analysis. An obvious example of a possible nuisance factor is the batch proportion of cases and controls in a case–control study. In a blocked design, samples should be measured with the same ratio of cases and controls within every batch as within the whole population involved in the analysis. In experiments where such blocking does not occur, any apparent between-batches changes between controls and cases could be due to batch effects rather than true differences between cases and controls. Other less obvious nuisance factors have been described in the literature, such as effects of gender [7]. A good rule for the design of a high throughput laboratory experiment is to block what you can and randomize what you cannot block [8]. Therefore, both blocking and randomization should be employed.

Replication is a method which acknowledges that there are sources of variability both between runs and (potentially) within runs and thus that replication is required to account for this. A replicate is a complete repetition of the same experimental conditions, beginning with the initial setup. Replicates in high-throughput glycomics may be achieved through two types of technical replicates: technical replicates of biological samples of interest (in future denoted as just *replicates*) and technical replicates as a special, usually in-house, sample to be used within all batches (in future denoted as *standards*). The importance of replicates comes from the idea that if everything in the experiment went perfectly then values for replicates should be the same. It is important to stress here that replicates are performed within an experiment and do not denote a special type of replication where the whole experiment is replicated in a larger sample size. Changes in results between replicates indicate the level of variability of the instrument (if samples were pooled before entering the instrument) or different internal (procedural) steps (if samples were pooled before a specific step) and can point to non-systematic changes possibly revealing previously unobserved nuisance factors.

2 Materials

For a proper experimental design it is important to obtain as much detailed knowledge of the study and information on known nuisance factors as possible. Thus, experimental design should be derived in collaboration between wet- and dry-labs.

Some of the known *nuisance factors* that are generally applicable to all human glycomics analysis are:

1. Age [7].
2. Gender [9].
3. Geographical location (Continent, State, Region, …).

Other known nuisance factors are more dependent on the underlying study and data on these are often hard to obtain. Examples are:

1. Case–control designation.
2. Batches in sample acquisition (e.g., samples could have been acquired village by village introducing possibly high genetic/location bias).

3. Sample acquisition dates.
4. Number of freeze–thaw cycles (e.g., newly obtained samples vs. old samples thawed many times).
5. Information on sample acquisition centers (e.g., studies combining samples from different hospitals).

3 Methods

Although choice of the most appropriate study design is highly dependent on the available data, the main ideas can be presented through four different approaches (with additional information given in **Notes 3–6**).

3.1 Cohort Study Where No Additional Information Is Available

This is an example of a study where a laboratory is asked to analyze glycosylation of a protein in a cohort study where the only data that can be shared are samples and sample names. Since there are no additional data on samples blocking cannot be applied. The following procedure can be used:

1. Decide if replicates are needed based on previous observations (e.g., systematic or non-systematic error).
2. Decide on the number of replicates and standards needed in the study. This decision should be based on cost–benefit analysis taking into account that larger numbers of replicates and standards increase time and budget costs while decreasing error.
3. Randomly assign standards to plates.
4. Select replicates randomly.
5. Randomly assign replicates to plates.
6. Randomly assign other samples to plates (*see* **Note 1**).

3.2 Case–Control Study Where No Additional Information Is Available

This is an example of a study where a laboratory is asked to analyze the glycosylation of a protein for a case–control cohort where the only data that can be shared are samples, sample names and case–control designation. Since there are additional data on samples blocking can be applied. The following procedure can be used:

1. Decide if replicates are needed based on previous observations (e.g., systematic or non-systematic error).
2. Decide on the number of replicates and standards needed in the study. This decision should be based on cost–benefit analysis taking into account that larger numbers of replicates and standards increase time and budget costs while decreasing error.
3. Randomly assign standards to plates.
4. Select replicates randomly with case–control ratio preserved as within the whole cohort (*see* **Note 2**).

5. Randomly assign replicates to plates.
6. Randomly assign other samples to plates with the case–control ratio preserved in plates as within the whole cohort (*see* **Note 2**). This can be achieved by randomly selecting appropriate number of samples from cases first and then appropriate number of samples from controls. The approach should be repeated plate by plate (*see* **Note 1**).

3.3 Cohort Study Where Age and Gender Data Is Available

This is an example of a study where a laboratory is asked to analyze the glycosylation of a protein for a cohort where the only data that can be shared are samples and sample names together with age and gender. Since there are additional data on samples blocking can be applied. The following procedure can be used:

1. Decide if replicates are needed based on previous observations (e.g., systematic or non-systematic error).
2. Decide on the number of replicates and standards needed in the study. This decision should be based on cost–benefit analysis taking into account that larger numbers of replicates and standards increase time and budget costs while decreasing error.
3. Randomly assign standards to plates.
4. Select replicates randomly with gender ratio and age distribution preserved as within the whole cohort (*see* **Note 2**).
5. Randomly assign replicates to plates.
6. Randomly assign other samples to plates with the gender ratio and age distribution preserved in plates as within the whole cohort (*see* **Note 2**). This can be achieved by randomly selecting appropriate number of samples from females first and then the appropriate number of samples from males. The approach should be repeated plate by plate (*see* **Note 1**). Since the random selection of samples from males and females could result in different age distributions the procedure can be repeated until more balanced results are obtained.

3.4 Case–Control Study Where Age and Gender Data Is Available

This is an example of a study where a laboratory is asked to analyze the glycosylation of a protein for a cohort where the only data that can be shared are samples, sample names, and case–control designation together with age and gender. Since there are additional data on samples blocking can be applied. The following procedure can be used:

1. Decide if replicates are needed based on previous observations (e.g., systematic or non-systematic error).
2. Decide on the number of replicates and standards needed in the study. This decision should be based on cost–benefit analysis taking into account that larger numbers of replicates and standards increase time and budget costs while decreasing error.

3. Randomly assign standards to plates.
4. Select replicates randomly with gender ratio, case–control ratio, and age distribution preserved as within the whole cohort (*see* **Note 2**).
5. Randomly assign replicates to plates.
6. Randomly assign other samples to plates with gender ratio, case–control ratio, and age distribution preserved in plates as within the whole cohort (*see* **Note 2**). This can be achieved by randomly selecting appropriate number of samples from joint distributions of male/case, male/control, female/case, and female/control groups. The approach should be repeated plate by plate (*see* **Note 1**). Since the random selection of samples from aforementioned four groups could result in different age distributions, the procedure can be repeated until more balanced results are obtained.

4 Notes

1. If a change in experimental design (plate layout) happens for a reason (e.g., not enough sample in a vial) consult the person who has derived the initial plate/experimental design. In the case of a missing sample a new one can sometimes be found conforming to the current design (blocking, randomization).
2. A perfect (equal) distribution between plates is hard to achieve when controlling (blocking) many factors. Sometimes it is even impossible to achieve it. Therefore, "good enough" (in an expert view) designs should be used.
3. Appropriate software tools are of great use in deriving experimental designs since designs derived by hand can be quite time consuming.
4. If there is a plate with many samples missing or not measured well enough (seen from the consequent quality control) this plate should be taken into consideration for exclusion from the study since its distribution (case–control, gender, age) of nuisance factors could be different from the rest of the experiment.
5. Try to avoid repeating samples that did not pass quality control on a new plate without consulting the person who derived the initial design or at least looking at the distribution of nuisance factors of the failed samples. It could be that these samples could have a completely different distribution from the initial design and could therefore introduce problems in later data analysis.
6. More information on the theory of Experimental Design can be found in books specialized for the topic [8].

References

1. Tzoulaki I, Ebbels TM, Valdes A, Elliott P, Ioannidis JP (2014) Design and analysis of metabolomics studies in epidemiologic research: a primer on -omic technologies. Am J Epidemiol 180:129–139
2. Gallo V, Egger M, McCormack V, Farmer PB, Ioannidis JP, Kirsch-Volders M, Matullo G, Phillips DH, Schoket B, Stromberg U, Vermeulen R, Wild C, Porta M, Vineis P, STROBE Statement (2011) STrengthening the Reporting of OBservational studies in Epidemiology—Molecular Epidemiology (STROBE-ME): an extension of the STROBE Statement. PLoS Med 8:e1001117
3. Bossuyt PM, Reitsma JB, Bruns DE, Gatsonis CA, Glasziou PP, Irwig L, Lijmer JG, Moher D, Rennie D, de Vet HC, Kressel HY, Rifai N, Golub RM, Altman DG, Hooft L, Korevaar DA, Cohen JF, STARD Group (2015) STARD 2015: an updated list of essential items for reporting diagnostic accuracy studies. BMJ 351:h5527
4. Altman DG, McShane LM, Sauerbrei W, Taube SE (2012) Reporting Recommendations for Tumor Marker Prognostic Studies (REMARK): explanation and elaboration. PLoS Med 9:e1001216
5. Janssens AC, Ioannidis JP, van Duijn CM, Little J, Khoury MJ, GRIPS Group (2011) Strengthening the reporting of genetic risk prediction studies: the GRIPS statement. Eur J Clin Invest 41:1004–1009
6. Fisher RA (1935) The design of experiments. Oliver and Boyd, Edinburgh
7. Krištić J, Vučković F, Menni C, Klarić L, Keser T, Beceheli I, Pučić-Baković M, Novokmet M, Mangino M, Thaqi K, Rudan P, Novokmet N, Šarac J, Missoni S, Kolčić I, Polašek O, Rudan I, Campbell H, Hayward C, Aulchenko Y, Valdes A, Wilson JF, Gornik O, Primorac D, Zoldoš V, Spector T, Lauc G (2014) Glycans are a novel biomarker of chronological and biological ages. J Gerontol A Biol Sci Med Sci 69(7):779–789
8. Anderson MJ, Whitcomb PJ (2015) DOE simplified: practical tools for effective experimentation. CRC Press, New York
9. Knezevic A, Gornik O, Polasek O, Pucic M, Redzic I, Novokmet M, Rudd PM, Wright AF, Campbell H, Rudan I, Lauc G (2010) Effects of aging, body mass index, plasma lipid profiles, and smoking on human plasma N-glycans. Glycobiology 20:959–969

Chapter 3

High-Throughput Analysis of the IgG *N*-Glycome by UPLC-FLR

Maja Pučić-Baković

Abstract

As biological and clinical relevance of glycosylation is becoming more apparent, interest in large scale studies of the glycome is growing. Glycans attached to immunoglobulin G (IgG) were shown to be essential for its function and IgG glycosylation was shown to change with various processes, making IgG one of the most studied glycoproteins. Many approaches including liquid chromatography, capillary gel electrophoresis, and mass spectrometry were developed to study IgG glycosylation. Generation of high-quality glycomics data in a high-throughput fashion requires reproducible and robust sample preparation and accurate and reliable quantitative analysis. This chapter presents a protocol for an optimized and high-throughput IgG *N*-glycan release, fluorescent labeling and cleanup, and analysis of fluorescently labeled IgG *N*-glycans by hydrophilic interaction liquid chromatography (HILIC) on an ultra performance liquid chromatography (UPLC) system with fluorescence (FLR) detection.

Key words Immunoglobulin G, *N*-glycan analysis, UPLC-FLR, High-throughput

1 Introduction

IgG is the most abundant antibody class in the human blood (approx. 10 mg/mL) and a major effector molecule of the humoral immune response. All four subclasses of human IgG (IgG1–4) have a single covalently attached *N*-glycan at the highly conserved asparagine 297 in the C_H2 domain of both heavy chains within the Fc region of the molecule [1]. In addition, 20 % of human IgG molecules also contain *N*-glycans within the Fab region [2]. The Fc *N*-glycans are essential structural components of the IgG molecule and can have a profound influence on its effector functions [2, 3]. Alterations of IgG *N*-glycome were shown to accompany many physiological parameters (e.g., age, sex, pregnancy) and pathological conditions (e.g., inflammatory and autoimmune diseases) [1, 4], leading to a growing interest in large-scale studies of the glycome. Several methods for high-throughput analysis of IgG *N*-glycome have been developed recently, with analysis of released

Gordan Lauc and Manfred Wuhrer (eds.), *High-Throughput Glycomics and Glycoproteomics: Methods and Protocols*, Methods in Molecular Biology, vol. 1503, DOI 10.1007/978-1-4939-6493-2_3, © Springer Science+Business Media New York 2017

glycans by liquid chromatography or capillary gel electrophoresis, and mass spectrometric analysis of glycopeptides currently being the methods of choice [5, 6]. Mass spectrometric analysis of tryptic IgG glycopeptides allows discrimination of different human IgG subclasses based on peptide differences, and additionally provides Fc specific glycoprofiling [7]. In contrast, analysis of released IgG glycans provides no information on the glycosylation site and thus includes glycans from both Fc and Fab regions of all the IgG subclasses [5].

To enable high-throughput analysis of IgG *N*-glycome, reproducible and robust sample preparation and accurate and reliable quantitative analysis is of utmost importance. Satisfying needs for generation of high-quality glycomics data while dealing with time constraints can sometimes be quite challenging. It is, therefore, crucial to thoroughly optimize and validate the whole protocol. This chapter presents a protocol for an optimized high-throughput sample preparation [8] and analysis of fluorescently labeled IgG *N*-glycans by HILIC on an UPLC-FLR system [9]. Following denaturation of IgG purified from human plasma or serum, IgG *N*-glycans are released by overnight incubation with Peptide-*N*-Glycosidase F (PNGase F). The released *N*-glycans are subsequently labeled with 2-aminobenzamide (2-AB) and purified from excess of label and reducing agent by hydrophilic interaction liquid chromatography solid phase extraction (HILIC-SPE). To achieve the best high-throughput of sample preparation, 96-well plates are used throughout. Separation of 2-AB labeled IgG *N*-glycans into 24 chromatographic peaks is achieved by HILIC within 25 min analytical run. This separation method enables identification of 33 IgG *N*-glycans with most of them being separated as a single chromatographic peak [9].

2 Materials

Personal protective equipment (laboratory coat and non-powdered gloves) should be used throughout. Prepare all solutions using ultrapure water (prepared by purifying deionized water to attain a sensitivity of 18 MΩ·cm at 25 °C) and high-purity chemicals. Prepare and store all reagents at room temperature (unless indicated otherwise). Diligently follow all waste disposal regulations when disposing of waste materials.

2.1 Denaturation

1. 96-well collection plate (1 or 2 mL well volume).
2. 1.33 % or 2 % (w/v) Sodium dodecyl sulphate (SDS; *see* **Note 1**): to prepare 100 mL of 1.33 % or 2 % (w/v) SDS add about 80 mL of water to a 100 mL glass beaker. Weigh 1.33 or 2 g of SDS and transfer to the beaker. Add a magnetic stirring bar

to the beaker, cover the beaker with Parafilm and put on a magnetic stirrer. When completely dissolved (*see* **Note 2**), transfer into a 100 mL volumetric flask and make up to 100 mL with water. Transfer the solution into a 100 mL glass bottle. Store at 37 °C.

3. Adhesive seal for the 96-well collection plate.
4. Oven at 65 °C.
5. 4 % Igepal CA-630: to prepare 100 mL of 4 % Igepal CA-630 add about 80 mL of water into a 100 mL glass volumetric flask. Pipette 4 mL of Igepal CA-630 using wide-orifice pipette tip (*see* **Note 3**) and add to the flask. Cover the flask with Parafilm and put in an oven at approx. 60 °C to help dissolving. When completely dissolved, make up to 100 mL with water. Transfer the solution into a 100 mL glass bottle. Store at 4 °C.
6. Plate shaker.

2.2 Deglycosylation

1. 5× phosphate buffered saline (PBS), pH 6.6–6.8: we prepare 5× PBS by diluting 10× PBS with water. For 1 L of 10× PBS add about 800 mL of water to a 1 L glass beaker. Weigh 80.028 g of NaCl, 13.832 g of Na_2HPO_4, 2.964 g of KH_2PO_4 and 1.976 g of KCl and transfer to the beaker. Add a magnetic stirring bar to the beaker, cover the beaker with Parafilm, and put on a magnetic stirrer. When completely dissolved (*see* **Note 2**), transfer into a 1 L volumetric flask and make up to 1 L with water. Filter the buffer through a bottle top 0.2 μm Supor filter into a clean 1 L brown glass bottle (*see* **Note 4**). Check the pH of the buffer (pH 6.6–6.8). To prepare 100 mL of 5× PBS add 50 mL of 10× PBS into a 100 mL volumetric flask and make up to 100 mL with water. Transfer the buffer into a 100 mL brown glass bottle (*see* **Note 4**) and store at 4 °C.
2. PNGase F (Promega, 10 U/μL). Store at 4 °C.
3. Adhesive seal for the 96-well collection plate.
4. Incubator at 37 °C.

2.3 Fluorescent 2-AB Labeling and HILIC-SPE Cleanup

1. 2-AB labeling solution (*see* **Note 5**): 50 μL of 30 % acetic acid (HAc) in dimethyl sulfoxide (DMSO), 0.96 mg 2-aminobenzamide (2-AB) and 2.24 mg 2-picoline borane complex (PB) per sample. Upscale as required (prepare a master mix with 10 % extra so as not to run out). First prepare 30 % HAc in DMSO solution. Then weigh out the required mass of 2-AB and PB in separate tubes (*see* **Note 6**). Next, add the HAc/DMSO mixture to the 2-AB and mix well by vortexing until completely dissolved. Add the entire volume to the PB and mix well by vortexing until completely dissolved. Prepare fresh each time.

2. Plate shaker.
3. 70 % ethanol in water: prepare fresh by diluting ethanol, absolute, ≥99.8 %.
4. Acetonitrile: HPLC grade or higher. Store at 4 °C (*see* **Note 7**).
5. 96 % acetonitrile: prepare fresh by diluting acetonitrile (HPLC grade or higher) with water. Store at 4 °C (*see* **Note 7**).
6. 1 mL AcroPrep™ 0.2 μm GHP membrane 96-well filter plate (Pall Corporation).
7. PCR plate (200 μL well volume) with plastic caps.
8. Multi-well plate vacuum manifold and vacuum pump.

2.4 UPLC-FLR Analysis

1. Solvent A: 100 mM ammonium formate, pH 4.4: prepare by diluting 2 M ammonium formate, pH 4.4, stock solution with water (*see* **Note 8**). To prepare 500 mL of the stock solution, first add 300 mL of ice cold water to a 500 mL glass beaker (*see* **Note 9**). Then add 37.75 mL of formic acid and a magnetic stirring bar to the beaker and stir on a magnetic stirrer. Insert an electrode of the pH meter into the beaker and check the pH which should be <2. Adjust the pH to 4.2 by adding cca. 75 mL of 25 % ammonia solution. This causes a rapid rise in temperature, so the ammonia solution must be added slowly in small amounts (*see* **Note 9**). Leave at room temperature overnight and finally adjust the pH to 4.4 with 25 % ammonia solution. Transfer the solution to a 500 mL volumetric flask and make up to 500 mL with water. Filter the solution through a 0.2 μm PES membrane into a brown glass bottle (*see* **Note 4**) and store at room temperature for 1 month. To prepare 100 mM ammonium formate, pH 4.4, transfer 50 mL of 2 M ammonium formate into a 1 L volumetric flask and make up to 1 L with water. Transfer the prepared solution into a brown glass bottle (*see* **Note 4**). Prepare fresh once a week.
2. Solvent B: LC-MS grade acetonitrile.
3. 2-AB labeled dextran ladder standard.
4. Acquity UPLC Glycan BEH Amide column, 130 Å, 1.7 μm, 2.1 mm × 100 mm (Waters).
5. Waters Acquity UPLC H-class system consisting of a quaternary solvent manager, sample manager and a fluorescence (FLR) detector set with excitation and emission wavelengths of 250 and 428 nm, respectively.

3 Methods

Carry out all procedures at room temperature unless otherwise specified.

3.1 Denaturation

1. Place 1.33 % or 2 % SDS, 4 % Igepal CA-630, and 5× PBS at laboratory countertop and allow to reach room temperature.
2. Take 96-well collection plate with your IgG samples. Recommended amount of each IgG sample for this protocol is 50–200 μg. The protocol works with both dissolved and lyophilized IgG. If using dissolved IgG, the total volume of the IgG sample should be 10 μL.
3. Denature the samples by adding 30 μL of 1.33 % SDS (for lyophilized samples) or 20 μL of 2 % SDS (for dissolved samples) to each sample and mix by pipetting. Make sure to completely dissolve the lyophilized samples.
4. Seal the plate with adhesive seal and incubate for 10 min at 65 °C.
5. Take the plate out of the oven and allow to cool to room temperature for 30 min.
6. Remove the adhesive seal and add 10 μL of 4 % Igepal CA-630 to each sample. Mix by pipetting.
7. Place the plate on the plate shaker to gently shake for 15 min.

3.2 Deglycosylation

1. Prepare the PNGase F solution by mixing 0.12 μL of PNGase F and 10 μL of 5× PBS per sample. Upscale as required (prepare a master mix with 5 % extra so as not to run out).
2. Add 10 μL of PNGase F solution to each samples and mix by pipetting.
3. Seal the plate with adhesive seal and incubate overnight at 37 °C.

3.3 Fluorescent 2-AB Labeling and HILIC-SPE Cleanup

1. Prepare the 2-AB labeling solution.
2. Take the plate out of the incubator, remove the adhesive seal and add 25 μL of 2-AB labeling solution to each sample (*see* **Note 5**). Mix well by pipetting.
3. Seal the plate with an adhesive seal and place on the plate shaker to gently shake for 10 min.
4. Incubate the plate for 2 h at 65 °C.
5. Take the plate out of the oven and allow to cool to room temperature for 30 min.
6. To remove free label and reducing agent, cleanup of 2-AB labeled *N*-glycans is performed by HILIC-SPE using 1 mL AcroPrep™ 0.2 μm GHP membrane 96-well filter plate and Multi-well plate vacuum manifold with vacuum pump (*see* **Note 10**).
7. Precondition the filter plate by adding 200 μL of 70 % ethanol to each well. Vacuum to waste.

8. Add 200 μL of water to each well of the filter plate and vacuum to waste.
9. Add 200 μL of 96 % acetonitrile to each well of the filter plate and vacuum to waste.
10. Prepare the 2-AB labeled *N*-glycan samples for cleanup by adding 700 μL of acetonitrile to each sample. Mix well by pipetting and transfer the entire sample volume to the filter plate.
11. Incubate for 2 min and then vacuum to waste.
12. Add 200 μL of 96 % acetonitrile to each well of the filter plate and vacuum to waste.
13. Repeat the previous **step 3** times.
14. Place the filter plate on top of a plate stand and add 200 μL of 96 % acetonitrile to each well of the filter plate.
15. Centrifuge the filter plate on top of a plate stand for 5 min at 1000 rpm. Make sure the centrifuge is balanced.
16. Place the filter plate on top of the PCR plate.
17. Add 90 μL of water to each well of the filter plate, place the plates on the plate shaker to gently shake for 15 min.
18. Centrifuge the plates for 5 min at 180 × *g* to collect the first fraction of the eluate into the PCR plate.
19. Repeat **steps 17** and **18** to collect the second fraction of the eluate.
20. Seal the plate with plastic caps and store at −20 °C.

3.4 UPLC-FLR Analysis

1. These instructions assume the use of a Waters Acquity UPLC H-class system consisting of a quaternary solvent manager, sample manager and a FLR detector set with excitation and emission wavelengths of 250 and 428 nm, respectively.
2. Prepare solvent A by diluting 2 M ammonium formate, pH 4.4, stock solution with water. Solvent B is LC-MS grade acetonitrile. Prime the solvents for 5 min if you are changing solvents with freshly prepared ones or after the system has been idle overnight.
3. Set the temperature of sample manager at 10 °C and the column temperature at 60 °C.
4. Turn on the lamp of the FLR detector (*see* **Note 11**).
5. Samples and dextran ladder standard are injected in 80 % LC-MS grade acetonitrile. Take an aliquot of the sample or dextran ladder standard and make up to 10 μL with water (if necessary), then add 40 μL of acetonitrile. We recommend to use only a small percentage of your samples in the first run in order to test how much must be loaded to get a good chromatogram.

Usually, to produce a good chromatogram it is enough to have 2-AB labeled *N*-glycans released from about 5 μg of IgG.

6. Equilibrate the system to starting conditions (0.4 mL/min, 25 % solvent A, 75 % solvent B) by increasing the flow rate by 0.1 mL from 0.1 mL/min to 0.4 mL/min. Make sure the pressure ripple delta value is ≤10 psi before each increase.
7. After the system has been equilibrated to starting conditions and FLR detector has been stabilized, run blank, dextran ladder standard and samples using the running method (*see* Table 1). Inject 10 μL of blank (water), 10 μL of dextran ladder standard and 20 μL of samples. Blank is used to make sure the column is well conditioned and the FLR detector stabilized and working optimally. Dextran ladder standard is used to make sure the system is working properly (absence of any peak distortion like tailing, fronting, split peaks, etc.) and to calibrate the UPLC runs against any day-to-day or system-to-system changes. Run dextran ladder standard after each 12–16 samples. We do not recommend to run more than 32 samples in a single sample set.
8. Calibration and allocation of glycan units (GU) [10]: the GU values are calculated by fitting a fifth order polynomial distribution curve to the dextran ladder (usually GU 3–11 with the running method in Table 1). This curve is then used to convert retention time of individual chromatography peaks to GU values (Waters Empower 3 software can be used to calculate the GU values). This allows comparison with reference values in the GlycoBase database (available at https://glycobase.nibrt.ie) for preliminary structure assignment [11].
9. Quantitation: integrate obtained chromatograms using an automatic processing method with a traditional integration

Table 1
IgG glycoprofiling by UPLC-FLR: running method

Time (min)	Flow (mL/min)	%A	%B
0	0.4	25	75
25	0.4	38	62
26	0.4	100	0
28	0.4	100	0
29	0.4	25	75
34	0.4	25	75
35	0.4	0	100
45	0	0	100

Run time 34 min. A: 100 mM ammonium formate, pH 4.4; B: LC-MS grade acetonitrile

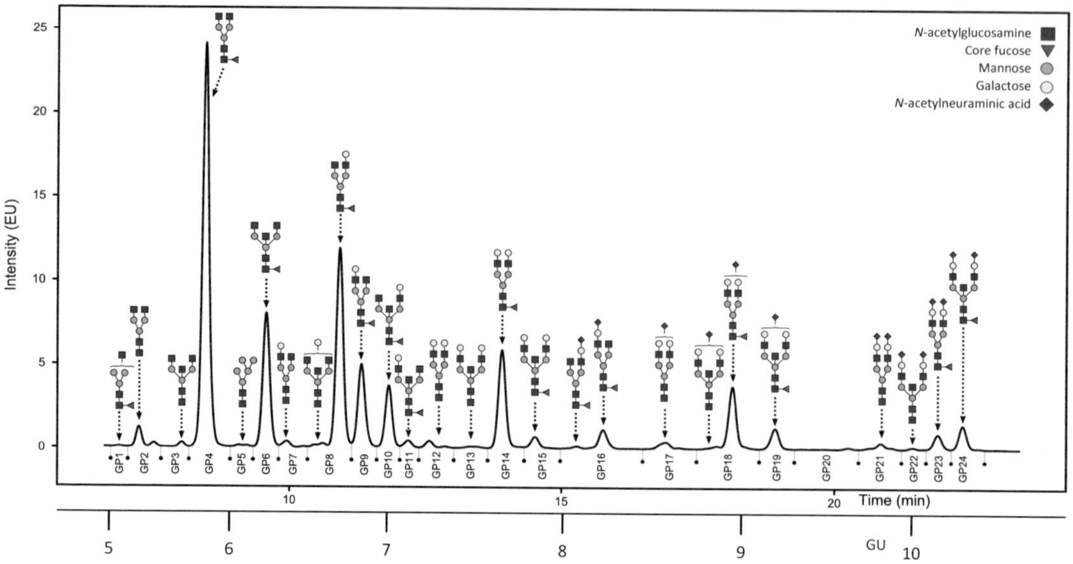

Fig. 1. A representative HILIC-UPLC chromatogram of 2-AB labeled IgG *N*-glycans separated into 24 chromatography peaks. Only the most abundant *N*-glycans in each chromatography peak are shown

algorithm (Waters Empower 3 software) and then manually check and correct (if necessary) each chromatogram to maintain the same intervals of integration for all the samples. We integrate 24 chromatography peaks for IgG *N*-glycan profile (Fig. 1). Within those peaks we are able to identify 33 IgG *N*-glycan structures: 30 complex, 1 oligomannose and 2 hybrid *N*-glycans [9]. Relative abundance of *N*-glycans in each peak is expressed as percentage of total integrated area.

4 Notes

1. Use 1.33 % SDS with lyophilized IgG samples and 2 % SDS with dissolved IgG samples. If using dissolved IgG, the total sample volume should be 10 μL.
2. Raise the temperature on the magnetic stirrer to approx. 60 °C to help dissolving.
3. Igepal CA-630 has very high viscosity and we recommend use of wide-orifice pipette tips. If these are not available, cut off the end of the regular pipette tip to make the orifice wider.
4. Use brown bottles to keep microorganisms from growing.
5. To avoid exposing hazardous chemicals to coworkers, prepare the labeling solution and label the samples in the fume hood.
6. PB should be stored at 4 °C. Before weighing, allow the PB to reach room temperature.

7. We find that the use of acetonitrile cooled to 4 °C has a positive effect on cleanup reproducibility and peak intensities. To save time, place the solutions at 4 °C the day before and allow them to cool overnight.
8. To avoid exposing hazardous chemicals to coworkers, prepare the 2 M ammonium formate, pH 4.4, stock solution in the fume hood.
9. Adding 25 % ammonia solution to diluted formic acid causes rapid rise in the temperature. To slow down the temperature rise use ice cold water and slowly add small amounts of 25 % ammonia solution. Do not let the temperature go above cca. 28 °C because ammonia will start to evaporate. Care should be taken to bring the solution to room temperature before final adjustment of pH.
10. Make sure the vacuum does not exceed 2 inHg and do not let the membrane become overly dry. We recommend to turn off the suction between vacuuming steps (i.e., while pipetting the next solution).
11. For optimal results, wait for 30–60 min for the FLR detector to stabilize before acquiring data.

References

1. Huhn C, Selman MHJ, Ruhaak LR, Deelder AM, Wuhrer M (2009) IgG glycosylation analysis. Proteomics 9:882–913
2. Jefferis R (2005) Glycosylation of recombinant antibody therapeutics. Biotechnol Prog 21: 11–16
3. Mimura Y, Church S, Ghirlando R, Ashton PR, Dong S, Goodall M, Lund J, Jefferis R (2001) The influence of glycosylation on the thermal stability and effector function expression of human IgG1-Fc: properties of a series of truncated glycoforms. Mol Immunol 37: 697–706
4. Arnold JN, Wormald MR, Sim RB, Rudd PM, Dwek RA (2007) The impact of glycosylation on the biological function and structure of human immunoglobulins. Annu Rev Immunol 25:21–50
5. Huffman JE, Pučić-Baković M, Klarić L, Hennig R, Selman MH, Vučković F, Novokmet M, Krištić J, Borowiak M, Muth T, Polašek O, Razdorov G, Gornik O, Plomp R, Theodoratou E, Wright AF, Rudan I, Hayward C, Campbell H, Deelder AM, Reichl U, Aulchenko YS, Rapp E, Wuhrer M, Lauc G (2014) Comparative performance of four methods for high-throughput glycosylation analysis of immunoglobulin G in genetic and epidemiological research. Mol Cell Proteomics 13:1598–1610
6. Zauner G, Selman MHJ, Bondt A, Rombouts Y, Blank D, Deelder AM, Wuhrer M (2013) Glycoproteomic analysis of antibodies. Mol Cell Proteomics 12:856–865
7. Wuhrer M, Stam JC, van de Geijn FE, Koeleman CAM, Verrips CT, Radboud RJEM, Hokke CH, Deelder AM (2007) Glycosylation profiling of immunoglobulin G (IgG) subclasses from human serum. Proteomics 7:4070–4081
8. Trbojević Akmačić I, Ugrina I, Štambuk J, Gudelj I, Vučković F, Lauc G, Pučić-Baković M (2015) High-throughput glycomics: optimization of sample preparation. Biochemistry (Mosc) 80:934–942
9. Pučić M, Knežević A, Vidič J, Adamczyk B, Novokmet M, Polašek O, Gornik O, Šupraha-Goreta S, Wormald MR, Redžić I, Campbell H, Wright A, Hastie ND, Wilson JF, Rudan I, Wuhrer M, Rudd PM, Josić D, Lauc G (2011) High throughput isolation and glycosylation analysis of IgG-variability and heritability of the IgG glycome in three isolated human populations. Mol Cell Proteomics 10:M111.010090
10. Royle L, Radcliffe CM, Dwek RA, Rudd PM (2006) Detailed structural analysis of N-glycans released from glycoproteins in SDS-PAGE gel bands using HPLC combined with exoglycosidase array digestions. Methods Mol Biol 347: 125–143
11. Campbell MP, Royle L, Radcliffe CM, Dwek RA, Rudd PM (2008) GlycoBase and autoGU: tools for HPLC-based glycan analysis. Bioinformatics 24:1214–1216

Chapter 4

High-Throughput Analysis of IgG Fc Glycopeptides by LC-MS

David Falck, Bas C. Jansen, Noortje de Haan, and Manfred Wuhrer

Abstract

This chapter contains a nanoscale liquid chromatography–mass spectrometry method for the glycoform profiling of the conserved Fc *N*-glycosylation site of monoclonal and polyclonal immunoglobulin G (IgG). It describes in detail LaCyTools, a program for automated data (pre-)processing of the obtained LC-MS data. The minimal sample preparation necessary is explained as well as an optional method for affinity purification of (polyclonal) antibodies from serum or plasma.

After (optional) affinity purification, the pure IgG is cleaved with trypsin. The tryptic glycopeptides are separated almost exclusively on their peptide backbone. This ensures similar response factors for all glycoforms in the MS detection and allows the collection of separate glycoform profiles for different IgG isoforms or allotypes. LaCyTools automatically performs label-free (relative) quantitation of the obtained data after minimal manual input and additionally calculates several quality criteria which can be used for data curation at the level of both individual analytes and entire LC-MS runs.

Key words Liquid chromatography–mass spectrometry (LC-MS), Glycopeptide analysis, Immunoglobulin G, High-throughput glycosylation profiling, Glycoform profiling, Bottom-up proteomics, Glycoproteomics, Automated data analysis, Label-free quantification, Quality scores

1 Introduction

Bottom-up proteomics is among the most well established and most frequently used approaches for the analysis of protein glycosylation [1, 2]. It relies on a combination of proteolytic cleavage, separation, and mass spectrometry. Proteases cleave the glycoprotein into (glyco-)peptides which are more accessible to the subsequent analytical techniques. Separation by liquid chromatography (LC) or, less frequently, capillary electrophoresis reduces the sample complexity by distributing analytes and matrix over the time scale. Finally, mass spectrometry (MS) provides the resolution necessary to select the analytes of interest out of the remaining complexity and high resolution (HR) mass analysers, e.g., time-of-flight (TOF) or orbitrap, have become the state of the art. The largest

Gordan Lauc and Manfred Wuhrer (eds.), *High-Throughput Glycomics and Glycoproteomics: Methods and Protocols*, Methods in Molecular Biology, vol. 1503, DOI 10.1007/978-1-4939-6493-2_4, © Springer Science+Business Media New York 2017

advantage of bottom-up proteomics over released glycan- or intact protein methods is that it readily provides site specific glycosylation profiles [1]. For immunoglobulin G (IgG), this means distinguishing between the conserved Fc *N*-glycosylation site and possible Fab glycosylation which is of utmost importance due to the very different functional implications of Fab versus Fc glycosylation changes [1, 3, 4]. While IgG Fab glycosylation is mainly affecting antigen binding, Fc *N*-glycosylation regulates the effector functions of IgG and its plasma half-life through modulation of receptor interaction [4, 5]. Prominent examples of this modulation include interaction with the Fcγ receptors [6] and the Fcn receptor [7]. Middle-down approaches, such as electrospray ionization-quadrupole TOF-mass spectrometry (ESI-qTOF-MS) after reduction and IdeS cleavage into Fab and Fc portions, can also make that distinction, but are generally less sensitive and may miss important information on low abundant glycoforms, most prominently afucosylated glycoforms [8].

Classical bottom-up proteomics approaches often incorporate extensive (glyco-)peptide separation usually in the form of reversed-phase LC. Gradients of up to 3 h are not uncommon. While a wealth of information on proteoforms can be extracted from such complex datasets, they do not provide the throughput required for the study of large clinical cohorts which may include thousands of samples. Affinity purification provides a convenient means of reducing the sample complexity. IgG can be purified using protein G, protein A or anti-Fc antibodies, but also using immobilized antigens to obtain a relevant subpopulation. However, even with only one (type of) protein in the sample, complexity is still too high to achieve a comprehensive proteoform analysis in high throughput. Therefore, the method presented in this chapter focuses on the glycosylation of IgG. Glycan profiles of IgG isoforms can be distinguished on account of their different amino acid sequences in the peptide moiety of the glycopeptides and interferences from Fab glycosylation are excluded. Targeted are specifically polyclonal human or recombinant, monoclonal human IgG, but adaptation to other mammalian IgG can be reasonably straightforward. The method can address both monoclonal as well as polyclonal antibodies as the tryptic glycopeptides of the different IgG isoforms are separated. The IgG3 glycopeptides present an exception as they coelute either with IgG2 or IgG4, because the two dominant allotypes of IgG3 have the same amino acid composition as either IgG2 or IgG4 [9].

Data extraction and analysis can be a tedious and time consuming step in any method, especially in the field of glycoproteomics where computational automation is in its infancy. However, untargeted automated data extraction can be performed, for example using Bruker DataAnalysis which is capable of performing automated peak detection and integration [10]. Using untargeted extraction for IgG glycosylation analysis can result in some glycoforms not being detected in some of the measurements, which may compromise comparability. Consequently, a targeted package that integrates robustly a set of user-specified features would be advantageous. For proteomics such packages exist, for example the Skyline package that is designed for label-free quantitation of selected reaction monitoring proteomic experiments [11]. However, the field of glycoproteomics lacks such targeted software packages. Therefore, this method uses a newly developed freely available software package, LaCyTools, for high-throughput label-free quantitation. This package is capable of calculating various quality criteria values, that allow for robust spectral and analyte curation. All quality criteria available in LaCyTools, except for the alignment residuals, are based on calculations published previously [12]. In conclusion, the entire method enables the high-throughput sample preparation, measurement and data integration of monoclonal and polyclonal IgG Fc *N*-glycosylation.

2 Materials

Water should be deionized and cleared of organic contaminants by an appropriate device. Reagents should be of at least analytical grade. Solvents are preferably of LC-MS grade, but certain HPLC grades might also be sufficient.

2.1 Affinity Purification

1. 2 mL Protein G (ProtG) Sepharose beads (GE Healthcare, Uppsala, Sweden) per 96-well plate (*see* **Note 1**).
2. Phosphate-buffered saline (PBS): Dissolve 5.7 g/L $Na_2HPO_4 \cdot 2H_2O$, 0.5 g/L KH_2PO_4, and 8.5 g/L NaCl in water, store at 4 °C.
3. 100 mM formic acid (FA): Add 189 μL concentrated formic acid to 49.8 mL water, store at 4 °C.

Additional equipment: 15 mL tube, 96-well filter plate (0.7 mL/well, PE frit, Orochem, Naperville, IL),centrifuge, vortex, Multiscreen vacuum manifold (Millipore, Amsterdam, The Netherlands), adhesive tape, plate shaker (Heidolph Titramax 100; Heidolph, Kelheim, Germany), two polypropylene 96-well V-bottom plates, vacuum concentrator (RVC2–25COplus, Martin Christ Gefriertrocknungsanlagen, Osterode am Harz, Germany).

2.2 Protease Treatment

1. 25 mM ammonium bicarbonate buffer pH 7.8 (ABC): Dissolve 99 mg of ammonium bicarbonate in 50.0 mL water. Prepare freshly or at least readjust pH. Test and, if necessary, adjust

the pH with formic acid and ammonia or sodium hydroxide. Store in the fridge, but do not use for more than a week.

2. 100 mM FA: *see* Subheading 2.1.
3. 20 mM acetic acid: Add 11.4 μL concentrated acetic acid to 10.0 mL water.
4. Modified and tosyl-phenylalanyl-chloromethyl-ketone (TPCK) treated, porcine trypsin (trypsin): Can be stored in 10 μL aliquots at −80 °C for 6 month, if dissolved at 4 °C in a 20 mM acetic acid solution at 10 μg/μL. Directly before use dilute to 25 ng/μL in ice-cold ABC and keep on ice until use.

2.3 LC-ESI-MS Setup

1. The liquid chromatography (LC) setup consists of a gradient pump, an isocratic (loading) pump, a programmable autosampler, a column oven with at least one switch valve. The gradient pump should be able to handle 400 bar backpressure, while a 100 bar pressure limit is sufficient for the loading pump. The setup is fitted with a Dionex Acclaim PepMap100 C18 5 mm × 300 μm (trap column, Thermo Fisher Scientific, Breda, The Netherlands), and an Ascentis Express C18 nanoLC column 50 mm × 75 μm with 2.7 μm fused core particles (analytical column, Supelco, Bellefonte, PA).
2. As an electrospray ionization interface, we had excellent results with the CaptiveSpray™ and nanoBooster™ technologies (Bruker, Daltonics, Bremen, Germany). Settings: end plate offset 500 V, capillary voltage 1200 V, nanoBooster™ pressure 0.2 bar, dry gas flow 3.0 L/min, dry temperature 180 °C.
3. Mass spectrometer (MS): Maxis Impact HD quadrupole-time-of-flight mass spectrometer (q-TOF; Bruker Daltonics). Profile spectra should be collected in order to make use of all LaCyTools features. General settings: mass range m/z 550–1800; spectral acquisition frequency 1 Hz, focus setting active; Collision cell settings: collision energy 7.0 eV, transfer time 110 μs, pre-pulse storage 21 μs (*see* **Notes 2** and **3**).
4. Solvent A consists of an aqueous 0.1 % trifluoroacetic acid (TFA) solution (v/v).
5. Solvent B is 95 % acetonitrile (ACN) in water (v/v).
6. Pure ACN is used as nanoBooster™ dopant.
7. Trap column wash solution A: 50 % isopropanol–50 % water–0.1 % TFA (v/v/v).

2.4 Hardware and Software for Processing

Data processing requires a computer running either a 32- or 64-bit operating system, containing at least 16GB of memory for larger measurements. Furthermore, python 2.7 is required with the numpy, scipy, matplotlib, and tkinter libraries installed [13–16]. We suggest using a complete python package that contains all the required libraries, such as Anaconda or Python(x,y) [17, 18]. Lastly, the LaCyTools package can be downloaded from the web: http://www.github.com/tarskin/lacytools.

3 Methods

3.1 Affinity Purification

1. Spin down 2 mL of ProtG beads in a 15 mL tube (3000 × *g*, 10 s; *see* **Note 1**).
2. Carefully pipette the supernatant from the beads.
3. Add 2 mL of PBS to the beads.
4. Vortex thoroughly and repeat spinning and removal of supernatant.
5. Resuspend the beads in 4.67 mL PBS.
6. Using a 200 μL multichannel pipette, apply 50 μL of slurry to each well of a 96-well filter plate (*see* **Note 4**).
7. Wash the beads three times with 200 μL PBS using the vacuum manifold (*see* **Note 5**).
8. Add 100 μL PBS and leave it on the beads.
9. Add 2 μL of serum or plasma (*see* **Note 6**).
10. Place the filter plate on a polypropylene 96-well V-bottom plate and seal wells with tape.
11. Incubate the filter plate on a plate shaker for 1 h at 1000 rpm.
12. Wash the samples three times with 200 μL of PBS using the vacuum manifold.
13. Wash the samples three times with 200 μL of MQ using the vacuum manifold.
14. Remove the filter plate from the vacuum manifold, tap the bottom carefully on lint-free paper to remove an excess of liquid, and place the filter plate on a new polypropylene 96-well V-bottom plate (*see* **Note 7**).
15. Add 100 μL 100 mM formic acid to the beads in each well and seal the wells with tape.
16. Incubate on the plate shaker for 5 min at 1000 rpm (*see* **Note 8**).
17. Remove the tape and elute the IgGs from the filter plate by centrifugation for 1 min at 100 × *g*.
18. Dry the samples using a vacuum concentrator at 60 °C to complete dryness, which will take approximately 2–3 h.

19. Samples can be digested directly or sealed with adhesive tape and stored at −20 °C.

3.2 Denaturation of Pure Antibody Samples (Alternative to Subheading 3.1)

1. Add 5 μg of antibody (sample) in an arbitrary volume to a 100 mM FA solution, but maintain a minimum final concentration of 75 mM FA (*see* **Notes 9** and **10**). Vortex.
2. Incubate for 15 min at room temperature (RT).
3. Dry in a centrifugal vacuum concentrator set to 60 °C until complete dryness (*see* **Note 11**).

3.3 Protease Treatment

1. Add 40 μL (1 μg) of the TPCK treated trypsin solution. Incubate for 5 min on a shaker at RT and low rpm (e.g., 400 rpm)
2. Incubate at 37 °C for 18 h (*see* **Note 12**).
3. Samples can be stored short term in a cooled (4 °C) autosampler or long-term at −20 °C

3.4 LC-MS

1. Inject 250 nL of sample into a flow of 25 μL/min of solvent A (*see* **Note 13**). The glycopeptides are trapped on the trap column.
2. Switching to the gradient pump flow of 900 nL/min, the glycopeptides are eluted onto the analytical column where they are separated (*see* **Notes 14** and **15**). Both trap and analytical column are kept at 30 °C. For more details on the LC method *see* Table 1. Changes in solvent B concentrations occur as linear gradients.
3. Nebulization and ionization is achieved in the CaptiveSpray™ ESI source assisted by ACN-doped nebulizing gas from the nanoBooster™.
4. Using highly optimized ion transfer settings, the glycopeptides are declustered and desolvated at high efficiency and with minimal fragmentation (between 0.1 % and 1 % compared to the expected parent peak).
5. Over a period of 5 min, every second, an MS spectrum is collected in full spectrum analysis mode from m/z 550 to m/z 1800. It is important to save the MS data as profile spectra, especially for accurate mass determination and quantification by MS peak area integration (*see* **Note 16**). Glycoforms derived from the same IgG isoform will mostly co-elute. In this way, three distinct "clusters" of glycopeptides will be visible in the separation (*see* Fig. 1).

3.5 Data (Pre-) processing with LaCyTools

1. Convert measured LC-MS runs into mzXML file format (*see* **Note 17**). Non-peak-picked data is essential for meaningful quality criteria values and for the proper functioning of the area integration feature at a later stage in the analysis.

Table 1
LC method

Time [min]	Solvent B [%]	Other actions
0	1	
0.001	3	
1		Valve switch: trap and analytical column in line with gradient pump
2	6	
3		MS data acquisition switched on
4.5	18	
5	30	
6		Valve switch: trap column back to loading pump
6+		Three injections of 5 μL washing solution[a]
7	30	
8	1	MS data acquisition switched off
10.9	1	End

[a]The injection loop is filled consecutively with 2.5 μL solvent B and 2.5 μL trap column wash solution A

2. Identify the retention times of the center of each IgG subclass (*see* **Note 18**).
3. Create a simple text file that will be used for alignment. This file should contain the exact mass and (desired) retention time of each feature that will be used for alignment which are to be put in a row divided by tabs. A minimum of three features with unique retention times is required for proper alignment (*see* **Note 19**).
4. Create an excel file that contains seven side-by-side columns for each analyte. The first two need to be filled with a composition and a retention time, respectively. In the following columns a mass window, time window, minimum charge for extraction or maximum charge for extraction can be specified optionally per analyte, if this is to deviate from the standard settings. In the last column, calibrants have to be marked with an X. For the optional columns, the default value, which is set in LaCyTools, will be used for values that are left blank. Finally, this file has to be saved as a tab-delimited text file.
5. Copy the newly made alignment file and reference file in the same folder that contains the LaCyTools file.
6. Adapt the parameters (*see* **steps 7–9**) that will be used by LaCyTools where necessary, by opening the program in a suit-

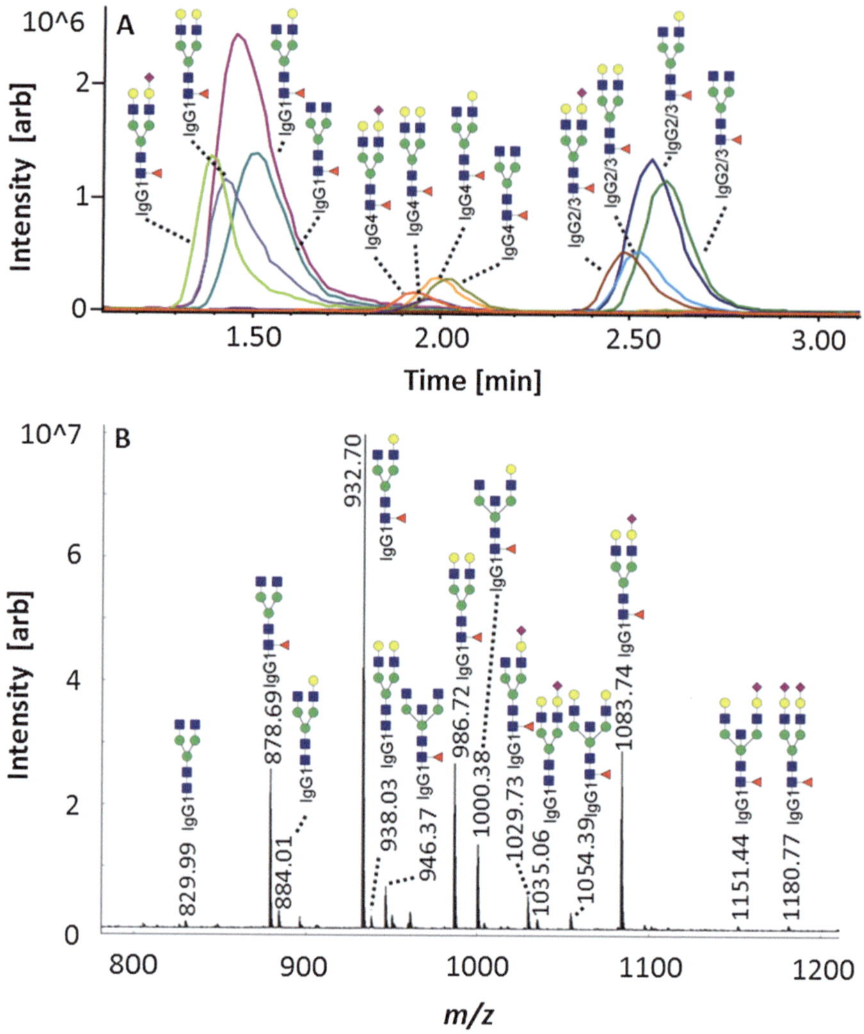

Fig. 1 LC-MS separation of tryptic Fc-glycopeptides of a polyclonal IgG standard. (**a**) Typical extracted ion chromatograms (EIC) of polyclonal human serum IgG. As the glycopeptides are separated mainly based on their peptide sequence, clusters of glycopeptides can be observed per IgG subclass. Of these clusters, summed mass spectra can be obtained, including all glycoforms present on the specific subclass. (**b**) A summed mass spectrum of the IgG1 Fc-glycopeptide cluster of the polyclonal IgG standard is shown. The mass region displayed contains the triply charged analytes. IgG1 = EEQYNSTYR; IgG2/3 = EEQFNSTFR; IgG4 = EEQFNSTYR; *green circle* = mannose; *yellow circle* = galactose; *blue square* = N-acetylglucosamine; *red triangle* = fucose and *purple diamond* = N-acetylneuraminic acid

able editor (*see* **Note 20**). There are parameters for alignment, calibration and extraction. All analytes are assumed to be protonated molecules ($[M + xH]^{x+}$). Alternatively, a different charge carrier can be specified in the program.

7. Alignment parameters:

All parameters listed below apply to the alignment which is performed on the mzXML files. Subsequently, LaCyTools creates an aligned_inputfile.mzXML file that will be used for the later steps of the workflow.

(a) ALIGNMENT_TIME_WINDOW: This parameter specifies the deviation (±) in seconds from the retention time, given in the alignment file, which the program will accept as the peak maximum for a feature. Change this value so that the peak maximum of the respective alignment feature falls within the resulting time window in all measured runs.

(b) ALIGNMENT_MASS_WINDOW: This parameter specifies the deviation (±) in Thomson around the exact mass, given in the alignment file, which the program will use to make an extracted-ion chromatogram. Change this value so that the accurate mass of every alignment feature (before calibration) falls within the resulting mass window in all measured runs.

(c) ALIGNMENT_S_N_CUTOFF: This parameter specifies the minimum signal-to-noise ratio (S/N) that a feature has to have, to be included in the alignment. Change this value to 27 so that only runs that have good signals are aligned (default is 9, the limit of quantitation, *see* **Note 21**).

(d) ALIGNMENT_MIN_PEAK: This parameter specifies the minimum number of features that have to pass the S/N cutoff for the program to execute alignment. A minimum of three alignment features should be used, as a lower number will result in over fitting. The higher the number of well represented alignment features the more confident the alignment (*see* **Note 22**).

8. Calibration parameters:

All parameters listed below apply to the calibration, which is performed on the basis of a sum spectrum. A sum spectrum is created by summing all spectra that fall within the calculated time window (*see* **step 9(a)**). In detail, a sum spectrum is created per unique retention time that is set in the extraction file. We advise to group (partially) co-eluting glycoforms together by giving them the same retention time in the reference file, e.g., by assigning one retention time to all IgG1 glycans. Thereby, these glycoforms will yield a single sum spectrum.

(a) SUM_SPECTRUM_RESOLUTION: This parameter refers to the number of datapoints per 1 Thomson that will be present in the sum spectrum which is created per unique retention time in the reference file. Change this value so that it is somewhat lower than the actual number of datapoints per 1 Thomson in the high mass range of the measured spectra (*see* **Note 23**).

(b) CALIB_MASS_WINDOW: This parameter specifies the deviation (±) in Dalton around the exact mass, calculated from the composition given in the reference file which the program is allowed to use to determine the accurate mass of a calibrant. This value will be charge state corrected in the program, by dividing the specified number by the charge states used for extraction. For example, specifying a window of 0.3 Da for a triply charged analyte results in a mass window of ±0.1 Thomson. Change this value so that the accurate mass of the current analyte falls within the resulting mass window in all sum spectra (*see* **Note 24**).

(c) CALIB_S_N_CUTOFF: This parameter specifies the minimum S/N that a calibrant has to have, to be included in the calibration. Change this value to 27 so that only clear signals are used for calibration (default is 9, the limit of quantitation, *see* **Note 25**).

(d) CALIB_MIN_PEAK: This parameter specifies the minimum number of calibrants that have to pass the S/N cutoff for the program to consider calibration. A minimum of four calibration features is required, as a lower number will result in overfitting. The quality of the calibration will improve by using as many well defined features as possible (*see* **Note 26**).

9. Extraction parameters:

All parameters listed below apply to the extraction, which is performed on the basis of a calibrated sum spectrum (created as described in **step 8**).

(a) TIME_WINDOW: This parameter specifies the deviation (±) in seconds around the retention time of an analyte, given in the reference file, which will be used for the sum spectrum creation. Change this value so that the width at 10 % peak height (FW_{10}) of the chromatographic peak of all analytes falls within the resulting time window in all aligned runs (*see* **Note 27**).

(b) EXTRACTION_TYPE: This parameter specifies the method of extraction that will be performed by LaCyTools. The most robust method is area integration (*see* **Note 28**).

(c) MASS_WINDOW: This parameter specifies the deviation (±) in Thomson around the exact mass, calculated by LaCyTools, which will be used for integration of the area in the sum spectra. Change this value so that the FW_{10} of the MS peak is included in the window (*see* **Note 29**).

(d) MIN_CHARGE: This parameter specifies the lowest charge state that will be extracted for each analyte. Change this parameter so that it matches to your experimental setup (default is 2, *see* **Note 30**).

(e) MAX_CHARGE: This parameter specifies the highest charge state that will be extracted for each analyte. Change this parameter so that it matches to your experimental setup (default is 3, *see* **Note 30**).

(f) MIN_TOTAL: This parameter specifies the minimum percentage of the total theoretical isotopic distribution that LaCyTools will extract. Change this parameter so that it is as high as possible without overlap of analytes (default is 0.99, *see* **Note 31**).

(g) BACKGROUND_WINDOW: This parameter specifies the deviation (±) in Thomson around the exact mass of the monoisotopic peak of an analyte, which the program will use to determine the background and noise. Change this value based on the data density in the spectrum (a value of around 10 is good for relatively clean spectra, while a value of around 20 is suggested for crowded spectra, *see* **Note 32**).

10. Run the LaCyTools program by double clicking the LaCyTools.py file.
11. Click on the batch process menu.
12. Click on the "Alignment File" button and select the newly created file containing the features for alignment.
13. Tick the "Calibration" checkbox.
14. Click on the "Reference File" button and select the newly created reference file containing all analytes that will be used for extraction.
15. Click on the "Batch Directory" button and select the folder that contains the mzXML files that are to be processed.
16. Click on the "Output Format" button and select "Analyte Intensity (Background subtracted)," "alignment residuals," "QC," "PPM error," and "Signal to Noise" as minimum outputs (*see* **Note 33**).
17. Click the "Run Batch Process" button, starting the automated processing (*see* Fig. 2).

4 Notes

1. 4 mL of the bead slurry contains approximately 2 mL beads. To determine the volume of the beads, spin the suspension down (3000 × *g*, 10 s) in a graduated tube.
2. Although the specific settings will only be useful to scientists working with the same or closely related mass spectrometers, it stresses the importance of optimizing declustering and ion transfer. Minimising the dependence of the ion transfer efficiency on m/z is essential for a correct relative quantitation (Transfer time and pre-pulse storage settings are crucial in our example). The fine tuning of parameters related to decluster-

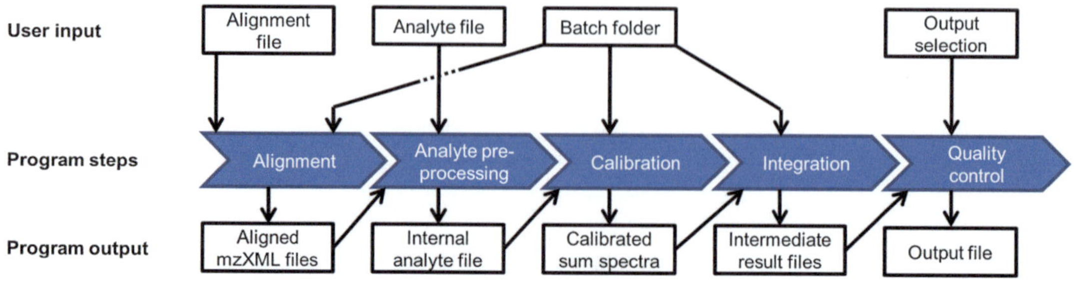

Fig. 2 The data processing workflow illustrates the sequential steps that are performed by the LaCyTools package, as well as the in- and output files per step of the program. Furthermore, the inputs are separated into user inputs and input generated by the program itself (e.g., the calibrated sum spectra are both an output of the calibration step as well as an input of the integration step). Afterwards, the resulting output file can be viewed and/or processed with any statistical package such as Microsoft Excel or IBM SPSS [27, 28]

ing and ion energy ensures an optimal compromise between high ion yield and low unwanted fragmentation.

3. *The m/z* range given is generally enough to detect the $[M + 3H]^{3+}$ and $[M + 2H]^{2+}$ ions of naturally occurring glycoforms. However, in some cases, for example in mAbs, larger glycan structures can occur. Though detection will likely succeed in the $[M + 3H]^{3+}$ ion the omission of the $[M + 2H]^{2+}$ ion will significantly impact quantitation. Larger structures may also attract more charges and the signals of their higher charged ions ($[M + 4H]^{4+}$) are more likely to fall into the optimal detection range. Dependent on the detected glycoforms, adjustments of the detection range may be needed to register all major charge states. However, the experimenter should be aware of the limitation of the optimal range for ion transition in their specific setup and should include this parameter in the decision about which *m/z* range and charge states to include.
4. To prevent the beads from clogging the pipette tips, cut off the ends of the tips. To keep the beads equally distributed in the suspension, pipette the slurry up and down three times before transferring it to the wells.
5. The vacuum should never exceed –4 inHg (–14 kPa), as this will cause damage to the beads. In addition, a slow flow will maximize the interaction between the beads and the liquid during washing.
6. 2 μL human plasma or serum contains approximately 20 μg IgG.
7. Tapping the plate on a lint-free paper to remove the lagging flow trough of the washing steps is a critical step. Without this step the elution will be contaminated.

8. This step can be extended with 15 min. This can be useful, for example, when two plates are prepared with a 15 min difference.
9. The steps in this section are meant to be applied, if the affinity purification in Subheading 3.1 is not required. This is the case for reasonably pure material, for example a therapeutic formulation of a mAb. If the affinity purification has been performed, Subheading 3.2 can be skipped.
10. The dilution factor with FA should be adjusted to the matrix intensity of the sample. For especially matrix intensive samples, it is recommended to make sure that the pH is around 3.
11. Drying time can depend strongly on the sample volumes and matrix intensity of the sample.
12. Shorter incubation times for tryptic digestion may be feasible, but biases of trypsin for certain IgG glycoforms can interfere with the profiling, if the digestion is incomplete [19]. Optimization of tryptic digestion protocols for IgG have been discussed extensively in the literature [20, 21]. Next to biases in the glycosylation profile, intact IgG can also damage the LC-MS system by blocking filters, tubings, columns, and even the ESI needle.
13. The limit of detection for abundant glycoforms (>10 %) expressed as concentration sensitivity with respect to the whole antibody is around 0.25 μg/mL. This high sensitivity of the system means that it is suitable to operate also with lower amounts of IgG such as they occur, for example, in the study of affinity-purified autoantibodies [22]. In these cases, we have successfully used injection volumes up to 5 μL (full loop injection) with only minor losses in performance.
14. Clusters of glycopeptides elute according to their peptide backbone, with only minimal separation of the glycoforms, in the order IgG1, IgG4, and IgG2/3. A2G1F peaks are separated by ca. 30 s and show a full width at half maximum of just under 10 s.
15. Special attention has to be paid to the integrity of the nanoLC flow system. Minor leakages and void volumes (especially post-column) can have great impact on the performance. Other frequently observed reasons for performance decreases are contamination of the MS transfer capillary (in bad cases including the ion optics) and deterioration of the trap column and analytical column. Cleaning of the MS transfer capillary is performed every one or two months, trap columns are exchanged roughly every two to three months, and the analytical column has been observed to be usable for up to ca. 8000 injections. However, frequency of use, but especially sample quality can have a large influence on the maintenance scheme.

16. In principle, the IgG tryptic glycopeptides can be detected with any MS. However, we recommend use of a high resolution (HR) instrument, such as the increasingly popular quadropole–time-of-flight hybrid (qTOF) MS or orbitrap-based instruments. Next to the improved speed and spectral quality in full spectrum acquisition mode, HR-MS allows to make use of all LacyTools features, including accurate mass determination for analyte quality assessment.
17. There are several methods to convert LC-MS data into other file formats. We suggest using the freely available Proteowizard package, containing the MSConvertGUI [23].
18. There are many packages available capable of processing open file formats that can be used to create an extracted-ion chromatogram. We suggest using the freely available MZMine2 or mMass [24, 25].
19. LaCyTools will align any LC-MS run that fulfils the user specified criteria for alignment. In case of a retention order shift, the LC-MS run might still be aligned but the alignment will be of poor quality. The final result file of the program includes an output "Alignment Residuals" that lists the root-mean-square of the residual retention time difference for all alignment features. This enables the fast exclusion of LC-MS runs showing low separation quality.
20. There are many good editors available for source code, some are even capable of running many different languages. We recommend the freely available Geany [26].
21. A value of 9 for the minimum S/N ratio for a biopharmaceutical feature is rather low. We advise to increase this value to at least 27, being three times the limit of quantitation, to get reproducible results with low variance.
22. In general, alignment will be better when using more features. However, we advise to add only features that have a high S/N ratio relative to your overall data quality. Performing the alignment exclusively with features that are well above the minimum S/N ratio will yield better alignment than the use of more features of a lower quality.
23. Setting this parameter higher than the number of data points per Thomson present in the raw data will result in a sum spectrum where there are "imaginary" 0 intensity data points present between "real" data points. Furthermore, a too low number will result in a poor accuracy of the mass error calculation and loss of resolution during the extraction. Lastly, a value that is close to the actual number of data points per Thomson in the input spectra can result in peak splitting due to an accurate mass shift during the measurement.

24. The calibration quality depends greatly on the determination of the accurate mass. Detecting the accurate mass is only possible if at least the FWHM of each calibrant feature for all LC-MS runs falls within the resulting mass window.
25. In general, this calibration parameter behaves similar as its alignment counterpart (*see* **Note 22**). Therefore, we advise to use only features that have a high S/N ratio.
26. The calibration quality is directly dependant on the mass range of specified calibrants. In the case of natural human glycosylation, we advise to use A2G0, FA2G0, FA2G1, FA2G2, FA2BG1, and FA2G2Sas calibrants, if the data quality permits, to ensure a good coverage of the mass range. Analyte charge states with an exact mass smaller than the lowest and larger than the highest calibrant m/z will have a larger mass error. Remember that the program uses all charge states specified in the reference file for calibration.
27. Please choose settings which ensure that the FW_{10} of the chromatographic peak falls within the time window in all runs. Otherwise, if early parts of early eluting analyte peaks and/or late parts of late eluting ones are not added to the sum spectrum, this can lead to a bias towards glycans that elute in the center of the extraction window.
28. Area integration by LaCyTools is performed by first multiplying the intensity of each data point with the data point width (in Thomson) before summing the multiplied intensity values (default option; option 2). Furthermore, LaCyTools is also capable of reporting the maximum intensity (option 0) or the total intensity (option 1) observed for a given feature. Total intensity is acquired by summing the intensity of all data points within the specified mass window (*see* **Note 29**). Maximum intensity is the intensity of the data point which had the highest intensity. Both of these methods are highly sensitive to spikes, potentially skewing the data, while area integration remains robust in spite of spikes.
29. The correct value of this parameter depends on which extraction method is used by LaCyTools (*see* **Note 28**). The default method, area integration, is ill suited for features that are not baseline separated in the sum spectrum, and such features will be overestimated in the final results. Therefore, in such a case, we suggest to use the maximum intensity. However, in that case this value should be changed so that the peak maximum falls within the mass window in all sum spectra.
30. Any change in the experimental setup that influences the ionization of analytes should be considered for this parameter. For example, the usage of a NanoBooster™ will in general increase the charge state for all analytes compared to conventional ESI.

31. The theoretical isotopic distribution is calculated based on the carbon, hydrogen, nitrogen, oxygen and sulphur content of each analyte. We suggest lowering the value to 0.95 in the scenario where the latter isotopes, of the majority of the analytes, are not visible above the background and noise. Note that any deviation from 0.99 may incur an error in the relative quantitation of two analytes, as different relative amounts of the isotope distribution will be integrated.
32. A too large value for this parameter will lead to underestimating the background value. In contrast, a too small value will lead to the program identifying a feature or interference as the background value, leading to overestimation of the background and noise values.
33. The Analyte Intensity (background subtracted) will yield the profile of the measured glycoforms. The other values are required for effective spectral and analyte curation. Firstly, the alignment residuals can be used to reject bad LC runs (a much higher than average value here can for instance be caused by a retention order shift of features that were used for alignment). Secondly, the PPM Error, QC and signal-to-noise values can be used for analyte and spectral curation (PPM errors above 10, QC values above 0.05 and SN values below 9 are all indicators of a poor quality feature).

References

1. Beck A, Wagner-Rousset E, Ayoub D, et al. (2013) Characterization of therapeutic antibodies and related products. Anal Chem 85:715–736
2. Dotz V, Haselberg R, Shubhakar A, et al. (2015) Mass spectrometry for glycosylation analysis of biopharmaceuticals. Trends Anal Chem 73:1–9
3. Schwab I, Nimmerjahn F (2014) Role of sialylation in the anti-inflammatory activity of intravenous immunoglobulin—F(ab′)(2) versus Fc sialylation. Clin Exp Immunol 178(Suppl 1): 97–99
4. Bondt A, Rombouts Y, Selman MH, et al. (2014) Immunoglobulin G (IgG) Fab glycosylation analysis using a new mass spectrometric high-throughput profiling method reveals pregnancy-associated changes. Mol Cell Proteomics 13:3029–3039
5. Jones AJ, Papac DI, Chin EH, et al. (2007) Selective clearance of glycoforms of a complex glycoprotein pharmaceutical caused by terminal N-acetylglucosamine is similar in humans and cynomolgus monkeys. Glycobiology 17: 529–540
6. Mimura Y, Sondermann P, Ghirlando R, et al. (2001) Role of oligosaccharide residues of IgG1-Fc in Fc gamma RIIb binding. J Biol Chem 276:45539–45547
7. Sesarman A, Vidarsson G, Sitaru C (2010) The neonatal Fc receptor as therapeutic target in IgG-mediated autoimmune diseases. Cell Mol Life Sci 67:2533–2550
8. Reusch D, Haberger M, Falck D, et al. (2015) Comparison of methods for the analysis of therapeutic immunoglobulin G Fc-glycosylation profiles-Part 2: mass spectrometric methods. MAbs 7:732–742
9. Vidarsson G, Dekkers G, Rispens T (2014) IgG subclasses and allotypes: from structure to effector functions. Front Immunol 5: 520
10. Bruker Daltonics. https://www.bruker.com/products/mass-spectrometry-and-separations/ms-software/compass/overview.html. Accessed 11 Mar 2015

11. Schilling B, Rardin MJ, MacLean BX, et al. (2012) Platform-independent and label-free quantitation of proteomic data using MS1 extracted ion chromatograms in skyline: application to protein acetylation and phosphorylation. Mol Cell Proteomics 11:202–214
12. Jansen BC, Reiding KR, Bondt A, et al. (2015) MassyTools: a high throughput targeted data processing tool for relative quantitation and quality control developed for glycomic and glycoproteomic MALDI-MS. J Proteome Res 14: 5088–5098
13. Van Rossum G, Drake FL Jr (1995) Python reference manual. Centrum voor Wiskunde en Informatica, Amsterdam
14. Van Der Walt S, Colbert SC, Varoquaux G (2011) The NumPy array: a structure for efficient numerical computation. Comput Sci Eng 13:22–30
15. Oliphant TE (2007) Python for scientific computing. Comput Sci Eng 9:10–20
16. Hunter JD (2007) Matplotlib: A 2D graphics environment. Comput Sci Eng 9:90–95
17. Python(x,y). http://python-xy.github.io/. Accessed 21 Aug 2015
18. Anaconda. https://www.continuum.io/downloads. Accessed 14 July 2015
19. Falck D, Jansen BC, Plomp R, et al. (2015) Glycoforms of immunoglobulin G based biopharmaceuticals are differentially cleaved by trypsin due to the glycoform influence on higher-order structure. J Proteome Res 14: 4019–4028
20. Switzar L, Giera M, Niessen WM (2013) Protein digestion: an overview of the available techniques and recent developments. J Proteome Res 12:1067–1077
21. Hustoft HK, Reubsaet L, Greibrokk T, et al. (2011) Critical assessment of accelerating trypsination methods. J Pharm Biomed Anal 56: 1069–1078
22. Rombouts Y, Ewing E, van de Stadt LA, et al. (2015) Anti-citrullinated protein antibodies acquire a pro-inflammatory Fc glycosylation phenotype prior to the onset of rheumatoid arthritis. Ann Rheum Dis 74:234–241
23. Kessner D, Chambers M, Burke R, et al. (2008) ProteoWizard: open source software for rapid proteomics tools development. Bioinformatics 24:2534–2536
24. Pluskal T, Castillo S, Villar-Briones A, et al. (2010) MZmine 2: modular framework for processing, visualizing, and analyzing mass spectrometry-based molecular profile data. BMC Biochem 11:395
25. Strohalm M, Kavan D, Novák P, et al. (2010) mMass 3: a cross-platform software environment for precise analysis of mass spectrometric data. Anal Chem 82:4648–4651
26. Geany. http://www.geany.org/. Accessed 15 Oct 2015
27. IBM SPSS Statistics. http://www-01.ibm.com/software/analytics/spss/products/statistics/. Accessed 3 Nov 2015
28. Microsoft Excel. https://products.office.com/en-us/excel. Accessed 3 Nov 2015

Chapter 5

Sialic Acid Derivatization for the Rapid Subclass- and Sialic Acid Linkage-Specific MALDI-TOF-MS Analysis of IgG Fc-Glycopeptides

Noortje de Haan, Karli R. Reiding, and Manfred Wuhrer

Abstract

Matrix-assisted laser desorption/ionization (MALDI)-time-of-flight (TOF)-mass spectrometry (MS) is a highly suitable method for the rapid analysis of IgG glycopeptides, providing a wealth of structural information. A limitation of this approach is that it generates a bias when analyzing sialylated species due to the labile nature of sialic acid glycosidic linkages. One way to overcome this problem is by chemical derivatization of the sialic acids. The method presented here results in both the stabilization of the sialic acids, as well as the differentiation of α2,3- and α2,6-linked sialic acids by mass. Described in this chapter are the isolation of IgG from plasma or serum, tryptic digestion of the samples, derivatization, and finally MALDI-TOF-MS measurement and data analysis.

Key words Immunoglobulin G (IgG), Fragment crystallizable (Fc)-glycosylation, Glycopeptide, Subclass-specific, Site-specific, Sialic acid, *N*-acetylneuraminic acid, *N*-glycolylneuraminic acid, Stabilization, Linkage-specific, Solid-phase extraction (SPE), Cotton hydrophilic interaction liquid chromatography (HILIC)Matrix assisted laser desorption/ionization (MALDI), Time-of-flight (TOF), Mass spectrometry (MS), Biopharmaceuticals, Clinical cohorts

1 Introduction

As glycosylation is a highly relevant co-translational and posttranslational protein modification that influences the physical and chemical properties of a protein, fast and accurate tools for the analysis of protein glycosylation are of great significance [1, 2]. Such tools find their application in both the analysis of human protein glycosylation for biomarker research, as well as for the design and control of biopharmaceutical glycosylation features.

Immunoglobulin G (IgG) is the most abundant immunoglobulin in the human body and contains one glycosylation site on each heavy chain in its fragment crystallizable (Fc)-region [3]. The heterogeneity at this glycosylation site influences the interac-

Gordan Lauc and Manfred Wuhrer (eds.), *High-Throughput Glycomics and Glycoproteomics: Methods and Protocols*, Methods in Molecular Biology, vol. 1503, DOI 10.1007/978-1-4939-6493-2_5, © Springer Science+Business Media New York 2017

tion of the antibody with antibody binding receptors and consequentially modulates the immune response [4, 5]. Not surprisingly, multiple studies have shown correlations between Fc glycosylation features and physiological states or disease activity. For example, galactosylation shows to decrease with age and with increased disease activity for rheumatoid arthritis and Crohn's disease, while the level of bisecting *N*-acetylglucosamine on digalactosylated glycans has been shown to increase with Crohn's disease [6–8]. Furthermore, IgG is one of the most frequently used antibodies for pharmaceutical purposes. Proper design of therapeutic antibody glycosylation has been found to greatly enhance antibody efficacy and plasma half-life, while toxicity is minimized [5, 9, 10]. With respect to sialylation, α2,6-linked *N*-acetylneuraminic acid on intravenous IgG (IVIG) has been reported to enhance anti-inflammatory activity, while α2,3-linked *N*-acetylneuraminic acid has not [11].

For the high-throughput analysis of IgG glycosylation, matrix-assisted laser desorption/ionization (MALDI)-time-of-flight (TOF)-mass spectrometry (MS) is a method of choice [12, 13]. However, due to the labile nature of sialic acids and their propensity for deprotonation, MALDI-TOF-MS causes biases in the observed abundances of sialylated species [12]. Chemical derivatization of the sialic acids can be performed to prevent these problems, which has been described in a linkage-specific manner for released glycans, as well as for tryptic IgG glycopeptides [14–18]. By studying released glycans, information about the protein is lost, while for the glycosylation analysis of IgG it can be advantageous to retain the protein-, site-, and subclass-specificity. Trypsin is a particularly suitable proteolytic enzyme for the generation of IgG glycopeptides, as it results in relatively small glycopeptides that are still specific for three of the four different subclasses of IgG that are present in human plasma. IgG1, 2, and 4 differ in the peptide sequences of the tryptic Fc glycopeptides, while IgG3 shows the same peptide sequence as either IgG2 or IgG4 depending on the IgG3 allotype [6, 19]. Using the carboxylic acid activator 1-ethyl-3-(3-(dimethylamino)propyl)carbodiimide (EDC) and the catalyst 1-hydroxybenzotriazole (HOBt) in the presence of the nucleophiles ethanol or dimethylamine, sialic acids are stabilized for measurements in reflectron positive mode MALDI-TOF-MS. In addition, the sialic acids react differentially depending on the linkage to their neighboring galactose. However, besides the sialic acid derivatization, also the carboxylic acids on the peptide moiety will be modified. Based on their surroundings, the carboxylic acids can either form a lactam or lactone with accompanying water loss, or react with the added nucleophile. In contrast to ethanol used for released glycans, dimethylamine showed more promise as a nucleophile for IgG glycopeptides as reactions performed fully selective on the carboxylic acids in the peptide moiety [17]. Under the optimized conditions, α2,3-linked sialic acids lactonize leading to water loss (−18.011 Da), whereas α2,6-linked sialic acids react with the added dimethylamine to form a dimethylamide (+27.047 Da). This results in a mass difference between α2,3- and α2,6-linked sialic acids of 45.058 Da, which enables their discrimination without the use of fragmentation. In addition, all IgG glycopeptides will have uniformly modified peptide moieties, carrying two dimethylamides and one lactam.

In this chapter the steps required for the sialic acid linkage-specific glycosylation analysis of IgG glycopeptides are described extensively. Covered by this chapter are the affinity capturing of IgG from a complex mixture, tryptic digestion, linkage-specific sialic acid derivatization with dimethylamine, cotton hydrophilic interaction liquid chromatography (HILIC) solid-phase extraction (SPE) [20], and measurement by MALDI-TOF-MS as well as subsequent data analysis.

2 Materials

The materials, chemicals, and reagents are listed for the preparation and measurement of one 96-well sample plate. The number of samples and the volumes used can be upscaled or downscaled according to needs. To facilitate the protocol it is recommended to use a 12- or 8-channel pipette for sample handling and purification by cotton HILIC SPE.

Ultrapure deionized water (MQ, ≥18 MΩ at 25 °C) is required for the preparation of all solutions.

2.1 IgG Capturing from Plasma or Serum

IgG capturing is additionally described in Chapter 4 (High-throughput analysis of IgG Fc glycopeptides by LC-MS). The procedure is reprinted here for completeness of the protocol.

1. 2 mL Protein G (ProtG) Sepharose beads (GE Healthcare, Uppsala, Sweden) (*see* **Note 1**).
2. 150 mL phosphate-buffered saline: dissolve 855 mg $Na_2HPO_4{\cdot}2H_2O$, 75 mg KH_2PO_4, and 1.275 g NaCl in 150 mL MQ, store at 4 °C.
3. 15 mL 100 mM aqueous formic acid: add 60 μL formic acid to 14.94 mL MQ in a glass bottle, store at 4 °C.

Additional equipment: 15 mL tube, 96-well filter plate (0.7 mL/well), centrifuge, vortex, vacuum manifold, adhesive tape, plate shaker, two polypropylene 96-well V-bottom plates, vacuum concentrator.

2.2 Tryptic Protein Digestion

When the IgG samples have been isolated from plasma or serum as described above, the formic acid solution can be omitted in this step. However, when no acid denaturation was performed before, it should be implemented before proteolytic digestion.

1. 100 mM aqueous formic acid: add 199 μL formic acid to 49.80 mL MQ in a glass bottle, store at 4 °C.
2. 2 mL 50 mM sodium bicarbonate buffer (pH 7.9) (*see* **Note 2**): dissolve 8.4 mg sodium bicarbonate in 2 mL MQ. Adjust the pH by titrating the solution with either 2 M HCl or 2 M NaOH. Do not use the solution longer than 1 week.
3. 20 mM acetic acid: add 57 μL acetic acid to 49.94 mL MQ in a glass flask, store at 4 °C.
4. 20 μL 10 mg/mL L-1-tosylamide-2-phenylethyl chloromethyl ketone (TPCK)-treated porcine trypsin: while keeping the trypsin on ice, dissolve 200 μg in 20 μL ice-cold 20 mM acetic acid. The solution can be stored for 6 months at −80 °C.

Additional equipment: vacuum concentrator, pH paper, adhesive tape, plate shaker.

2.3 Glycopeptide Derivatization

1. Dimethylamidation reagent: 250 mM EDC hydrochloride (Fluorochem, Hadfield, UK), 500 mM HOBt monohydrate (Sigma-Aldrich, Steinheim, Germany) and 250 mM dimethylamine (40 % in water, Sigma-Aldrich) in dimethylsulfoxide (DMSO): Weigh 95.85 mg EDC and transfer to a 2 mL Eppendorf tube, weigh 135.12 mg HOBt and transfer to another 2 mL Eppendorf tube. Add 1436.7 μL DMSO to the EDC and 500 μL to the HOBt. Mix both solutions (*see* **Note 3**) and finally add 63.3 μL 40 % dimethylamine in water (*see* **Notes 4** and **5**). Mix thoroughly.

Additional equipment: polypropylene 96-well V-bottom plate, adhesive tape, 60 °C stove.

2.4 Cotton HILIC SPE

1. 12 mL 100 % ACN.
2. 96 handcrafted cotton tips (*see* **Note 6**): cut 3 mm of cotton rope (180 μg, Pipoos, Utrecht, Netherlands) (*see* **Note 7**) and put one strand (out of the eight strands in the rope) into a 20 μL pipette tip. Apply air pressure on top of the tip (50 kPa pressure difference) (*see* **Note 8**).
3. 50 mL 85 % ACN: in a glass bottle, mix 7.5 mL MQ with 42.5 mL ACN.
4. 50 mL 85 % ACN, 1 % TFA: in a glass bottle, mix 7.5 mL MQ with 42.5 mL ACN and 500 μL TFA.

Additional equipment: solvent reservoirs for multichannel pipetting, polypropylene 96-well V-bottom plate, adhesive tape.

2.5 MALDI Target Spotting

1. 100 μL 5 mg/mL 4-chloro-α-cyanocinnamic acid matrix (Cl-CCA) : weigh 1 mg Cl-CCA, add 70 μL ACN and 30 μL MQ, mix well and store at −20 °C (*see* **Note 9**).
2. 1 μL external calibration standard, for example peptide calibration standard (Bruker Daltonics, Bremen, Germany) (*see* **Note 10**).
3. Polished steel 384 TF MALDI target (Bruker Daltonics).

3 Methods

3.1 IgG Capturing from Plasma or Serum

IgG capturing is additionally described in Chapter 4. The procedure is reprinted here for completeness of the protocol.

1. Spin down 2 mL of ProtG beads in a 15 mL tube (3000 × *g*, 10 s).
2. Carefully pipette the supernatant from the beads.
3. Add 2 mL of PBS to the beads.
4. Vortex thoroughly and repeat spinning and removal of supernatant.
5. Resuspend the beads in 4.67 mL PBS.
6. Using a 200 μL multichannel pipette, apply 50 μL of slurry to each well of a 96-well filter plate (*see* **Note 11**).
7. Wash the beads three times with 200 μL PBS using the vacuum manifold.
8. Add 100 μL PBS and leave it on the beads.
9. Add 2 μL of serum or plasma (*see* **Note 12**).
10. Place the filter plate on a polypropylene 96-well V-bottom plate and seal wells with tape.
11. Incubate the filter plate on a plate shaker for 1 h at 1000 rpm.
12. Wash the samples three times with 200 μL of PBS using the vacuum manifold.
13. Wash the samples three times with 200 μL of MQ using the vacuum manifold.
14. Remove the filter plate from the vacuum manifold, tap the bottom carefully on lint-free paper to remove an excess of liquid, and place the filter plate on a new polypropylene 96-well V-bottom plate (*see* **Note 13**).
15. Add 100 μL 100 mM formic acid to the beads in each well and seal the wells with tape.
16. Incubate on the plate shaker for 5 min at 1000 rpm (*see* **Note 14**).
17. Remove the tape and elute the IgGs from the filter plate by centrifugation for 1 min at 100 × *g*.

18. Dry the samples using a vacuum concentrator for 3 h at 60 °C.
19. Samples can be digested directly or sealed by adhesive tape and stored at −20 °C.

3.2 Tryptic Protein Digestion

When no acid denaturation has been performed on the samples before, **steps 1–3** should be implemented before proteolytic digestion. When the IgG was isolated as described in Subheading 3.1, one can start with **step 4**.

1. Mix 1 μL of the IgG samples with 20 μL 100 mM formic acid (*see* **Note 15**).
2. Incubate the samples 15 min at room temperature.
3. Dry the samples using a vacuum concentrator for 3 h at 60 °C.
4. Mix 20 μL 10 mg/mL TPCK trypsin with 2 mL sodium bicarbonate buffer.
5. Add 20 μL of the trypsin solution to each sample.
6. Seal the sample plate and incubate it on the plate shaker for 10 min at 600 rpm (*see* **Note 16**).
7. Select some random samples and take 0.5 μL to check the pH to be between 7 and 9 using pH paper.
8. Incubate the samples for 17 h at 37 °C.
9. When not used immediately, seal the plate with adhesive tape and store at −20 °C.

3.3 Glycopeptide Derivatization

1. Add 20 μL dimethylamidation reagent to each well of a clean 96-well V-bottom plate (*see* **Note 17**).
2. Add 1 μL tryptic digest to each well, pipette up and down in the reagent three times (*see* **Note 18**).
3. Seal the plate with adhesive tape and incubate for 3 h at 60 °C.

3.4 Cotton HILIC Solid-Phase Extraction

1. To make the samples cotton HILIC compatible, add 113.3 μL ACN to each well (*see* **Note 19**).
2. Pipet 10 μL MQ in each well of a new polypropylene 96-well V-bottom plate (the elution plate).
3. Equilibrate the cotton tips by pipetting 15 μL of MQ and discarding it in a waste bucket. Repeat this two times (*see* **Note 20**).
4. Condition the tips by pipetting three times 15 μL of 85 % ACN and discard the solvent (*see* **Note 21**).
5. Load the samples by slowly pipetting up and down 20 times in the reaction mixture (*see* **Note 22**).

6. Wash the samples by pipetting three times 15 μL of 85 % ACN with 1 % TFA and discard the flow-through.
7. Wash the samples by pipetting three times 15 μL of 85 % ACN and discard the flow-through.
8. Elute the derivatized glycopeptides by pipetting the MQ in the prepared elution plate up and down 5 times.
9. The eluted samples cannot be stored for a long time as the lactonized sialic acids will hydrolyze over time (*see* **Note 23**).

3.5 MALDI Target Spotting

1. For each sample, apply 1 μL Cl-CCA matrix on the MALDI target (*see* **Note 24**).
2. Add 1 μL sample to each matrix droplet (*see* **Note 25**).
3. Prepare one sample with the external calibration standard (*see* **Note 26**).
4. Let the spots dry by air (*see* **Note 27**).

3.6 MALDI-TOF (/TOF)-MS(/MS) Measurement

1. To measure the derivatized glycopeptides, the MALDI-TOF-MS should be operated in positive reflectron mode, covering an analysis window between m/z 1000 and 4000 (*see* **Note 28**).
2. Calibrate the instrument using the external calibration standard.
3. Adjust the laser power until high intensity signals are obtained that still show a resolution well enough to have a baseline separation of the isotopes.
4. Acquire spectra in a random walk pattern to prevent bias by potential heterogeneity in the spot crystallization (*see* **Note 29**).
5. Typical IgG glycopeptide spectra obtained by the method can be found in Fig. 1, showing biological differences between the IgG glycosylation of two individuals. Be aware that sialylated species are either decreased in mass or increased in mass, depending on the linkage of the sialic acid. Instead of 291.095 Da, α2,6-linked *N*-acetylneuraminic acids show a mass increment of 318.142 Da, while *N*-acetylneuraminic acids in α2,3-linkage show a mass increment of 273.085 Da. The difference of 45.058 Da between the differently linked species enables discrimination of the linkage variants without the use of fragmentation. *N*-glycolylneuraminic acids also form dimethylamides and lactones as described for the *N*-acetylneuraminic acids. This results in a mass increment of 334.138 Da for α2,6-linked species and 289.080 Da for α2,3-linked species. For more examples of spectra obtained for IgG samples derived from various sources and with different sialic acid structures and linkages, see De Haan et al. *Anal Chem.* 2015 [17].

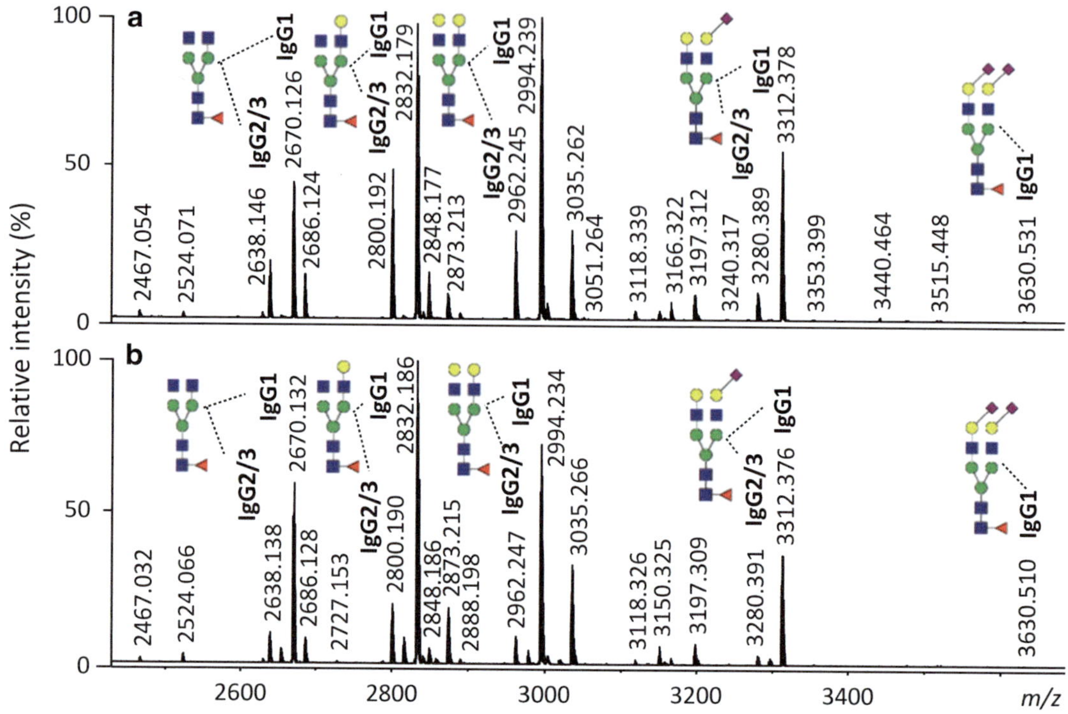

Fig. 1 Reflectron positive MALDI-TOF-MS spectra of human plasma IgG glycopeptides after linkage-specific glycopeptide derivatization of a (**a**) 25-year-old female and (**b**) 40-year-old male. Glycosylation can be studied separately for IgG1 and IgG2/3 and shows to differ between the samples with regard to levels of, for example, galactosylation and sialylation. IgG1 = EEQYNSTYR and IgG2/3 = EEQFNSTFR, both modified on the peptide sequence by two dimethylamidation reactions and one loss of water. *Green circle* = mannose; *yellow circle* = galactose; *blue square* = *N*-acetylglucosamine; *red triangle* = fucose; *right pointing purple diamond* = α2,6-linked *N*-acetylneuraminic acid

6. Besides sialic acids, the peptide moiety of all glycopeptides will be derivatized during the reaction as well, as three carboxylic acids are present on the amino acid sequences derived from IgG (two at the glutamic acid residues and one at the C-terminus of the peptide). The C-terminus becomes dimethylamidated, as does the intrachain glutamic acid. The N-terminal glutamic acid loses water by forming a lactam with the N-terminal amine. In all, the modifications on the peptide will lead to a mass increase of 36.084 Da for all glycopeptides (*see* **Note 30**).
7. The method is able to discriminate between glycosylation present on the different subclasses of human IgG. The peptide sequences of the different subclasses are: IgG2/3 = EEQFNSTFR($[M+H]^+ = 1157.522$ Da); IgG4 = EEQFNSTYR ($[M+H]^+ = 1173.517$ Da); IgG1 = EEQYNSTYR ($[M+H]^+ = 1189.512$ Da), resulting in a mass increment of 15.995 Da for each subclass (*see* **Notes 31** and **32**).

Table 1
Potential by-products and suggestions to prevent them

	Caused by	Prevented by
Δm for all glycopeptides (Da)		
−45.058	Exchange of the formation of dimethylamide for the loss of water on the peptide.	Increasing the concentrations EDC, HOBt and dimethylamine in the reagent, or using more reagent for the sample.
−5.065	Incomplete dimethylamidation on the peptide, with proton to sodium exchange on the remaining carboxylic acid.	Same as above.
Δm for α2,3-sialylated species (Da)		
+17.027	Amidation of the sialic acid.	Preventing ammonia from being present in the samples or reagents. Do not use ammonia buffers.
+18.011	Undermodification of the sialic acid.	Reduce time between sample preparation and measurement.
+45.058	Dimethylamidation of the sialic acid. This is usually seen for less than 5 % for all α2,3-linked sialic acids.	Decrease dimethylamine concentration during the reaction.
Δm for α2,6-sialylated species (Da)		
−28.031	Amidation with ammonia of the sialic acid.	Preventing ammonia from being present in the samples or reagents.
−27.047	Undermodification of the sialic acid.	Increasing the concentrations EDC, HOBt and dimethylamine in the reagent, or using more reagent for the sample.

8. Be sure the reaction was performed well by checking the mass spectra. Table 1 lists some potential by-products and suggestions how to prevent them.
9. Fragmentation spectra can be obtained by laser- or collision-induced fragmentation, and are informative for both the glycan composition as well as the peptide sequence. The most intense fragment observed for both the sialylated as well as the non-sialylated glycopeptides originates from the peptide containing the $^{0,2}X$ cross-ring fragment of the reducing end *N*-acetylglycosamine. For additionally observed glycan fragments, *see* Table 2.

Table 2
Mass differences with respect to the precursor of commonly observed glycan fragments observed in MS/MS and their explanation

Δmass (Da)	Glycan fragment
-146.06	Fucose
-318.14	α2,6-linked *N*-acetylneuraminic acid
-334.14	α2,6-linked *N*-glycolylneuraminic acid
-365.13	*N*-acetyllactosamine (LacNAc)
-638.22	LacNAc carrying an α2,3-linked *N*-acetylneuraminic acid
-654.21	LacNAc carrying an α2,3-linked *N*-glycolylneuraminic acid
-683.28	LacNAc carrying an α2,6-linked *N*-acetylneuraminic acid
-699.27	LacNAc carrying an α2,6-linked *N*-glycolylneuraminic acid

4 Notes

1. 4 mL of the bead slurry contains approximately 2 mL beads. To determine the volume of the beads, spin the suspension down (3000 × *g*, 10 s) in a graduated tube.
2. Although ammonium bicarbonate buffer is used regularly in tryptic digestion protocols, it is of particular importance to use a sodium bicarbonate buffer in this procedure. Ammonia is a nucleophilic molecule that interferes with the derivatization by reacting non-selectively with both α2,3- and α2,6-linked sialic acids. Even minor amounts of ammonia will result in by-products that are amidated with ammonia.
3. Only limited amounts of EDC can be dissolved in DMSO. Adding the specified amount of DMSO to the EDC may result in incomplete solvation of the EDC. However, after adding the HOBt solution, a clear light yellow solution is expected.
4. As both EDC and dimethylamine have a strong odor, it is recommended to work in a flow cabinet while preparing and handling the reagent.
5. The reagent can be stored at −20 °C for at least 1 month. However, freeze-thaw cycles may cause a rapid decrease of the reagent quality and it is therefore recommended to store freshly prepared reagent in small aliquots. As DMSO has a theoretical melting point of +19 °C, also storage in the fridge causes the reagent to become solid.
6. 96-well HILIC SPE formats are commercially available and have the potential to be of use in this protocol, however care

should be taken that the equipment and HILIC material used are compatible with DMSO.

7. Cotton can be bought from various suppliers; similar glycopeptide profiles were obtained using different brands of cotton rope or cotton pads [20].
8. Always wear gloves when preparing cotton tips to prevent contamination. Cotton rope can readily be cut using a surgical knife, and may for tip loading be handled using fine tweezers.
9. This solution should stay a liquid while stored at –20 °C. As soon as the solution crystallizes at –20 °C, contaminations are likely present.
10. Essentially every analyte mixture containing well defined molecules covering the mass range of interest (m/z 1000–4000) is suitable for the calibration of the instrument.
11. To prevent the beads from clogging the pipette tips, cut off the final ends of the tips. To keep the beads equally distributed in the suspension, pipette the slurry up and down three times before transferring it to the wells.
12. 2 μL human plasma or serum contains approximately 20 μg IgG.
13. Tapping the plate on a lint-free paper to remove the lagging flow trough of the washing steps is a critical step. Without this step the elution will be contaminated.
14. This step can be extended with 15 min. This can be for example of use when two plates are prepared with a 15 min difference.
15. In general, add 20 additional volumes of 100 mM formic acid to the IgG samples. The final concentration of formic acid should never be below 75 mM.
16. As the samples were dried before, this step is essential to get the proteins solubilized and well mixed with the trypsin.
17. This can be done rapidly by using a repetition pipette.
18. The sample–reagent ratio described here is 1:20. No loss in reaction performance was observed when a 1:10 ratio was used.
19. When the volume of reagent deviates from the protocol, adjust the ACN fraction to be 85 % compared to the reagent fraction.
20. Perform initial pipetting very slowly to prevent air bubble formation.
21. During this step, the pieces of cotton may come off loose from the tip. This situation can be prevented by pipetting up slowly and pipetting down firmly until the first stop of the pipette. Should the cotton still loosen, it can be forced back in place by slowly surrounding it in liquid and pipetting down firmly until the first stop of the pipette.

22. When using larger sample volumes it may be necessary to pipette up and down more often to enable the whole sample to interact with the cotton. As an alternative, 200 μL tips can be used for cotton HILIC SPE.
23. In the presence of water, and in slightly acidic or alkaline conditions, lactones are not stable and will be hydrolyzed over time. This may lead to an underestimation of the α2,3-linked sialic acids in samples that are not directly measured. The dimethylamidated α2,6-linked species do not suffer from this instability as amides are more resistant to hydrolysis. The ideal situation is to spot and analyze the samples within one day.
24. This can be done rapidly by using a repetition pipette. Make sure the pipette can handle volumes as small as 1 μL.
25. Samples can be spotted row by row using a multichannel pipette. It is recommended to apply the samples to the matrix as rapid as possible. The optimal circumstances are to spot one row of matrix, apply the samples and then spot the next row of matrix. In this way the evaporation of the organics in the matrix solution is minimal and the crystallization will be uniform.
26. Preferably position the calibration spot close to the samples measured, as the calibration may be slightly inconsistent over the MALDI target.
27. It is advised to cover the plate with a tilted lid, to allow an air flow over the samples while limiting sample contamination.
28. Glycopeptides preferable ionize as $[M + H]^+$, therefore best results are obtained when operating the MALDI-TOF-MS in positive ion mode. One of the advantages of the derivatization is the neutralization of the sialylated species, which obviates the ionization bias of these species. For high resolution spectra, the use of the reflectron essential.
29. With the Bruker UltraFlextreme, usually 10,000 shots are combined in a random walk pattern with a frequency of 2000 Hz. The laser power, number of shots combined and the frequency used are highly dependent on the instrument and should be optimized for each specific case.
30. When other nucleophiles are used in the reaction, like ethanol or methanol, the reaction on the peptide is observed to be variable. The exchange of one amidation/esterification with the external nucleophile for the internal loss of water was observed when ethanol or methanol was used or when dimethylamine was used in combination with lower concentrations EDC and HOBt. For the dimethylamidation conditions described here, this side reaction should be less than 1 % (observable as –45.058 Da for all glycopeptides).

31. Be aware that non-fucosylated glycoforms on IgG2/3 have exactly the same elemental composition as the fucosylated glycoforms of IgG4 and will therefore have overlapping signals in the mass spectra. For human plasma IgG, which is a mix of IgG1, 2, 3, and 4, this means that IgG4 signals cannot be accurately extracted and that only the fucosylated glycoforms of IgG2/3 can be properly extracted. The same overlap is present for fucosylated IgG4 and nonfucosylated IgG1 glycoforms. However, as the relative abundance of IgG1 is about 65 % and that of IgG4 about 4 % in healthy subjects [21], the influence of IgG4 on IgG1 is considered minor, and in most cases negligible.
32. Ionization of IgG1 and IgG2/3 glycopeptides was observed to differ from each other during MALDI-TOF-MS analysis. Therefore, it is advised to do relative quantification of the analytes for IgG1 and IgG2/3 separately.

Acknowledgements

This work was supported by the European Union Seventh Framework Programmes IBD-BIOM (grant number 305479) and HighGlycan (grant number 278535), as well as by the Netherlands Genomic Initiative Horizon Programme Zenith project (grant number 93511033). Additional financial support was provided by Hoffmann-la Roche.

References

1. Crocker PR, Paulson JC, Varki A (2007) Siglecs and their roles in the immune system. Nat Rev Immunol 7:255–266
2. Muramatsu T (2000) Essential roles of carbohydrate signals in development, immune response and tissue functions, as revealed by gene targeting. J Biochem 127:171–176
3. Vidarsson G, Dekkers G, Rispens T (2014) IgG subclasses and allotypes: from structure to effector functions. Front Immunol 5:520
4. Houde D, Peng Y, Berkowitz SA, et al. (2010) Post-translational modifications differentially affect IgG1 conformation and receptor binding. Mol Cell Proteomics 9:1716–1728
5. Okazaki A, Shoji-Hosaka E, Nakamura K, et al. (2004) Fucose depletion from human IgG1 oligosaccharide enhances binding enthalpy and association rate between IgG1 and FcgammaRIIIa. J Mol Biol 336:1239–1249
6. Bondt A, Selman MH, Deelder AM, et al. (2013) Association between galactosylation of immunoglobulin G and improvement of rheumatoid arthritis during pregnancy is independent of sialylation. J Proteome Res 12: 4522–4531
7. Dall'Olio F, Vanhooren V, Chen CC, et al. (2013) N-glycomic biomarkers of biological aging and longevity: a link with inflammaging. Ageing Res Rev 12:685–698
8. Trbojevic AI, Ventham NT, Theodoratou E, et al. (2015) Inflammatory bowel disease associates with proinflammatory potential of the immunoglobulin G glycome. Inflamm Bowel Dis 21:1237–1247
9. Chung CH, Mirakhur B, Chan E, et al. (2008) Cetuximab-induced anaphylaxis and IgE specific for galactose-alpha-1,3-galactose. N Engl J Med 358:1109–1117
10. Goetze AM, Liu YD, Zhang Z, et al. (2011) High-mannose glycans on the Fc region of therapeutic IgG antibodies increase serum clearance in humans. Glycobiology 21:949–959
11. Anthony RM, Nimmerjahn F, Ashline DJ, et al. (2008) Recapitulation of IVIG anti-

inflammatory activity with a recombinant IgG Fc. Science 320:373–376
12. Harvey DJ (2006) Analysis of carbohydrates and glycoconjugates by matrix-assisted laser desorption/ionization mass spectrometry: an update covering the period 1999-2000. Mass Spectrom Rev 25:595–662
13. Huffman JE, Pucic-Bakovic M, Klaric L, et al. (2014) Comparative performance of four methods for high-throughput glycosylation analysis of immunoglobulin G in genetic and epidemiological research. Mol Cell Proteomics 13:1598–1610
14. Reiding KR, Blank D, Kuijper DM, et al. (2014) High-throughput profiling of protein N-glycosylation by MALDI-TOF-MS employing linkage-specific sialic acid esterification. Anal Chem 86:5784–5793
15. Alley WR Jr, Novotny MV (2010) Glycomic analysis of sialic acid linkages in glycans derived from blood serum glycoproteins. J Proteome Res 9:3062–3072
16. Wheeler SF, Domann P, Harvey DJ (2009) Derivatization of sialic acids for stabilization in matrix-assisted laser desorption/ionization mass spectrometry and concomitant differentiation of alpha(2 -->3)- and alpha(2 --> 6)-isomers. Rapid Commun Mass Spectrom 23: 303–312
17. de Haan N, Reiding KR, Haberger M, et al. (2015) Linkage-specific sialic acid derivatization for MALDI-TOF-MS profiling of IgG glycopeptides. Anal Chem 87:8284–8291
18. Gomes de Oliveira AG, Roy R, Raymond C, et al. (2015) A systematic study of glycopeptide esterification for the semi-quantitative determination of sialylation in antibodies. Rapid Commun Mass Spectrom 29:817–1826
19. Einarsdottir HK, Selman MH, Kapur R, et al. (2013) Comparison of the Fc glycosylation of fetal and maternal immunoglobulin G. Glycoconj J 30:147–157
20. Selman MH, Hemayatkar M, Deelder AM, et al. (2011) Cotton HILIC SPE microtips for microscale purification and enrichment of glycans and glycopeptides. Anal Chem 83: 2492–2499
21. Shakib F, Stanworth DR (1980) Human IgG subclasses in health and disease. (A review). Part II. Ric Clin Lab 10:561–580

Chapter 6

The Use of Multiple Reaction Monitoring on QQQ-MS for the Analysis of Protein- and Site-Specific Glycosylation Patterns in Serum

L. Renee Ruhaak

Abstract

In recent years, high-throughput glycomics approaches have been developed and applied to either complete biofluids, cell lysates or tissues, or proteins isolated thereof. However, during such analyses the N-glycan are released from the protein backbone and therefore site- and protein-specific information is lost. There exists a need for high-throughput methods that allow quantification of site- and protein-specific glycosylation patterns from complex biological mixtures. We here describe the use of a multiple reaction monitoring mass spectrometry based method for the generation of glycopeptide profiles of the nine high abundance glycoproteins IgG, IgA, IgM, haptoglobin, alpha-1-antitrypsin, alpha-2-macroglobulin, alpha-1-acid glycoprotein, transferrin, and complement C3. We show that the sample preparation can be performed at the 96-well level, and using a 17-min gradient on a RP-UPLC-QQQ instrument, 96 samples can be analyzed within 3 days.

Key words N-glycopeptides, Quantitation, Serum, Multiple reaction monitoring, RP-LC-MS

1 Introduction

Over the past decade, a substantial body of research has been aimed towards the identification of differential glycosylation patterns in biofluids, with an emphasis on serum and plasma, that associate with certain health and disease states. Differential glycosylation patterns have been observed in a wide variety of physiological states, including, but not limited to multiple autoimmune diseases—including rheumatoid arthritis [1], lupus erythematosus [2], inflammatory bowel disease [3] and Crohn's disease, aging [4, 5], diabetes [6], and multiple types of malignancies [7–13]. However, the strategy employed in most of these studies is to release the glycans from the whole biofluids, thereby losing protein- and site-specific information.

Protein and site-specific glycosylation profiling methods are not yet readily available, and surely the identification of glycopeptides

Gordan Lauc and Manfred Wuhrer (eds.), *High-Throughput Glycomics and Glycoproteomics: Methods and Protocols*, Methods in Molecular Biology, vol. 1503, DOI 10.1007/978-1-4939-6493-2_6, © Springer Science+Business Media New York 2017

from more or less complex mixtures is much more complicated than peptide identification. However, several methods have been employed to provide glycosylation maps for individual proteins, either by released glycans (e.g., [14]), glycopeptides (e.g., [15]), or even intact glycoproteins (e.g., [16]). The best studied glycoprotein thus far is IgG [17, 18], while other glycoproteins are much less characterized, in part because the current analytical methods often rely on protein capturing, which is not easily achieved.

Multiple reaction monitoring (MRM), typically performed on triple quadrupole instruments, has recently been applied for the targeted quantification of proteins and metabolites in biofluids [19] and is being evaluated for its potential in a clinical setting [20]. MRM is a powerful technique, which demonstrates good sensitivity and specificity for quantitation. However, its use for glycopeptides has thus far been limited [15, 21–24], potentially due to a lack of good protein- and site-specific glycosylation maps. Methods to generate protein- and site-specific glycosylation maps using both nonspecific proteases and trypsin were recently developed [15]. The method for targeted glycosylation fingerprinting presented here is based on the maps generated using this method.

To determine a samples' protein- and site-specific glycosylation pattern using MRM, the sample is first diluted, followed by cysteine reduction and alkylation. The sample is then tryptically digested to generate peptides and glycopeptides. Without further sample purification, the sample can then be conducted for LC-QQQ-MS analysis, thus avoiding the need for protein capturing, at least for higher abundance proteins. MRM analysis is performed using transitions from the molecular glycopeptide ion to the oxonium ions at m/z 204 (HexNAc) or m/z 366 (HexNAcHex).

To allow for protein glycosylation fingerprinting irrespective of an individual's protein concentration, we not only determine the ion abundance of the glycopeptides, but also of non-glycosylated peptides. These can then be compared to a protein standard using a standard calibration curve to determine the absolute concentration of a specific protein, thus eliminating the effect of protein abundance differences in the glycosylation fingerprint. The method is described here for the nine highest abundance serum glycoproteins, but could theoretically be adjusted for every other protein, provided that a glycopeptide map is available or generated.

2 Materials

2.1 Tryptic Digestion

1. Standard proteins:
   (a) Immunoglobulin G (IgG), Sigma-Aldrich, I4506.
   (b) Immunoglobulin A (IgA), Sigma-Aldrich, I8260.
   (c) Immunoglobulin M (IgM), Sigma-Aldrich, I4036.

(d) Alpha-2-macroglobulin (A2MG), Sigma-Aldrich, M6159.
(e) Alpha-1-antitrypsin (A1AT), Sigma-Aldrich, A9024.
(f) Alpha-1-acid glycoprotein (AGP), Sigma-Aldrich, G9885.
(g) Haptoglobin (HP), Sigma-Aldrich, H3536.
(h) Transferrin (TF), Sigma-Aldrich, T3309.
(i) Complement C3 (C3), Athens Research & Technology, 16-16-030303.

2. Serum samples, preferably one standard serum and the samples of interest.
3. Ammonium bicarbonate (NH_4HCO_3, Sigma-Aldrich, 09830).
4. Dithiothreitol (DTT, Sigma-Aldrich, 43815).
5. Iodoacetamide (IAA, Sigma-Aldrich, I1149).
6. Trypsin (Promega, sequencing grade modified trypsin, V5117 (100 μg)).
7. Water bath.
8. 96-well microtiter plate (Eppendorf, 951020401).
9. Cover for 96-well plate (Phenix Research Products, SMX-PCR96).
10. Microbalance.
11. Falcon 15 ml tube (VWR, 21008-918).
12. Sonicator bath.
13. Incubator.

2.2 Mass Spectrometry Analysis Using LC-QQQ-MS/MS

1. UPLC (Agilent 1290 infinity UPLC consisting of autosampler, pump, and column oven, coupled to:
2. QQQ-MS (Agilent 6490) equipped with jetspray ionization source.
3. Agilent Eclipse plus C18 column (RRHD, 1.8 μm, 2.1 mm × 100 mm) with an Agilent Eclipse plus C18 pre-column (RRHD 1.8 μm, 2.1 × 5 mm) guard column.
4. Cover for 96-well plate (Phenix Research Products, SMX-PCR96).
5. Acetonitrile (HPLC grade).
6. Formic acid.

2.3 Data Analysis

1. Computer with at least 64-bit Windows 7 Professional SP1 and Microsoft Office.
2. Masshunter Quantitative analysis (Agilent, Santa Clara, CA).

3 Methods

To facilitate high-throughput analysis, both the sample preparation and analysis are performed in a 96-well microtiter plate. The procedure requires only one plate, in which sample preparation is performed and which is subsequently used as the sample container for LC-MS analysis. An overview of the sample preparation procedure is shown in Fig. 1. Briefly, a standard protein mixture is made, consisting of the standard glycoproteins to be studied. Then, samples, control samples, and the standard protein mixture are denatured, reduced, and alkylated prior to tryptic digestion to generate peptides and glycopeptides. Without further purification, the samples are loaded for LC-QQQ-MS analysis.

Typically, when a QQQ-MS is delivered, its parameters are optimized for proteomics or small molecules. However, since glycolytic bonds, particularly those that link sialic acids, are rather weak, the instrument parameters need to be tuned to reduce degradation and increase analyte sensitivity [25]. The parameters that were particularly optimized are: gas temperature, fragmentor voltage (the voltage placed at the exit of the capillary to transmit ions into the mass analyzer), and ion funnel rf amplitude voltages (The ion funnel consists of a series of concentric electrodes with an applied radio frequency (rf) to compress the ion cloud, and a DC

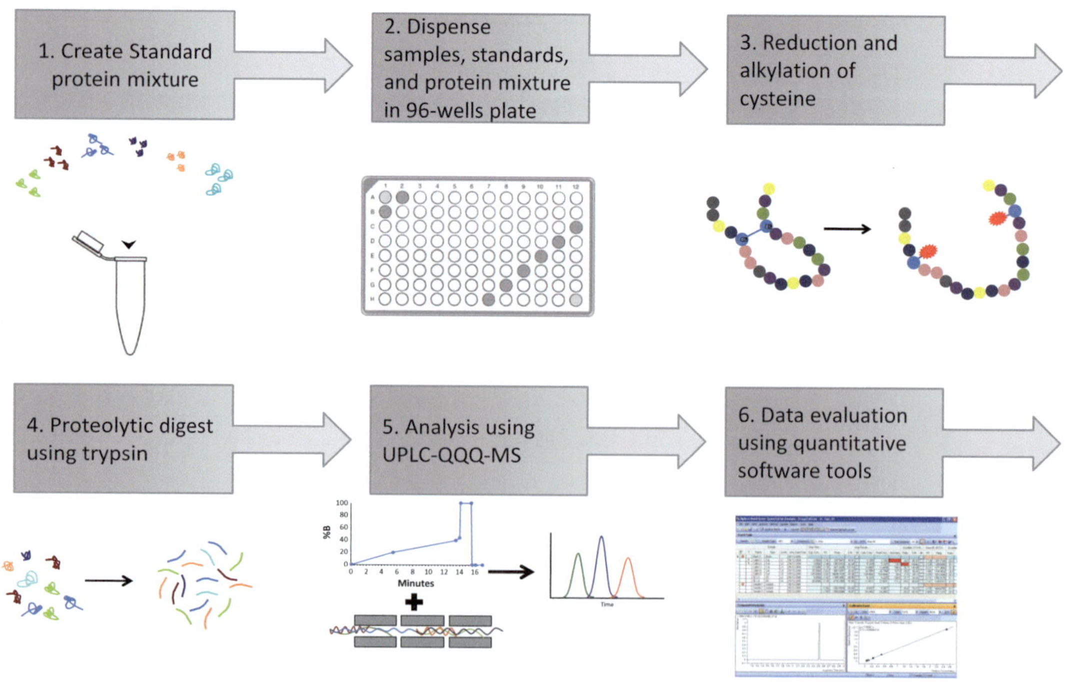

Fig. 1 Overview of the analytical pipeline for protein- and site-specific glycosylation profiling using multiple reaction monitoring

electrical field to transmit ions into the mass analyzers). An MS1 spectrum of the human milk oligosaccharide LNT prior and after instrument optimization, together with the effects of these instrument parameters on ion intensity and ion degradation is shown in Fig. 2. Upon instrument optimization, an MRM method, which contained transitions for nine glycoproteins was then developed based on literature and glycopeptide mapping (*see* **Note 1**).

3.1 Tryptic Digestion

1. Freshly prepare 10 ml of a 50 mM NH_4HCO_3 solution by dissolving 39.5 mg NH_4HCO_3 in water to a total volume of 10 ml in a 15 ml tube.
2. Prepare a standard protein mixture: weigh 100 μg IgG and 50 μg of each of the other standard proteins using a microbalance. Add all proteins to one eppendorf tube and add 100 μl water. Mix and dissolve well, by vortex for 5 min, then sonicate in a sonicator bath for 30 min and vortex again for 5 min.
3. Prepare 550 mM DTT solution dissolving 84.84 mg DTT in 1 ml water in an eppendorf tube.
4. Prepare 450 mM IAA solution by weighing 82.23 mg IAA and dissolving that in 1 ml water in an eppendorf tube. To prevent degradation, the IAA solution should be kept in the dark prior to use.
5. Dispense 2 μl of serum into each of the wells of a 96-well microtitration plate and add 87 μl of the 50 mM NH_4HCO_3 solution to each of the wells (*see* **Note 2**).
6. Cover the microtiter plate and mix on a shaker for 5 min.

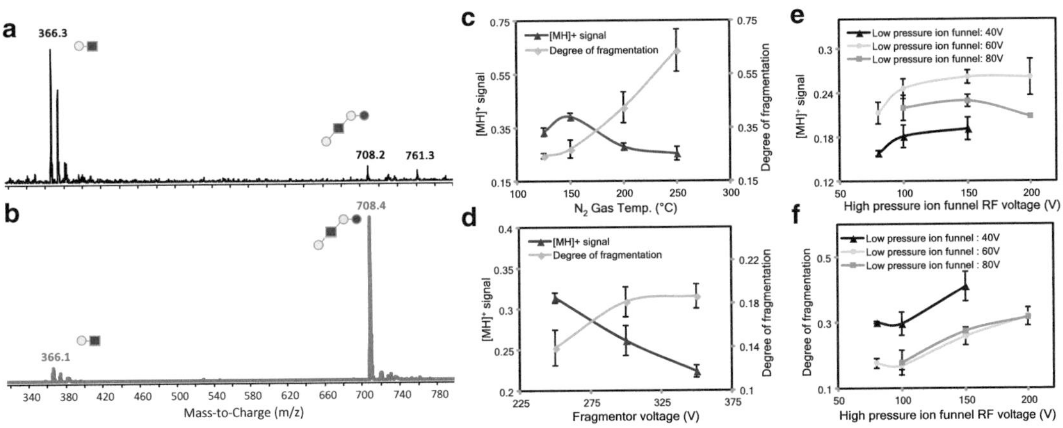

Fig. 2 Optimization of mass spectrometer parameters using a common tetrasaccharide from human milk. MS scans (**a**) using the manufacturers default parameters, which shows high levels of fragment ions at *m/z* 366, and (**b**) using the optimized instrument parameters with limited ion fragmentation. (**c**) Optimization of drying gas temperature. (**d**) Optimization of Fragmentor voltage. Optimization of the rf-amplitude, ion signal (**e**) and degree of fragmentation (**f**)

7. Add 2 μl of the 550 mM DTT solution to each of the wells of the microtiter plate, cover and mix on a shaker for 5 min (*see* **Note 3**).
8. Place the microtiter plate in a 60 °C water bath for 50 min to facilitate protein reduction and let cool to room temperature for 3 min. Centrifuge at 1000 × *g* for 20 s to remove droplets from the cover if necessary.
9. Add 4 μl of the 450 mM IAA solution to each of the wells of the microtiter plate, cover and mix on a shaker for 5 min. Then incubate for 1 h at room temperature in the dark to alkylate the proteins.
10. Add 300 μl of the 50 mM NH_4HCO_3 solution to one 100 μg vial of trypsin, mix and place on ice. Dispense 3 μl of the trypsin solution (*see* **Note 4**) to each of the wells of the microtiter plate, cover and mix for 5 min. Incubate the plate overnight in an incubator at 37 °C to digest the proteins (*see* **Note 5**).
11. Store the samples at −20 °C for up to 1 month prior to analysis.

3.2 QQQ-MS Analysis of Glycopeptides in Serum

1. HPLC gradient: the peptides and glycopeptides were separated using a 17-min binary gradient consisting of solvent A of 3 % acetonitrile, 0.1 % formic acid; solvent B of 90 % acetonitrile, 0.1 % formic acid in nanopure water (v/v) at a flow rate of 0.5 mL/min. The gradient starts at 0 %B, increases linearly till 20 %B over 5.5 min, increases further till 40 %B at 13.5 min and 44 %B at 14 min. The column is then washed at 100 %B for 30 s, upon which the column is re-equilibrated for 1.5 min, prior to the next injection.
2. Instrument parameters critical to reduce ion fragmentation:
   (a) Set gas temperature at 150 °C.
   (b) Set fragmentor voltage to 250 V.
   (c) Set the rf amplitude for ion funnel 1 to 100 V and ion funnel 2 to 60 V.
3. Transitions: Peptides and glycopeptides are monitored using QQQ-MS/MS in a dynamic mode (*see* **Note 6**). Transitions were generated based on literature and a previously generated glycopeptide map [15], developed using LC-QTOF-MS/MS analysis of tryptically digested protein standards (*see* **Note 1**). The transitions and retention times shown in Table 1 were used (Figs. 3 and 4).
4. Running the samples:
   (a) Always check after the first sample (which is a protein standard) if the retention times have shifted and adjust the dynamic range window as needed.

Table 1
Transitions monitored for the nine highest abundant serum proteins

No.	Peptide	Glycopeptide	Protein	Site	Glycan moiety	Precursor ion	Product ion	Ret time (min)
1	x		A1AT	–	–	444.8	718.4	4.6
2	x		A2MG	–	–	628.3	1071.5	5.2
3	x		AGP1	–	–	556.8	712.3	5.4
4	x		AGP1/2	–	–	497.8	764.4	6
5	x		AGP2	–	–	586.3	728.9	6.6
6	x		C3	–	–	701.6	928.6	9.05
7	x		HP	–	–	460.8	629.4	5.9
8	x		IgA1	–	–	466.3	415.8	4.7
9	x		IgA1/2	–	–	448.7	620.3	5.3
10	x		IgA2	–	–	756.9	475.3	5.9
11	x		IgG1	–	–	839.4	968.5	6.5
12	x		IgG1/2/3/4	–	–	418.2	506.3	4.7
13	x		IgG2	–	–	970.1	1100.6	8.6
14	x		IgG3	–	–	472.9	534.3	5.2
15	x		IgG4	–	–	634.7	425.2	8.9
16	x		IgM	–	–	573.0	734.9	5.8
17	x		TF	–	–	489.8	735.4	4.8
18		x	A1AT	107	H5N4F1S1	1151.6	366.1	12.2
19		x	A1AT	107	H5N4F0S2	1180.6	366.1	12.2
20		x	A1AT	107	H5N4F1S2	1209.8	366.1	12.2
21		x	A1AT	107	H6N5F0S3	1311.8	366.1	12.4
22		x	A1AT	107	H6N5F1S3	1341.0	366.1	12.4
23		x	A1AT	271	H5N4F0S2	991.2	366.1	9.3
24		x	A1AT	271	H5N4F1S2	1027.7	366.1	9.3
25		x	A1AT	271_MC	H5N4F0S2	1149.9	366.1	11.8
26		x	A1AT	271_MC	H5N4F1S2	1179.1	366.1	11.8
27		x	A1AT	70	H5N4F0S2	1078.5	366.1	14.6
28		x	A1AT	70	H5N4F1S2	1107.7	366.1	14.7
29		x	A2MG	70	H3N3F0S0	721.4	204.1	0.5
30		x	A2MG	70	H5N4F0S1	1130.5	366.1	0.5

(continued)

Table 1
(continued)

No.	Peptide	Glycopeptide	Protein	Site	Glycan moiety	Precursor ion	Product ion	Ret time (min)
31		x	A2MG	70	H5N4F1S1	1203.6	366.1	0.5
32		x	A2MG	70	H6N5F1S1	1386.1	366.1	0.5
33		x	A2MG	70	H5N4F0S2	1276.1	366.1	0.55
34		x	A2MG	70	H5N4F1S2	1349.1	366.1	0.5
35		x	A2MG	869	H5N2F0S0	1158.8	1206.9	7.9
36		x	A2MG	869	H6N2F0S0	1199.3	1206.9	7.9
37		x	A2MG	869	H7N2F0S0	1239.8	1206.9	7.9
38		x	A2MG	869	H5N4F0S1	1066.7	366.1	8.1
39		x	A2MG	869	H5N4F0S2	1124.9	366.1	8.1
40		x	A2MG	247	H5N2F0S0	1239.2	1314.2	9.7
41		x	A2MG	247	H5N4F0S1	1131.0	366.1	9.7
42		x	A2MG	247	H5N4F0S2	1189.2	366.1	10
43		x	A2MG	55	H5N4F0S1	1078.9	366.1	11.2
44		x	A2MG	55	H5N4F1S1	1115.4	366.1	11.2
45		x	A2MG	55	H5N4F0S2	1151.6	366.1	11.5
46		x	A2MG	55	H5N4F1S2	1188.2	366.1	11.5
47		x	A2MG	1424	H5N4F0S1	1020.3	366.1	12.8
48		x	A2MG	1424	H5N4F1S1	1056.8	366.1	12.8
49		x	A2MG	1424	H6N5F0S1	1111.6	366.1	12.8
50		x	A2MG	1424	H6N5F1S1	1148.1	366.1	12.8
51		x	A2MG	1424	H5N4F0S2	1093.1	366.1	12.8
52		x	A2MG	1424	H5N4F1S2	1129.6	366.1	12.8
53		x	A2MG	991	H5N4F0S2	1206.3	366.1	14.7
54		x	AGP1	103	H6N5F0S3	1213.3	366.1	1.9
55		x	AGP1	103	H6N5F1S3	1262.0	366.1	1.9
56		x	AGP1	103	H7N6F0S2	1238.0	366.1	1.9
57		x	AGP1	103	H7N6F0S3	1335.0	366.1	1.9
58		x	AGP1	103	H7N6F0S4	1074.3	366.1	1.9
59		x	AGP1	103	H7N6F1S2	1286.6	366.1	1.9
60		x	AGP1	103	H7N6F1S3	1383.7	366.1	1.9

(continued)

Table 1 (continued)

No.	Peptide	Glycopeptide	Protein	Site	Glycan moiety	Precursor ion	Product ion	Ret time (min)
61		x	AGP1	103	H7N6F1S4	1110.8	366.1	1.9
62		x	AGP1	103	H7N6F2S4	1147.3	366.1	1.9
63		x	AGP1	103	H8N7F0S3	1092.8	366.1	1.9
64		x	AGP1	103	H8N7F0S4	1165.6	366.1	1.9
65		x	AGP1	103	H9N8F0S4	1256.8	366.1	1.9
66		x	AGP1	93	H6N5F0S2	1122.5	366.1	5.8
67		x	AGP1	93	H6N5F1S2	1159.0	366.1	5.8
68		x	AGP1	93	H7N6F0S2	1213.8	366.1	5.8
69		x	AGP1	93	H7N6F1S2	1250.3	366.1	5.8
70		x	AGP1	93	H6N5F0S3	1195.3	366.1	5.8
71		x	AGP1	93	H6N5F1S3	1231.8	366.1	5.8
72		x	AGP1	93	H7N6F0S3	1286.6	366.1	5.8
73		x	AGP1	93	H7N6F1S3	1323.1	366.1	5.8
74		x	AGP1	93	H8N7F0S3	1102.5	366.1	5.8
75		x	AGP1	93	H8N7F1S3	1131.7	366.1	5.8
76		x	AGP1	93	H7N6F0S4	1087.7	366.1	5.8
77		x	AGP1	93	H7N6F1S4	1116.9	366.1	5.8
78		x	AGP1	93	H8N7F0S4	967.4	366.1	5.8
79		x	AGP1/2	56/56	H5N4F0S2	1001.2	366.1	1.7
80		x	AGP1/2	56/56	H6N5F0S2	1122.9	366.1	1.7
81		x	AGP1/2	56/56	H6N5F0S3	1219.9	366.1	1.7
82		x	AGP1/2	56/56	H6N5F1S3	1268.6	366.1	1.7
83		x	AGP1/2	72/72_MC	H7N6F0S2	1167.3	366.1	10.7
84		x	AGP1/2	72/72_MC	H6N5F0S3	1152.5	366.1	10.7
85		x	AGP1/2	72/72_MC	H6N5F1S3	1181.7	366.1	10.7
86		x	AGP1/2	72/72_MC	H7N6F0S3	1225.5	366.1	10.7

(continued)

Table 1
(continued)

No.	Peptide	Glycopeptide	Protein	Site	Glycan moiety	Precursor ion	Product ion	Ret time (min)
87		x	AGP1/2	72/72_MC	H7N6F1S3	1254.7	366.1	10.7
88		x	AGP1/2	72/72_MC	H7N6F0S4	1283.7	366.1	10.7
89		x	AGP1/2	72/72_MC	H7N6F1S4	1094.3	366.1	10.7
90		x	C3	85	H5N2F0S0	1158.3	1230.3	6.4
91		x	C3	85	H6N2F0S0	909.5	1230.3	6.4
92		x	C3	85	H7N2F0S0	950.0	1230.3	6.4
93		x	HP	184	H5N4F0S1	1149.4	366.1	7.8
94		x	HP	184	H5N4F1S1	1186.0	366.1	7.8
95		x	HP	184	H6N5F0S1	992.8	366.1	7.8
96		x	HP	184	H5N4F0S2	1222.2	366.1	7.8
97		x	HP	184	H5N4F1S2	1258.7	366.1	7.8
98		x	HP	184	H6N5F0S2	1051.0	366.1	7.8
99		x	HP	184	H6N5F1S2	1080.2	366.1	7.8
100		x	HP	184	H6N5F0S3	1109.2	366.1	7.8
101		x	HP	184	H6N5F1S3	1138.4	366.1	7.8
102		x	HP	207	H10N8F0S3	1116.4	366.1	3.9
103		x	HP	207	H10N8F0S4	1174.6	366.1	4
104		x	HP	207	H11N9F0S4	1247.7	366.1	4
105		x	HP	207	H11N9F1S4	1276.9	366.1	4
106		x	HP	207	H11N9F0S5	1305.9	366.1	4
107		x	HP	207	H11N9F1S5	1335.1	366.1	4
108		x	HP	207	H12N10F0S5	1378.9	366.1	4
109		x	HP	207	H12N10F1S5	1173.6	366.1	4
110		x	HP	241		599.0	658.4	7
111		x	HP	241	H5N4F0S1	1237.3	366.1	7.1
112		x	HP	241	H6N5F0S1	1019.5	366.1	7.1

(continued)

Table 1
(continued)

No.	Peptide	Glycopeptide	Protein	Site	Glycan moiety	Precursor ion	Product ion	Ret time (min)
113		x	HP	241	H5N4F0S2	1001.0	366.1	7.1
114		x	HP	241	H5N4F1S2	1383.0	366.1	7.1
115		x	HP	241	H6N5F0S2	1092.3	366.1	7.1
116		x	HP	241	H6N5F1S2	1128.8	366.1	7.1
117		x	HP	241	H7N6F0S2	1183.5	366.1	7.1
118		x	HP	241	H6N5F0S3	1165.0	366.1	7.1
119		x	HP	241	H6N5F1S3	1201.5	366.1	7.1
120		x	HP	241	H7N6F0S3	1256.3	366.1	7.1
121		x	HP	241	H7N6F1S3	1292.8	366.1	7.1
122		x	HP	241	H7N6F0S4	1063.5	366.1	7.1
123		x	IgA1/2	144/131	H3N5F0S0	1117.1	366.1	10.8
124		x	IgA1/2	144/131	H4N4F0S1	943.9	366.1	10.8
125		x	IgA1/2	144/131	H4N5F0S0	1157.6	366.1	10.8
126		x	IgA1/2	144/131	H4N5F0S1	1230.4	366.1	10.8
127		x	IgA1/2	144/131	H5N4F0S0	1147.3	366.1	10.8
128		x	IgA1/2	144/131	H5N4F0S1	976.3	366.1	10.8
129		x	IgA1/2	144/131	H5N4F0S2	1292.9	366.1	10.8
130		x	IgA1/2	144/131	H5N5F0S0	1198.1	366.1	10.8
131		x	IgA1/2	144/131	H5N5F0S1	1016.9	366.1	10.8
132		x	IgA1/2	144/131	H5N5F0S2	1075.1	366.1	10.8
133		x	IgA2	205	H4N5F1S0	923.5	366.1	4.1
134		x	IgA2	205	H5N4F1S0	909.8	366.1	4.1
135		x	IgA2	205	H5N4F1S1	1006.8	366.1	4.1
136		x	IgA2	205	H5N4F1S2	1103.8	366.1	4.1
137		x	IgA2	205	H5N5F1S0	977.5	366.1	4.1
138		x	IgA2	205	H5N5F1S1	1074.5	366.1	4.1
139		x	IgA2	205	H5N5F1S2	878.9	366.1	4.1
140		x	IgG1	180	H3N4F1S0	878.8	204.1	2.4
141		x	IgG1	180	H3N5F1S0	946.5	204.1	2.4
142		x	IgG1	180	H4N4F0S0	884.1	204.1	2.4

(continued)

Table 1 (continued)

No.	Peptide	Glycopeptide	Protein	Site	Glycan moiety	Precursor ion	Product ion	Ret time (min)
143		x	IgG1	180	H4N4F1S0	932.8	204.1	2.4
144		x	IgG1	180	H4N4F1S1	1029.8	204.1	2.4
145		x	IgG1	180	H4N5F0S0	951.7	204.1	2.4
146		x	IgG1	180	H4N5F1S0	1000.5	204.1	2.4
147		x	IgG1	180	H5N4F0S0	938.1	366.1	2.4
148		x	IgG1	180	H5N4F1S0	986.8	366.1	2.4
149		x	IgG1	180	H5N4F1S1	1083.8	366.1	2.4
150		x	IgG1	180	H5N5F1S0	1054.5	366.1	2.4
151		x	IgG2	176	H3N4F1S0	868.1	204.1	4
152		x	IgG2	176	H3N5F1S0	935.8	204.1	4
153		x	IgG2	176	H4N4F0S0	873.4	204.1	4
154		x	IgG2	176	H4N4F1S0	922.1	204.1	4
155		x	IgG2	176	H4N4F1S1	1019.1	204.1	4
156		x	IgG2	176	H4N5F0S0	941.1	204.1	4
157		x	IgG2	176	H4N5F1S0	989.9	204.1	4
158		x	IgG2	176	H5N4F1S0	976.1	366.1	4
159		x	IgG2	176	H5N4F1S1	1073.1	366.1	4
160		x	IgG2	176	H5N5F1S0	1043.8	366.1	4
161		x	IgG3/4	227/177	H4N4F1S1	1024.5	204.1	2.5
162		x	IgG3/4	227/177	H4N5F1S0	995.1	204.1	2.5
163		x	IgG3/4	227/177	H5N4F1S1	1078.4	366.1	2.5
164		x	IgG3/4	227/177	H3N5F1S0	941.1	204.1	2.5
165		x	IgG3/4	227/177	H4N4F1S0	927.4	204.1	2.55
166		x	IgM	46	H4N3F1S1	993.8	366.1	2.1
167		x	IgM	46	H4N4F0S1	1012.8	366.1	2.1
168		x	IgM	46	H5N4F1S1	1115.5	366.1	2.1
169		x	IgM	46	H5N4F1S2	909.6	366.1	2.1
170		x	IgM	46	H5N5F0S1	851.1	366.1	2.1
171		x	IgM	46	H5N5F0S2	923.9	366.1	2.1
172		x	IgM	46	H5N5F1S1	1183.2	366.1	2.1

(continued)

Table 1
(continued)

No.	Peptide	Glycopeptide	Protein	Site	Glycan moiety	Precursor ion	Product ion	Ret time (min)
173		x	IgM	46	H5N6F0S1	901.9	366.1	2.1
174		x	IgM	209	H4N5F1S1	1110.8	366.1	6.1
175		x	IgM	209	H5N4F1S1	1100.5	366.1	6.1
176		x	IgM	209	H5N4F1S2	1173.3	366.1	6.1
177		x	IgM	209	H5N5F1S1	1151.3	366.1	6.1
178		x	IgM	209	H5N5F1S2	1224.1	366.1	6.1
179		x	IgM	439	H5N2F0S0	1194.5	1284.7	7.3
180		x	IgM	439	H6N2F0S0	1248.5	1284.7	7.3
181		x	IgM	439	H7N2F0S0	1302.6	1284.7	7.3
182		x	IgM	439	H8N2F0S0	1356.6	1284.7	7.3
183		x	IgM	439	H9N2F0S0	1058.3	1284.7	7.3
184		x	IgM	439		1183.1	342.1	8.2
185		x	TF	432	H5N4F0S1	1131.1	366.1	6.6
186		x	TF	432	H5N4F0S2	921.4	366.1	6.6
187		x	TF	432	H5N4F1S2	957.9	366.1	6.6
188		x	TF	432	H6N5F0S2	1012.7	366.1	6.6
189		x	TF	432	H6N5F0S3	1085.4	366.1	6.6
190		x	TF	630	H5N4F0S1	1108.4	366.1	7.7
191		x	TF	630	H5N4F0S2	1181.1	366.1	7.7
192		x	TF	630	H5N4F1S2	1217.7	366.1	7.6
193		x	TF	630	H6N5F0S2	1018.1	366.1	7.7
194		x	TF	630	H6N5F0S3	1076.4	366.1	7.7
195		x	TF	630	H6N5F1S3	1105.6	366.1	7.7

(b) Do a dilution curve of the standard protein sample to generate a calibration curve for protein concentration determination (Fig. 3). The samples are not diluted prior to analysis, and the injection volume is 1 μl

3.3 Data Analysis

1. Perform data analysis as indicated in the MassHunter manual (*see* **Note 7**), in essence:

   (a) Generate a batch file based on the method file used for the analysis.

   (b) Load all data and have MassHunter integrate the generated data.

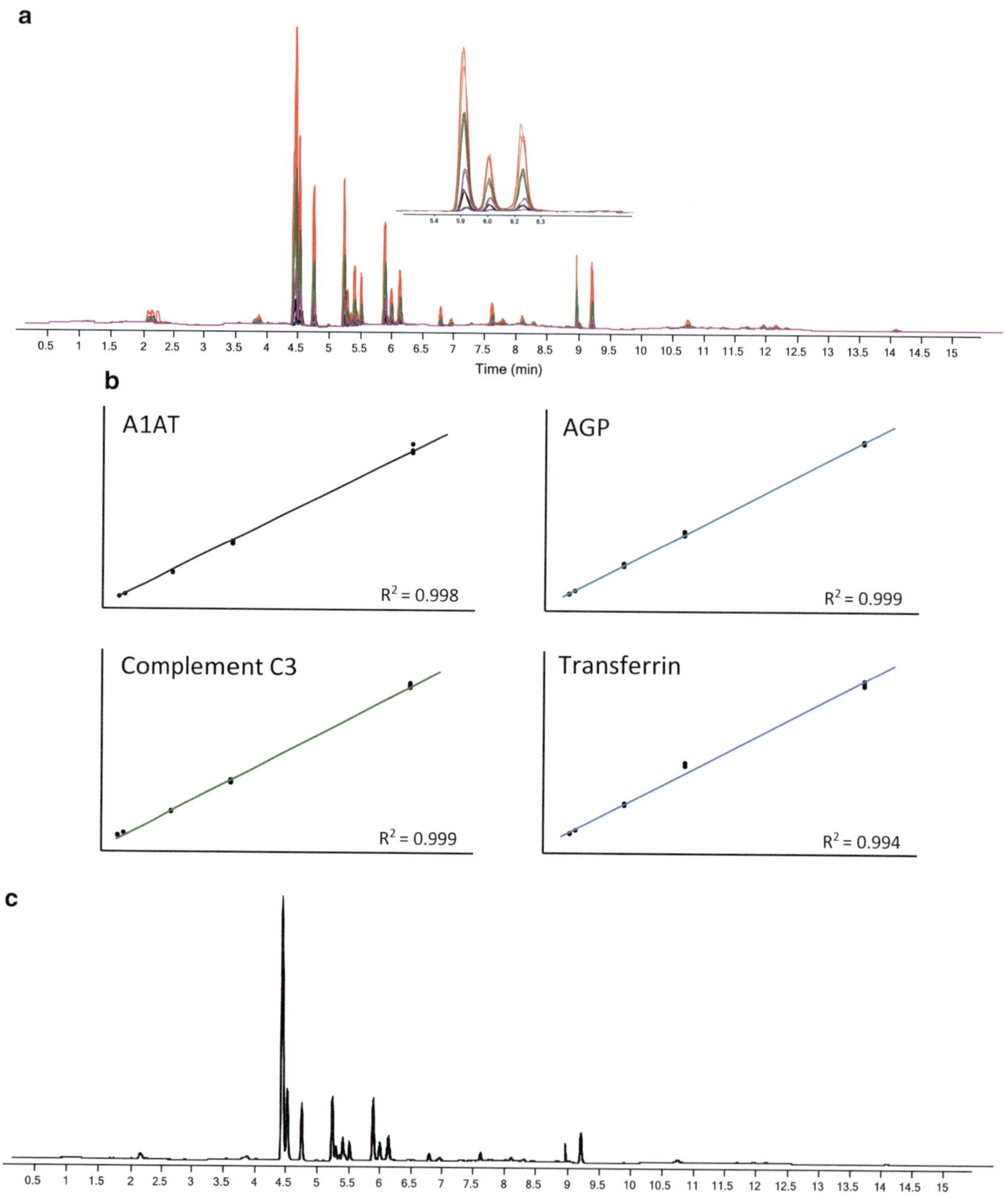

Fig. 3 Stability and accuracy of the MRM method. (**a**) Overlays of MRM chromatograms for the nine high abundance protein standards obtained from a dilution curve. (**b**) Calibration curves for four of the targeted proteins, showing good linearity over at least three orders of magnitude. (**c**) Overlay of chromatograms of ten standard serum samples, indicating good repeatability for serum samples as well

(c) Manually check the performance of the integration, particularly for lower abundant glycopeptides.

2. Export the data as .csv files to excel and manually perform the glycopeptide standardization according to the Table 2 (*see* **Note 8**). The data is now ready for further statistical analysis.

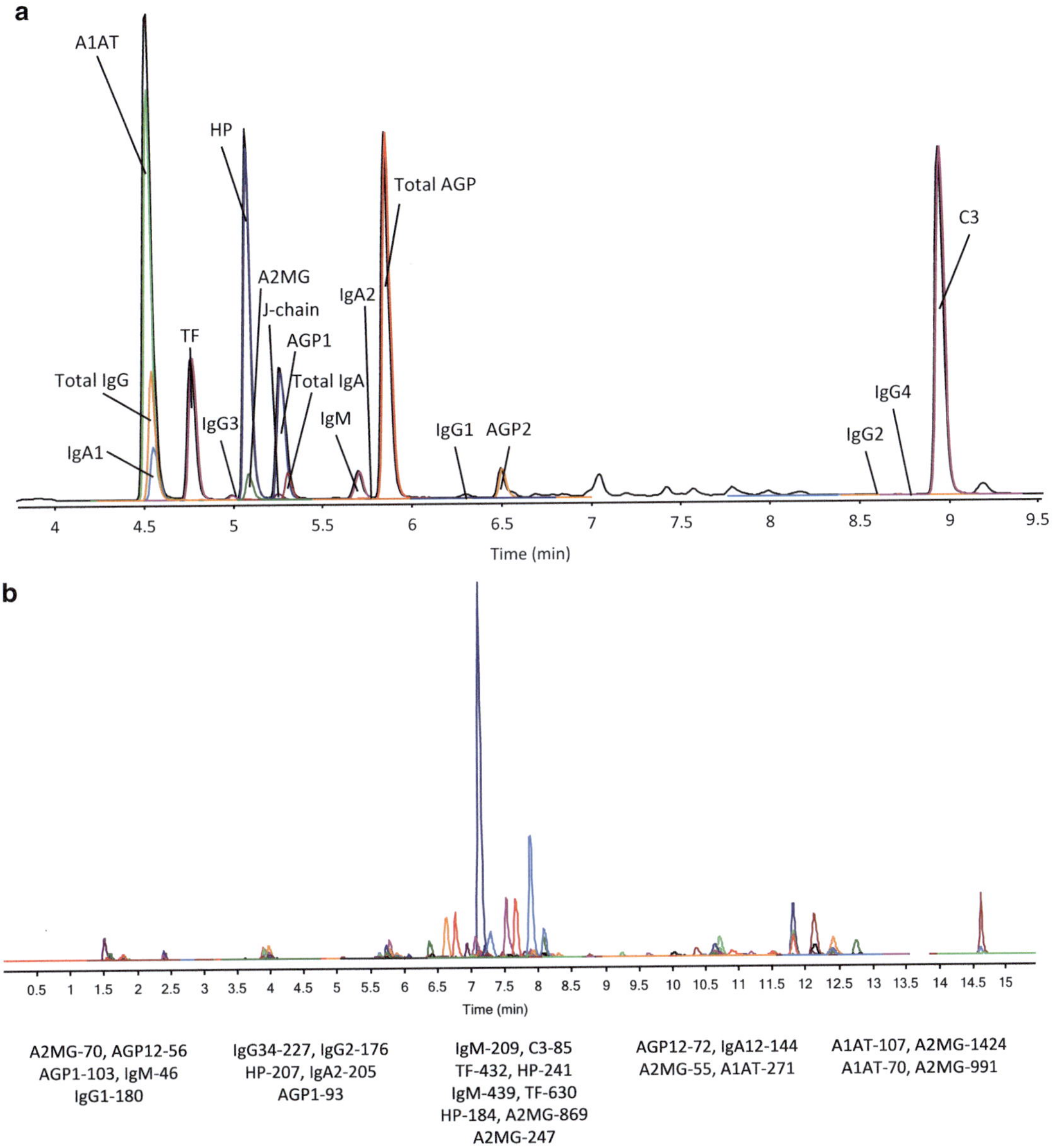

Fig. 4 Signal identification in the MRM chromatograms. (**a**) Chromatogram annotated with the peptide signals used for protein quantitation. The peptide signals are substantially more intense than the glycopeptide signals (**b**) Extracted chromatograms for the glycopeptides monitored, annotated with the peptide moieties of the glycopeptides. Since glycopeptides retain on a C18 stationary phase mostly based on their peptide moiety, all glycopeptides from one site of glycosylation elute around the same retention time window

4 Notes

1. Because MRM is a targeted technique, it is necessary to first generate a glycosylation map of the glycoprotein of interest. Glycosylation maps can be generated by QTOF-MS analysis of tryptic and nonspecific protease digests of standard proteins, as

Table 2
Peptides used for standardization to eliminate differences due to differential protein expression

Glycopeptides from site	Relative to peptide
A1AT(107)	A1AT
A1AT(271)	A1AT
A1AT(70)	A1AT
A2MG(70)	A2MG
A2MG(869)	A2MG
A2MG(247)	A2MG
A2MG(55)	A2MG
A2MG(1424)	A2MG
A2MG(991)	A2MG
AGP1(103)	AGP1
AGP1(93)	AGP1
AGP1/2(56/56)	AGP1/2
AGP1/2(72/72)	AGP1/2
C3(85)	C3
HP(184)	HP
HP(207)	HP
HP(241)	HP
IgA1/2(144/131)	IgA1/2
IgA2(205)	IgA2
IgG1(180)	IgG1
IgG2(176)	IgG2
IgG3/4(227/177)	IgG3 + IgG4
IgM(46)	IgM
IgM(209)	IgM
IgM(439)	IgM
TF(432)	TF
TF(630)	TF

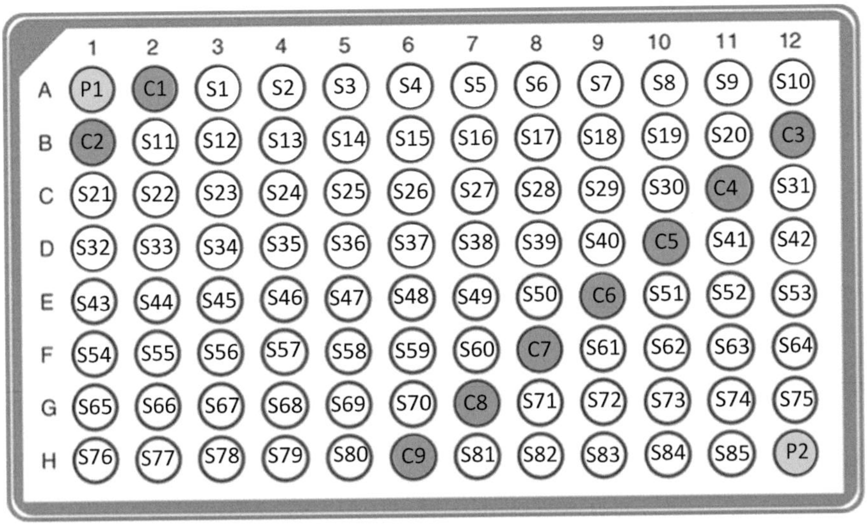

Fig. 5 Potential 96-well plate layout for glycosylation profiling of larger sample sets. The protein mixtures are indicated by P# with wells colored in *light grey*, the standard control samples are indicated with C# with wells colored in *dark grey* and the individual samples are indicated by S# with wells colored in *white*

was done for IgG, IgA, IgM, HP, TF, C3, A2MG, AGP, and A1AT for this study [15, 22].

2. When performing larger scale studies in a high-throughput fashion, it is important to develop a study design and analysis plan prior to the start of the experiments. An important aspect is the plate-layout, for which an example is given in Fig. 5. The protein standard is included in the first and last well of the 96-well plate, while a control sample with similar composition to the samples is included every ten samples. In this example no blank samples are included in the plate-layout, but including blank samples after each control sample could be considered.
3. In order for the protease digestion to be effective, proteins need to be denatured, reduced and alkylated so that their amino acid backbone is exposed and can be accessed by the protease enzymes. In this protocol, this is achieved by addition of DTT for cysteine reduction and iodoacetamide for cysteine alkylation. This results in a 57.05 Da mass increase of cysteine residues. However, other chemicals, such as TCEP can be used for reduction and alternatives such as acrylamide and iodoacetic acid are available for alkylation. Using this strategy samples may also be labeled isotopically.
4. While trypsin is the most commonly used protease, it may not be ideal for the generation of glycopeptides due to for example

multiply glycosylated peptides, as can be shown for one of the tryptic glycopeptides of haptoglobin (HP(207)). Therefore, other proteases, such as chymotrypsin, Lys-C and Lys-N may be evaluated for their generation of more specific glycopeptides. However, this will affect the transitions to be monitored.

5. In this protocol the proteins are digested using trypsin for 16 h at 37 °C. Alternatively, the reaction can be sped up dramatically in multiple ways, such as a microwave reactor or pressure cycling technology [26, 27]. Digestion surfactants, such as RapiGest [28] may also be used, but this will require sample cleanup prior to LC-MS analysis.
6. In MRM, multiple analytes or transitions are detected individually and multiple transitions are detected concurrently in duty cycles. In MRM, the cycle time is the time spent monitoring all transitions in one duty cycle, while the dwell time is the time spent acquiring a specific transition during each duty cycle. Increasing the cycle time will result in limited sampling and thus poor data quality, while shorter dwell times would result in a poorer signal-to-noise ratio, especially for lower abundance analytes.

   In a dynamic MRM method, a specific analyte is only monitored at the time it elutes, which greatly reduces the number of concurrent transitions and saves the duty cycle for the coeluting compounds only. Separation of glycopeptides on reverse phase stationary phases relies mainly on the retention of the peptide moiety and glycopeptides with the same peptide moiety typically elute therefore around the same retention time. To avoid long duty cycles, it is thus necessary to analyze at a specific time point only those glycopeptides that are eluting. In this method, the cycle time was fixed at 500 ms to avoid undersampling.
7. The data evaluation for the study presented here was performed using MassHunter Qualitative analysis, which is Agilent's commercially available software. Within the quantitative mass spectrometry community Skyline has been developed, which is an open access software tool for the evaluation of quantitative (MRM based) mass spectrometry data [29]. Most mass spectrometry vendors support this software tool, and data generated in this study could be evaluated using Skyline, however, to date all glycopeptides will have to be entered manually and the ease of use of Skyline for glycopeptide quantitation is therefore currently low.
8. Using MRM, absolute abundances are obtained. However, absolute ion abundances of glycopeptides are a result of both the absolute glycoprotein concentration and the glycosylation pattern. The glycoprotein concentrations can vary substantially, particularly in biological samples, and thus affect the absolute ion abundances of glycopeptides dramatically [22]. To assess

the glycosylation pattern independent of the protein concentration, transitions are monitored for non-glycosylated peptides and the absolute glycopeptide ion abundances are standardized to the peptide ion abundances, according to the equation shown below.

Degree of glycosylation = glycopeptide ion abundance/ peptide ion abundance

Table 2 lists the sites of glycosylation and their respective peptides used for normalization for the nine glycoproteins monitored in this method.

References

1. Parekh RB, Dwek RA, Sutton BJ, Fernandes DL, Leung A, Stanworth D, et al. (1985) Association of rheumatoid arthritis and primary osteoarthritis with changes in the glycosylation pattern of total serum IgG. Nature 316:452–457
2. Vuckovic F, Kristic J, Gudelj I, Teruel M, Keser T, Pezer M, et al. (2015) Association of systemic lupus erythematosus with decreased immunosuppressive potential of the IgG glycome. Arthritis Rheumatol 67:2978–2989
3. Theodoratou E, Campbell H, Ventham NT, Kolarich D, Pucic-Bakovic M, Zoldos V, et al. (2014) The role of glycosylation in IBD. Nat Rev Gastroenterol Hepatol 11:588–600
4. Ruhaak LR, Uh HW, Beekman M, Hokke CH, Westendorp RG, Houwing-Duistermaat J, et al. (2011) Plasma protein N-glycan profiles are associated with calendar age, familial longevity and health. J Proteome Res 10:1667–1674
5. Yamada E, Tsukamoto Y, Sasaki R, Yagyu K, Takahashi N (1997) Structural changes of immunoglobulin G oligosaccharides with age in healthy human serum. Glycoconj J 14: 401–405
6. Testa R, Vanhooren V, Bonfigli AR, Boemi M, Olivieri F, Ceriello A, et al. (2015) N-glycomic changes in serum proteins in type 2 diabetes mellitus correlate with complications and with metabolic syndrome parameters. PLoS One 10:e0119983
7. Ruhaak LR, Barkauskas DA, Torres J, Cooke CL, Wu LD, Stroble C, et al. (2015) The serum immunoglobulin G glycosylation signature of gastric cancer. EuPA Open Proteom 6:1–9
8. Kim K, Ruhaak LR, Nguyen UT, Taylor SL, Dimapasoc L, Williams C, et al. (2014) Evaluation of glycomic profiling as a diagnostic biomarker for epithelial ovarian cancer. Cancer Epidemiol Biomarkers Prev 23:611–621
9. Ruhaak LR, Miyamoto S, Lebrilla CB (2013) Developments in the identification of glycan biomarkers for the detection of cancer. Mol Cell Proteomics 12:846–855
10. Alley WR Jr, Vasseur JA, Goetz JA, Svoboda M, Mann BF, Matei DE, et al. (2012) N-linked glycan structures and their expressions change in the blood sera of ovarian cancer patients. J Proteome Res 11:2282–2300
11. Goldman R, Ressom HW, Varghese RS, Goldman L, Bascug G, Loffredo CA, et al. (2009) Detection of hepatocellular carcinoma using glycomic analysis. Clin Cancer Res 15:1808–1813
12. Bones J, Byrne JC, O'Donoghue N, McManus C, Scaife C, Boissin H, et al. (2011) Glycomic and glycoproteomic analysis of serum from patients with stomach cancer reveals potential markers arising from host defense response mechanisms. J Proteome Res 10:1246–1265
13. Arnold JN, Saldova R, Galligan MC, Murphy TB, Mimura-Kimura Y, Telford JE, et al. (2011) Novel glycan biomarkers for the detection of lung cancer. J Proteome Res 10: 1755–1764
14. Ruhaak LR, Koeleman CA, Uh HW, Stam JC, van Heemst D, Maier AB, et al. (2013) Targeted biomarker discovery by high throughput glycosylation profiling of human plasma alpha1-antitrypsin and immunoglobulin a. PLoS One 8:e73082
15. Hong Q, Ruhaak LR, Stroble C, Parker E, Huang J, Maverakis E, et al. (2015) A method for comprehensive glycosite-mapping and direct quantitation of serum glycoproteins. J Proteome Res 14(12):5179–5192
16. Behnken HN, Ruthenbeck A, Schulz JM, Meyer B (2014) Glycan analysis of Prostate Specific Antigen (PSA) directly from the intact glycoprotein by HR-ESI/TOF-MS. J Proteome Res 13:997–1001

17. Zauner G, Selman MH, Bondt A, Rombouts Y, Blank D, Deelder AM, et al. (2013) Glycoproteomic analysis of antibodies. Mol Cell Proteomics 12:856–865
18. Huhn C, Selman MH, Ruhaak LR, Deelder AM, Wuhrer M (2009) IgG glycosylation analysis. Proteomics 9:882–913
19. Picotti P, Aebersold R (2012) Selected reaction monitoring-based proteomics: workflows, potential, pitfalls and future directions. Nat Methods 9:555–566
20. Hoofnagle AN, Becker JO, Oda MN, Cavigiolio G, Mayer P, Vaisar T (2012) Multiple-reaction monitoring-mass spectrometric assays can accurately measure the relative protein abundance in complex mixtures. Clin Chem 58:777–781
21. Ruhaak LR, Lebrilla CB (2015) Applications of multiple reaction monitoring to clinical glycomics. Chromatographia 78:335–342
22. Hong Q, Lebrilla CB, Miyamoto S, Ruhaak LR (2013) Absolute quantitation of immunoglobulin G and its glycoforms using multiple reaction monitoring. Anal Chem 85: 8585–8593
23. Song E, Pyreddy S, Mechref Y (2012) Quantification of glycopeptides by multiple reaction monitoring liquid chromatography/tandem mass spectrometry. Rapid Commun Mass Spectrom 26:1941–1954
24. Sanda M, Pompach P, Brnakova Z, Wu J, Makambi K, Goldman R (2012) Quantitative liquid chromatography-mass spectrometry-multiple reaction monitoring (LC-MS-MRM) analysis of site-specific glycoforms of haptoglobin in liver disease. Mol Cell Proteomics 12:1294–1305
25. Hong Q, Ruhaak LR, Totten SM, Smilowitz JT, German JB, Lebrilla CB (2014) Label-free absolute quantitation of oligosaccharides using multiple reaction monitoring. Anal Chem 86(5):2640–2647
26. Alvarado R, Tran D, Ching B, Phinney BS (2010) A comparative study of in-gel digestions using microwave and pressure-accelerated technologies. J Biomol Tech 21:148–155
27. Pramanik BN, Mirza UA, Ing YH, Liu YH, Bartner PL, Weber PC, et al. (2002) Microwave-enhanced enzyme reaction for protein mapping by mass spectrometry: a new approach to protein digestion in minutes. Protein Sci 11:2676–2687
28. Yu YQ, Gilar M, Lee PJ, Bouvier ES, Gebler JC (2003) Enzyme-friendly, mass spectrometry-compatible surfactant for in-solution enzymatic digestion of proteins. Anal Chem 75: 6023–6028
29. MacLean B, Tomazela DM, Shulman N, Chambers M, Finney GL, Frewen B, et al. (2010) Skyline: an open source document editor for creating and analyzing targeted proteomics experiments. Bioinformatics 26: 966–968

Chapter 7

Analysis of Permethylated Glycan by Liquid Chromatography (LC) and Mass Spectrometry (MS)

Shiyue Zhou, Kerry M. Wooding, and Yehia Mechref

Abstract

The development of a reliable and high-throughput glycomic profiling strategy is in high demand due to the biological roles of glycans and their association with different diseases. Native analysis can be quite difficult because of the low ionization efficiency and microheterogeneity of glycans. In this chapter, the sample preparation protocols and LC-MS analysis of permethylated glycan strategies are introduced. Solid-phase permethylation is a fast, convenient, and high-yield method to stabilize sialic acid and improve glycan ionization efficiency and analysis in positive mode; this results in a more sensitive and reliable glycomic profiling strategy. Several modifications in the LC method are also mentioned in this chapter. Online purification simplifies sample preparation and reduces sample loss. Elevating the column temperature significantly improves the peak shape of permethylated glycans and results in isomeric separation. The identification and quantification of permethylated glycans can be achieved through high resolution MS and MS/MS experiments using a MRM method; both approaches are reliable, sensitive, and conducive to high-throughput glycomic studies.

Key words Quantitative glycomics, LC-MS, Permethylation, MRM, Online purification

1 Introduction

Protein glycosylation is a common protein posttranslational modification (PTM) because of its effect on protein function and its prevalence in numerous biological processes. For example, cell-surface glycans are lectin receptors influencing cell adhesion and signaling [1]; N-glycans on the Fc fragment of antibodies affect its stability and efficacy [2]. Glycans are also considered potential biomarkers for different diseases such as immune deficiencies [3–5], hereditary disorders, and various types of cancer [6–8].

Mass spectrometry (MS) is one of the most powerful analytical techniques in bioanalysis. Biomolecules are initially identified with high-resolution MS experiments. Subsequent tandem MS (MS/MS) experiments yield structural information such as protein sequence and glycan composition. Challenges still exist in the

Gordan Lauc and Manfred Wuhrer (eds.), *High-Throughput Glycomics and Glycoproteomics: Methods and Protocols*, Methods in Molecular Biology, vol. 1503, DOI 10.1007/978-1-4939-6493-2_7, © Springer Science+Business Media New York 2017

characterization and quantitation of glycans with MS. Unlike nucleotides and peptides, glycan biosynthesis is a template-free process resulting in the presence of highly branched, complex glycan structures with a wide dynamic range of intensity. The most prevalent sialic acid in mammalian cells, N-acetylneuraminic acid, is widely distributed in N- and O-glycans. The native form of N-acetylneuraminic acid is unstable during the ionization process and has a lower positive mode ionization efficiency that results in a quantitative measurement bias of sialylated glycans. All of these factors contribute to the difficulty in MS-based glycomics analysis even with the use of state-of-the-art modern MS instruments.

Permethylation is one of the most commonly utilized derivatization methods to address the stability and ionization issues mentioned above. During the permethylation process, all reactive glycan hydrogens are replaced with methyl groups. The most widely used glycan permethylation protocols are the method introduced by Ciccanu and Kerek in 1984 [9], which is based on the utilization of methyl iodide, DMSO, and solid sodium hydroxide. This method was further modified with the introduction of minuscule amounts of water to avoid potential oxidative degradation during the reaction [10]; however, side reactions occur when using small quantities of glycan. To overcome this issue, a solid-phase permethylation method has been developed [11]. In solid-phase permethylation, the reaction is processed in micro-spin columns or micro-reactors modified from the capillary. The spin column or capillary are packed with sodium hydroxide beads followed by the mixture of glycans and reaction reagents. In this method, both neutral and sialylated glycans are quickly and efficiently permethylated in a minimized reaction volume. After the reaction, liquid–liquid extraction is utilized to remove salts generated during permethylation.

Permethylated glycans are stable and readily ionize with either matrix-assisted laser desorption ionization (MALDI) or electrospray ionization (ESI). Numerous glycomics studies have employed the combination of permethylation derivatization and MALDI-MS but this approach is not able to detect lower abundant glycan structures. A more robust approach is required for the comprehensive analysis of trace glycans derived from complex biological samples. Coupling MS to separation techniques significantly improves the identification and quantitation of low abundant glycans by reducing competitive ionization and detector saturation. The increased hydrophobicity of permethylated glycans allows for efficient separation with reverse phase liquid chromatography (RPLC). A recent study comparing the glycomic profiling reproducibility of MALDI-MS, RPLC-ESI-MS, and offline RPLC-MALDI-MS [12] highlighted the benefits of employing an online separation step before the MS analysis of permethylated glycans; RPLC-ESI-MS identified the most glycans. The sensitivity of LC-MS analysis of permethylated glycans can be further improved by employing an online purification step [13]. A ten port valve with C18 trapping option is utilized for online glycan purification before separation on the analytical column. This method is more sensitive than offline liquid–liquid extraction and C18 solid-phase extraction (SPE) and yields a simplified sample preparation process making it suitable for high-throughput studies. Recently, we have demonstrated an increase in the separation efficiency of permethylated glycans on a C18 analytical column at elevated temperatures [14] (Fig. 1); this is especially useful for separating highly branched glycans. An optimized column temperature of 55 °C is sufficient to acquire this resolution without damaging the C18 column (*see* **Note 1**). The peak width under this condition is one-third to one fourth of the peak width obtained at ambient temperature separation; an improved separation efficiency would further increase sensitivity and enable isomer identification.

Most LC-MS analysis of permethylated glycans relies on high-resolution MS (HRMS) to resolve the close *m/z* of complex glycans. Multiple reaction monitoring (MRM) experiments on a triple quadrupole mass spectrometer is another strategy for the identification and quantitation of permethylated glycans [15]. In MRM mode, the first quadrupole (Q1) isolates and transfers the analyte of interest, the precursor ion, to the second quadrupole (Q2) where it undergoes collision induced dissociation (CID) to yield fragment ions. These product ions are then transferred to the third quadrupole (Q3) for mass analysis and subsequent abundance determination by a detector (Fig. 2). The quantitation of MRM is based on the intensities of the transitions. SRM/MRM technique has already been utilized in proteomics and glycoproteomics studies [16–19]. Recently, this strategy has been applied to the quantitative analysis of permethylated N-glycans [15]. This approach is quite sensitive as evidenced by the confident relative quantitation of 10 nL of human blood serum.

2 Materials

2.1 PNGase F Digestion

1. PNGase F (glycerol-free, 500,000 units/ml) (New England Biolabs, Ipswich, MA).
2. Ammonium bicarbonate buffer (50 mM solution, pH 7.5–8.0) (Sigma-Aldrich, St. Louis, MO).

2.2 Charcoal SPE Purification

1. Solution A: 85 % acetonitrile, 15 % water, 0.1 % formic acid.
2. Solution B: 5 % acetonitrile, 95 % water, 0.1 % formic acid.
3. Solution C: 40 % acetonitrile, 60 % water, 0.1 % formic acid.

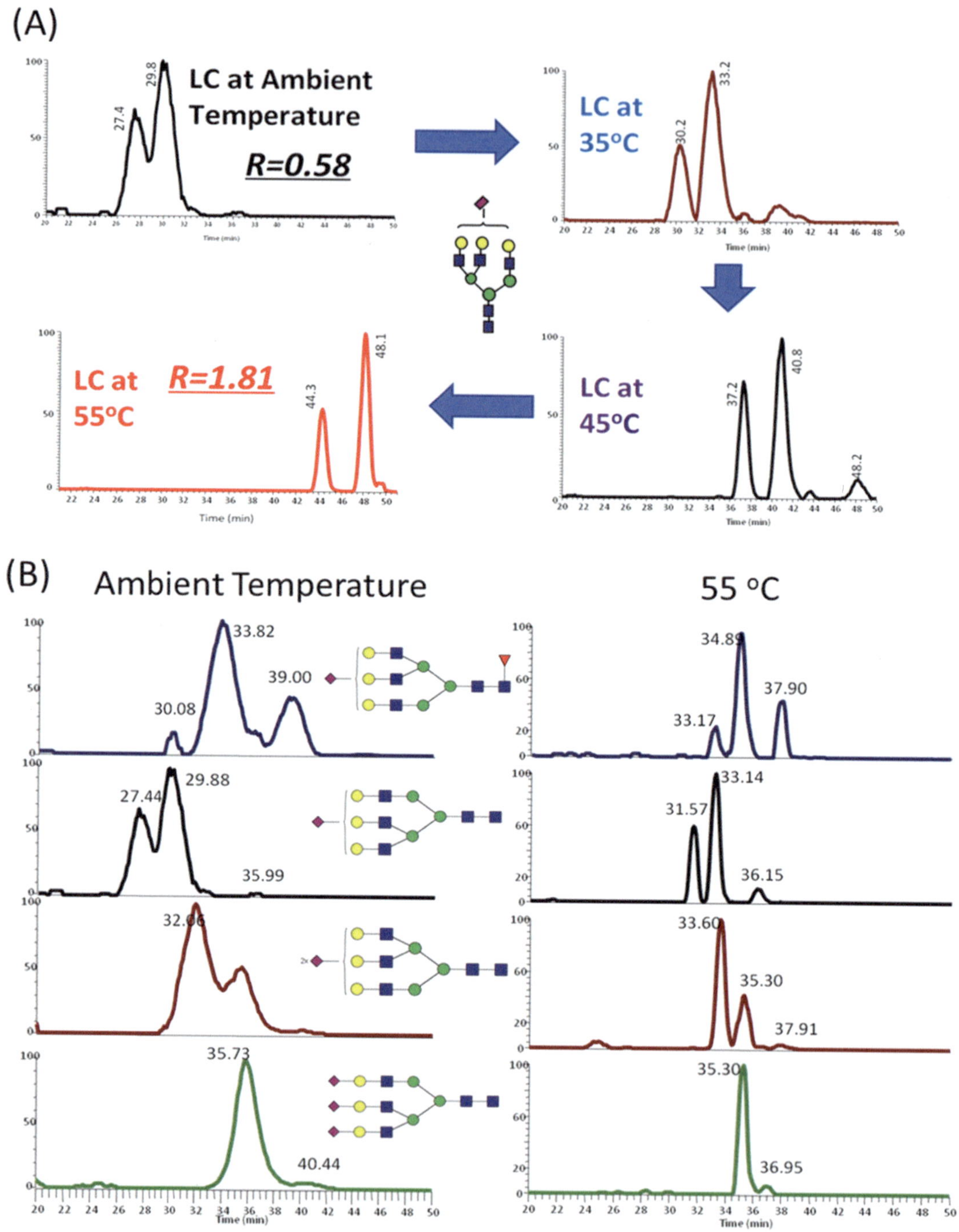

Fig. 1 Extracted ion chromatograph of glycans separated at different temperatures. (**a**) Increased resolution of isomeric peaks from ambient temperature to 55 °C separation; (**b**) Comparison of tri-antennary glycans separated at both ambient temperature and 55 °C

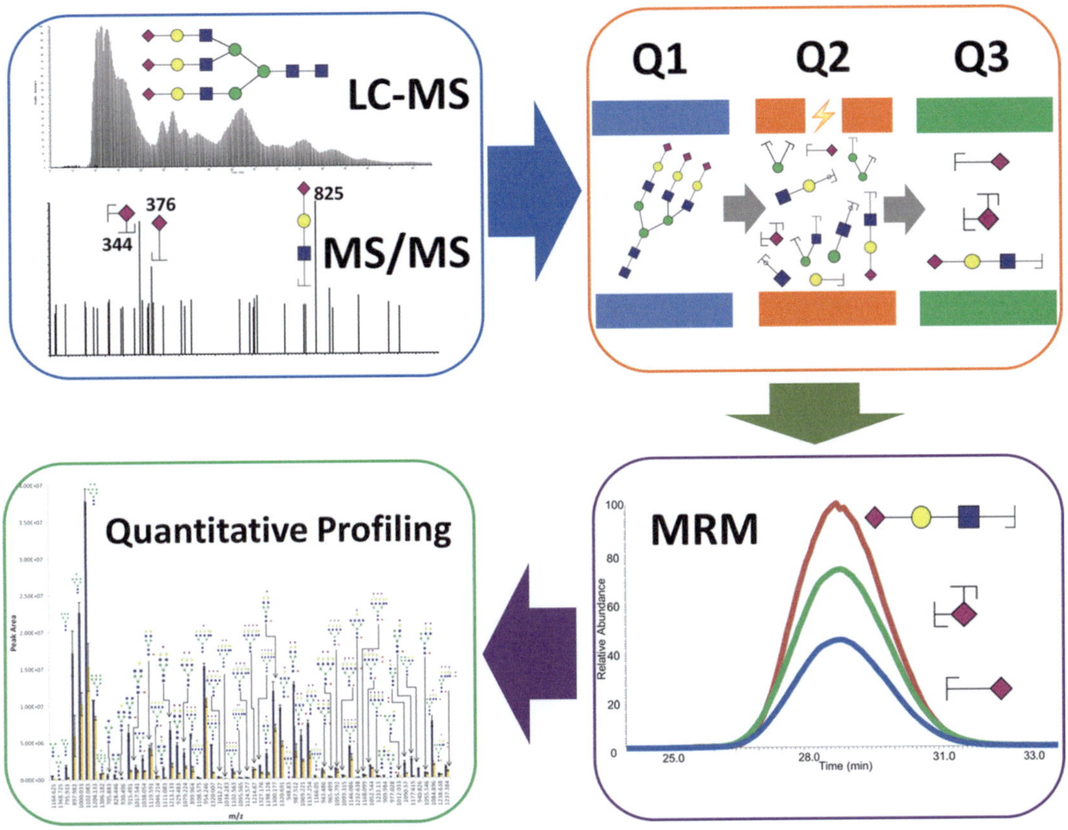

Fig. 2 Schematic representation of MRM analysis of permethylated glycans from transition selection to quantitative profiling

2.3 Reduction

1. Borane-ammonia complex, 97 % (Sigma-Aldrich, St. Louis, MO).
2. HPLC grade menthol.

2.4 Solid-Phase Permethylation

1. Dimethyl sulfoxide, >99.9 % (Sigma-Aldrich, St. Louis, MO).
2. sodium hydroxide beads, 20–40 mesh, 97 % (Sigma-Aldrich, St. Louis, MO).
3. Empty micro spin column, 5 μm frit (Harvard Apparatus, Holliston, MA).
4. Iodomethane contains copper as stabilizer, 99.5 % (Sigma-Aldrich, St. Louis, MO).

2.5 Liquid Chromatography

1. Solvent A (98 % water, 2 % acetonitrile with 0.1 % formic acid).
2. Solvent B (acetonitrile with 0.1 % formic acid).
3. C18 pre-column (3 μm particle size, 100 Å pore size, 75 μm i.d., 2 cm length), Acclaim Pepmap RSLC C18 column (2 μm particle size, 100 Å pore size, 75 μm i.d., 15 cm length) (Thermo Scientific, Sunnyvale, CA).
4. Ultimate 3000 Nano LC system (Thermo Scientific, Sunnyvale, CA).

2.6 Mass Spectrometer

1. LTQ Orbitrap Velos mass spectrometer (Thermo Fisher Scientific Inc., Waltham, MA).
2. Vantage mass spectrometer (Thermo Fisher Scientific Inc., Waltham, MA).

3 Methods

3.1 Release N-Glycan

N-glycans are released from glycoproteins using PNGase F. The protocol may vary depending on the sample matrix. Extracted glycoproteins and biofluids can be directly denatured and deglycosylated with PNGase F while cell lines and tissue samples require additional steps for sample homogenization and glycoprotein extraction.

1. Homogenize tissue or cell line samples by bead beater homogenizer at 4 °C. A cycle of 30 s of beating can be applied until no pellets can be observed in sample
2. Sonicate homogenized sample in an ice–water bath for 1 h.
3. Mix sample with 50 mM pH 7.5 ammonium bicarbonate (ABC) or PBS buffer in a 1:1 ratio.
4. Denature sample in 80 °C water bath for 30 min.
5. When the denatured sample is cooled down to room temperature, add excess PNGase F (~250 units) and mix well by a pipette tip. Incubate in 37 °C water bath for 18 h (*see* **Note 2**).

3.2 Purification of Released N-Glycan

3.2.1 Charcoal SPE Purification

1. Wash charcoal spin column with 400 μL of acetonitrile, spin down at 1.6 krpm for 2 min.
2. Wash charcoal spin column with 400 μL of 85 % acetonitrile with 15 % water, spin down at 1.6 krpm for 2 min.
3. Wash charcoal spin column with 400 μL of Solution B, spin down at 1.6 krpm for 2 min; repeat this step two times.
4. Adjust the sample volume to 400 μL of Solution A, load the sample into spin column and spin down at 1.6 krpm for 2 min
5. Wash the spin column with 400 μL of Solution A, spin down at 1.6 krpm for 2 min; repeat this step three times.
6. Elute the spin column with 400 μL of Solution C, spin down at 1.6 krpm for 2 min; repeat this step two times. Collect and dry the eluents.

3.2.2 Protein Precipitation

1. Add 500 μL of ice-cold 90 % ethanol into the sample.
2. Incubate the mixture in −20 °C freezer for 30 min.
3. Centrifuge at 14.8 krpm for 10 min, collect and dry the supernatant part.

3.2.3 Dialysis

1. Place dialysis membrane (MWCO: 500–1000 Da) into the homemade dialysis device.
2. Resuspend sample in 50 μL water and apply sample on the top of the dialysis membrane, the lower chamber is filled with cycled deionized water.
3. After 12 h, samples are collected from dialysis device, each well is rinsed with 100 μL of deionized water containing 0.1 % formic acid.

3.3 Reduce N-Glycan

1. Prepare fresh 10 mg/mL borane-ammonia aqueous solution.
2. Add 10 μL of borane-ammonia solution into purified and dried glycan sample.
3. Incubate the mixture in 60 °C water bath for 1 h.
4. Take the sample out and dry it in a speed vacuum.
5. Add 400 μL of methanol into the sample and dry it in a speed vacuum; repeat this step until there are no white pellets left in the dried sample

3.4 Solid-Phase Permethylation

3.4.1 Packing Sodium Hydroxide Beads into Empty Column

1. Sodium hydroxide beads are soaked in DMSO for packing into the reaction column. The mixture of sodium hydroxide and DMSO can be stored at room temperature for up to 1 month.
2. Use a 1 mL pipette to transfer the sodium hydroxide beads in DMSO mixture into an empty spin column. The pipette tip should be cut so that sodium hydroxide beads can be aspirated into the pipette tip. The packed sodium hydroxide bead bed volume should be ~2 cm in height.
3. Centrifuge the spin column at 1.8 krpm for 2 min (or longer time until no visible DMSO is left in the column), to force the DMSO out of the spin column.
4. Add 200 μL of DMSO to the spin column for washing beads, also, Centrifuge the column at 1.8 krpm for 2 min (or longer time until no visible DMSO is left in the column). The spin column is ready for the permethylation reaction (*see* **Note 3**).

3.4.2 Solid-Phase Derivatization

1. The purified and dried glycan is resuspended with 30 μL DMSO, 1.2 μL water, and 20 μL iodomethane (*see* **Note 4**).
2. The whole mixture is added to the prepared sodium hydroxide column and incubated at room temperature for 25 min; then add 20 μL of iodomethane into the column and incubate at room temperature for 10 min. Try to keep the column in a low-humidity environment during the incubation.
3. Place the spin column in a clean Eppendorf tube and centrifuge the spin column to elute the mixture.

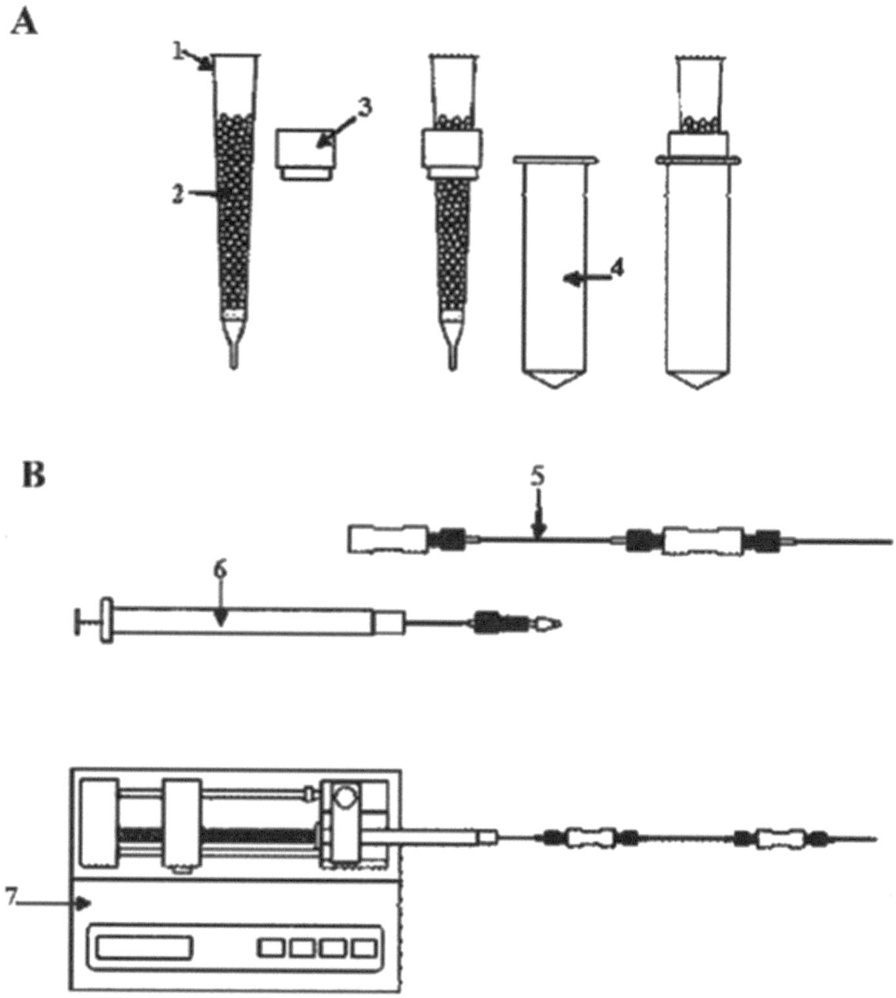

Fig. 3 Schematic representation of the different permethylation techniques. (**a**) Spin-column setup: **1.** spin column; **2.** NaOH beads; **3.** column holder; **4.** microcentrifuge tube. (**b**) 500 μm i.d. fused-silica capillary setup: **5.** 500 μm i.d. fused-silica capillary; **6.** syringe; **7.** syringe pump. This figure was reproduced from ref. 11 with permission

4. Dry the eluent in a speed vacuum, it may take longer than 10 h since DMSO is hard to evaporate (*see* **Note 5**).
5. After drying, there may be white sodium and iodide salts in the bottom of the Eppendorf tube. The dried mixture can be directly analyzed by LC-MS with an online trap. For MALDI or direct infusion ESI-MS, liquid–liquid extraction, or SPE should be used to remove salts. The schematic representation of glycan permethylation is shown in Fig. 3.

3.5 Liquid Chromatography Conditions

After permethylation, the increased hydrophobicity of the glycan allows for efficient separation on a C18 column. An elevated temperature setting improves the chromatographic resolution, reduces analytical column back pressure, and obtains a stable nanoESI spray. We have included an online purification step in this LC method to remove any by-products generated during the permethylation reaction.

1. Resuspend dried sample using 20 % acetonitrile with 0.1 % formic acid.
2. Set the column oven temperature to 55 °C and allow it to equilibrate for ~30 min before starting the analysis.
3. The multi-gradient method is as follows:

   0–10 min: Initial position of the 10-port valve is set to the 1_2 position. Loading pump has a 3 μL/min flow rate, 100 % A. Nano pump has 0.35 μL/min flow rate, 20 % B

   10–11 min: 10-port valve is switched to the 1_10 position. Nano pump has mobile phase B increased to 38 %.

   12–43 min: ramping of mobile phase B from 38 % to 60 %.

   44–45 min: ramping of mobile phase B from 60 % to 90 %.

   46–50 min: keep mobile phase B at 90 %.

   51 min: decrease mobile phase B from 90 % to 20 %.

   52–60 min: keep mobile phase B at 20 %.

3.6 Mass Spectrometer Parameters

3.6.1 High-Resolution Mass Spectrometer Parameters

A nanoESI source is utilized to interface a nanoLC to an LTQ Velos Orbitrap (*see* **Note 6**). The ESI spray voltage is set to 1.6 kV. Transfer tube temperature is set to 275 °C. The data is acquired in data-dependent acquisition (DDA) mode; the four most abundant precursor ions are isolated for MS^2 analysis by collision-induced dissociation (CID) and higher energy collision dissociation (HCD). The dynamic exclusion feature was utilized to ensure no repeated ion is selected during a 30-s moving window. MS acquisition is set to scan a range of 600–2000 m/z with the resolution set to 15,000. The CID MS^2 settings are the following: 3.0 m/z isolation width, 10 ms activation time, 30 % normalized collision energy, 0.25 activation Q value. The HCD MS^2 utilizes a 1.0 m/z isolation width with 40 % normalized collision energy.

3.6.2 MRM Tandem Mass Spectrometer Parameters

A nanoESI source with the same settings is utilized to interface the LC to a TSQ Vantage. DDA mode is utilized in this MRM experiment. The full scan is generated with an acquisition range of 300–1500 m/z in Q3. The scan time is set to 0.7 s with a peak width of 0.7 FWHM. The five most intense ions are selected from the full scan and subjected to collision induced dissociation. The normalized collision energy is set to 30–45 %, depending on the structure of the precursor ion. The intensities of three transitions were recorded for each precursor ion. The m/z values of three transitions are glycan structure dependent. The workflow for analyzing permethylated N-glycans is shown in Fig. 4.

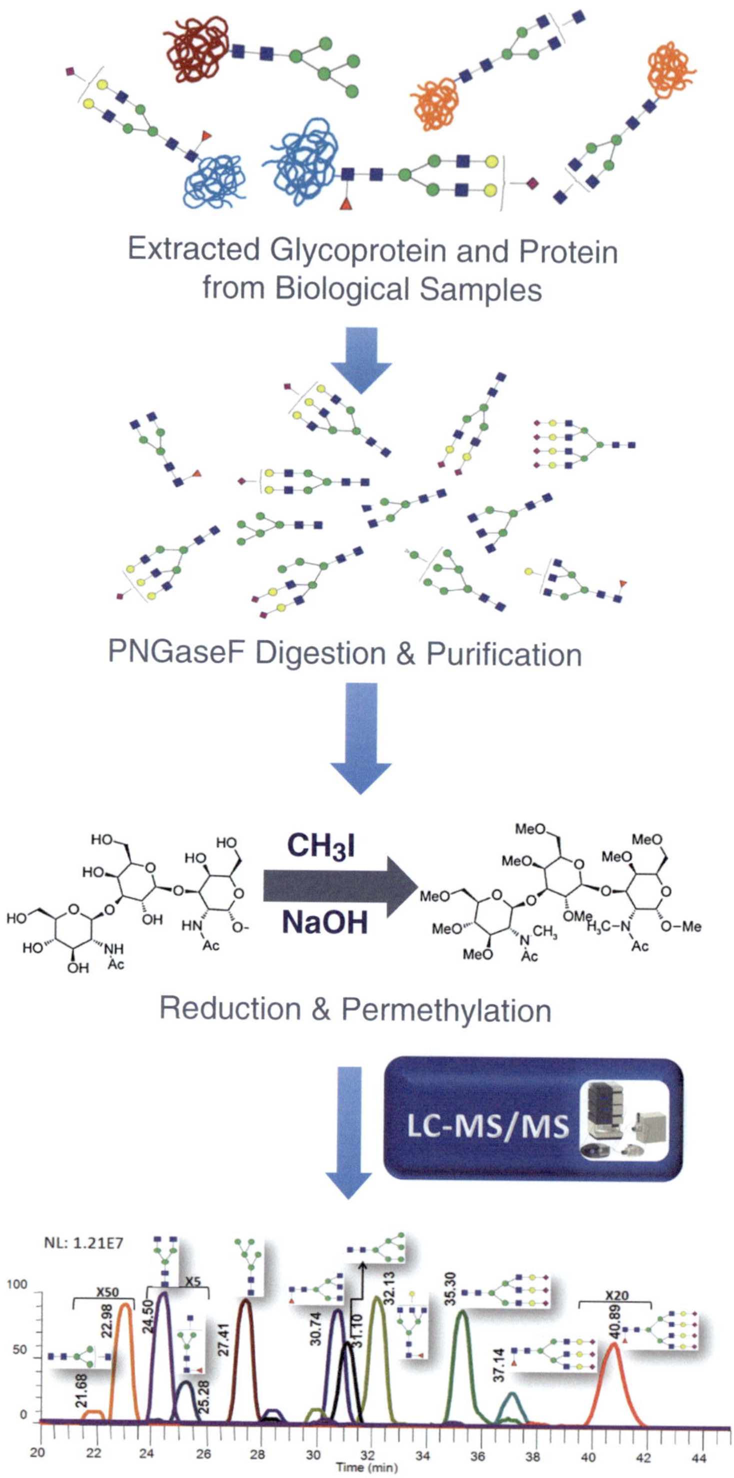

Fig. 4 Sample preparation work flow including glycan release, purification, reduction, and permethylation; extracted ion chromatogram (EIC) of LC-MS analysis of permethylated glycans is shown at the bottom

3.7 Data Processing

3.7.1 High-Resolution Mass Spectrometer

The relative quantitation of glycans in Orbitrap mass spectrometers is based on the ion intensities acquired in the full scan. For the manual data processing, Xcalibur Qual Browser is utilized to generate the extracted ion chromatogram (EIC) for all adducts and charge states of glycans. The mass tolerance is set to 10 ppm. The single full scan spectrums under EIC are manually checked to eliminate false positive results generated by Xcalibur. The accumulated peak area of EIC is recorded to represent the intensity of the corresponding glycan. Another software is available, and some have proven to permit rapid data processing. Here we use MultiGlycan [20, 21] to process the data generated in this method (Fig. 5). Algorithms were employed for the peak validation, and different adducts are merged to calculate the total glycan intensity. Fast and reliable annotation and quantitation of glycans can be achieved by using MulitGlycan (*see* **Note 7**).

3.7.2 MRM Tandem Mass Spectrometer

Xcalibur Qual Browser is used for processing the MRM experiment data. The intensities of three transitions were summed for each precursor ion to rebuild the EIC by selecting scan filters in the mass range option. The peak area of the EIC represents the intensity of a corresponding glycan.

4 Notes

1. We examined the effect of temperature (25–75 °C) on analytical column performance; there were no discernible improvement in resolution above 55 °C. Also of note, the C18 trap loses its trapping efficiency after approximately 20 runs at 75 °C. We selected 55 °C as an optimal temperature for LC-MS analysis of permethylated glycans.
2. The amount of PNGase F added will depend on the quantity of glycoproteins subjected to PNGase F treatment. For example, 60 Units will be used to release glycans from 50 μg of the glycoprotein. In the case of blood serum, 100 units of PNGase F are added to 10 μL sample.
3. Sodium hydroxide is hygroscopic, and the absorbance of water interferes with the reaction. Load the sample into the spin column as soon as the sodium hydroxide column is ready. The incubation should be performed in a low-humidity environment.
4. The volume of water, DMSO and iodomethane added are sufficient for up to 10 μg N-glycans. We recommend reducing these volumes by half when the glycan amount is less than 1 μg.
5. To ensure permethylation labeling efficiency, we recommend limiting each sample batch to no more than 24 samples.

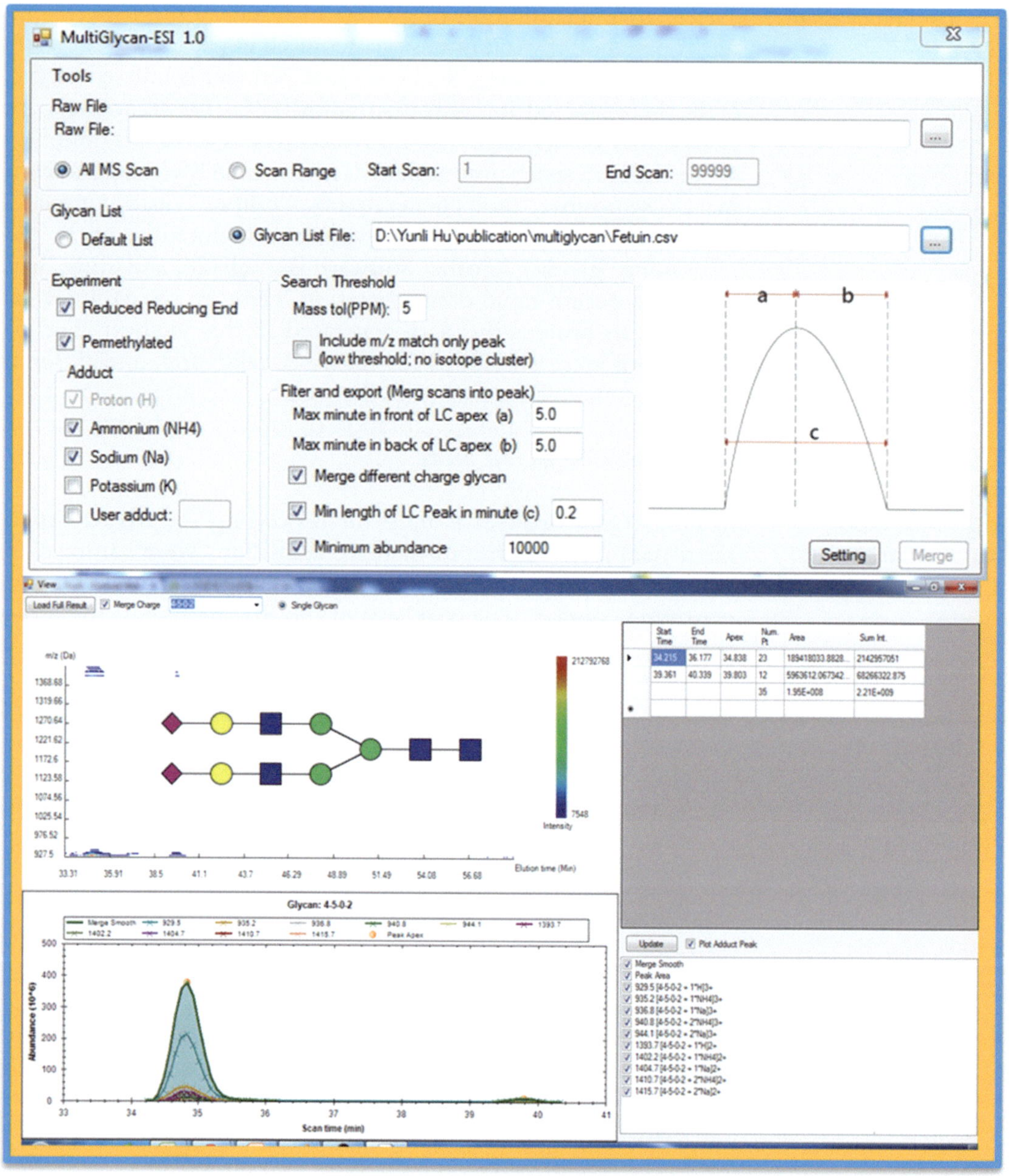

Fig. 5 Representative data from MultiGlycan-ESI-based automatic software annotation and quantitation method. This figure was reproduced from ref. 21 with permission

6. Any high resolution mass spectrometer can be used; usually a 10 ppm mass accuracy is sufficient to identify glycans through a full MS scan.
7. MultiGlycan is discussed in detail here because it has been specifically developed for analyzing permethylated glycans on

LC-MS. Other software also enables rapid and reliable glycomic data processing: SimGlycan [22, 23] (identification and quantitation of differently derivatized glycans and TMT labeled multiplex glycan data), GlycoWorkbench [24–26] (prediction of glycan fragment ions); Cartoonist [27] (automated annotation of MALDI-MS analyzed glycans). Glycomod [28], GlycoQuest (Bruker Daltonics Inc., Billerica, MA), Glyco-Peakfinder [29], and SysBioWare [30] are also routinely used for analyzing glycomics data.

References

1. Haslam SM, North SJ, Dell A (2006) Mass spectrometric analysis of N- and O-glycosylation of tissues and cells. Curr Opin Struct Biol 16(5):584–591
2. Raju TS (2008) Terminal sugars of Fc glycans influence antibody effector functions of IgGs. Curr Opin Immunol 20(4):471–478
3. Back NK et al. (1994) An N-glycan within the human immunodeficiency virus type 1 gp120 V3 loop affects virus neutralization. Virology 199(2):431–438
4. Kolchinsky P et al. (2001) Loss of a single N-linked glycan allows CD4-independent human immunodeficiency virus type 1 infection by altering the position of the gp120 V1/V2 variable loops. J Virol 75(7):3435–3443
5. Doores KJ et al. (2010) Envelope glycans of immunodeficiency virions are almost entirely oligomannose antigens. Proc Natl Acad Sci U S A 107(31):13800–13805
6. Dennis JW, Granovsky M, Warren CE (1999) Protein glycosylation in development and disease. Bioassays 21(5):412–421
7. Lowe JB, Marth JD (2003) A genetic approach to mammalian glycan function. Annu Rev Biochem 72:643–691
8. Mechref Y et al. (2012) Identifying cancer biomarkers by mass spectrometry-based glycomics. Electrophoresis 33(12):1755–1767
9. Ciucanu I, Kerek F (1984) A simple and rapid method for the permethylation of carbohydrates. Carbohydr Res 131(2):209–217
10. Ciucanu I, Costello CE (2003) Elimination of oxidative degradation during the per-O-methylation of carbohydrates. J Am Chem Soc 125(52):16213–16219
11. Kang P et al. (2005) Solid-phase permethylation of glycans for mass spectrometric analysis. Rapid Commun Mass Spectrom 19(23): 3421–3428
12. Hu Y, Mechref Y (2012) Comparing MALDI-MS, RP-LC-MALDI-MS and RP-LC-ESI-MS glycomic profiles of permethylated N-glycans derived from model glycoproteins and human blood serum. Electrophoresis 33(12):1768–1777
13. Desantos-Garcia JL et al. (2011) Enhanced sensitivity of LC-MS analysis of permethylated N-glycans through online purification. Electrophoresis 32(24):3516–3525
14. Zhou S, Hu Y, Mechref Y High-temperature LC-MS analysis of permethylated glycans derived from glycoproteins. Electrophoresis 37:1506–1513
15. Zhou S et al. (2015) Quantitation of permethylated N-glycans through multiple-reaction monitoring (MRM) LC-MS/MS. J Am Soc Mass Spectrom 26(4):596–603
16. Gil GC, Velander WH, Van Cott KE (2009) N-glycosylation microheterogeneity and site occupancy of an Asn-X-Cys sequon in plasma-derived and recombinant protein C. Proteomics 9(9):2555–2567
17. Ahn YH et al. (2009) Quantitative analysis of an aberrant glycoform of TIMP1 from colon cancer serum by L-PHA-enrichment and SISCAPA with MRM mass spectrometry. J Proteome Res 8(9):4216–4224
18. Kurogochi M et al. (2010) Sialic acid-focused quantitative mouse serum glycoproteomics by multiple reaction monitoring assay. Mol Cell Proteomics 9(11):2354–2368
19. Zhao Y et al. (2011) Fragmentation and site-specific quantification of core fucosylated glycoprotein by multiple reaction monitoring-mass spectrometry. Anal Chem 83(22):8802–8809
20. Yu CY et al. (2013) Automated annotation and quantification of glycans using liquid chromatography-mass spectrometry. Bioinformatics 29(13):1706–1707
21. Hu Y et al. (2015) Automated annotation and quantitation of glycans by liquid chromatography/electrospray ionization mass spectrometric analysis using the MultiGlycan-ESI

computational tool. Rapid Commun Mass Spectrom 29(1):135–142

22. Apte A, Meitei NS (2010) Bioinformatics in glycomics: glycan characterization with mass spectrometric data using SimGlycan. Methods Mol Biol 600:269–281
23. Meitei NS et al. (2015) Automating mass spectrometry-based quantitative glycomics using aminoxy tandem mass tag reagents with SimGlycan. J Proteomics 127(Pt A):211–222
24. Ceroni A et al. (2008) GlycoWorkbench: a tool for the computer-assisted annotation of mass spectra of glycans. J Proteome Res 7(4): 1650–1659
25. Damerell D et al. (2015) Annotation of glycomics MS and MS/MS spectra using the GlycoWorkbench software tool. Methods Mol Biol 1273:3–15
26. Damerell D et al. (2012) The GlycanBuilder and GlycoWorkbench glycoinformatics tools: updates and new developments. Biol Chem 393(11):1357–1362
27. Goldberg D et al. (2005) Automatic annotation of matrix-assisted laser desorption/ionization N-glycan spectra. Proteomics 5(4):865–875
28. Cooper CA, Gasteiger E, Packer NH (2001) GlycoMod—a software tool for determining glycosylation compositions from mass spectrometric data. Proteomics 1(2):340–349
29. Maass K et al. (2007) "Glyco-peakfinder"—de novo composition analysis of glycoconjugates. Proteomics 7(24):4435–4444
30. Vakhrushev SY, Dadimov Ď, Peter-Katalinic J (2009) Software platform for high-throughput glycomics. Anal Chem 81(9): 3252–3260

Chapter 8

High-Throughput Analysis of the Plasma *N*-Glycome by UHPLC

Barbara Adamczyk, Henning Stöckmann, Róisín O'Flaherty, Niclas G. Karlsson, and Pauline M. Rudd

Abstract

The understanding of glycosylation alterations in health and disease has evolved significantly and glycans are considered to be relevant biomarker candidates. High-throughput analytical technologies capable of generating high-quality, large-scale glycoprofiling data are in high demand. Here, we describe an automated sample preparation workflow and analysis of *N*-linked glycans from plasma samples using hydrophilic interaction liquid chromatography with fluorescence detection on an ultrahigh-performance liquid chromatography (UHPLC) instrument. Samples are prepared in 96-well plates and the workflow features rapid glycoprotein denaturation, enzymatic glycan release, glycan purification on solid-supported hydrazide, fluorescent labeling, and post-labeling cleanup with solid-phase extraction. The development of a novel approach for plasma *N*-glycan analysis and its implementation on a robotic platform significantly reduces the time required for sample preparation and minimizes technical variation. It is anticipated that the developed method will contribute to expanding high-throughput capabilities to analyze protein glycosylation.

Key words *N*-linked glycosylation, Plasma, Glycan analysis, Ultrahigh-performance liquid chromatography, Robotics, Automation, High-throughput

1 Introduction

The growing interest in glycomics has provided an impetus for the development of high-throughput workflows for sample preparation and analysis. The biological relevance of glycans has been exemplified in many processes including protein folding, secretion and degradation, cell signaling, immune response, to name but a few [1–4]. Changes in glycosylation play a pivotal role in many diseases, such as cancer, autoimmune and infectious diseases as well as congenital disorders of glycosylation [5, 6]. Therefore, exploiting disease-associated alterations in glycosylation represents an excellent target for the discovery and development of biomarkers. The identification of potential glycan-based targets can improve the diagnostic value of protein biomarkers currently used

Gordan Lauc and Manfred Wuhrer (eds.), *High-Throughput Glycomics and Glycoproteomics: Methods and Protocols*, Methods in Molecular Biology, vol. 1503, DOI 10.1007/978-1-4939-6493-2_8,

by increasing their sensitivity and specificity since many cancer associated proteins display clear shifts in their glycan profiles.

Glycomics has lagged behind other -omics disciplines, in part due to the complexity of glycans, which pose more significant analytical challenges compared to genomics or proteomics. The ongoing drive to elucidate these complex structures facilitates the development and great advancement of current glycomics technologies. Deciphering the glycome is a challenging task, in which glycoanalytical methods such as high or ultrahigh-performance liquid chromatography (HPLC, UHPLC), mass spectrometry (MS), capillary electrophoresis (CE), nuclear magnetic resonance (NMR), lectin and glycan arrays are aiding in the characterization of the complete glycome [7–13].

One of the most widely reported and well-known methods for glycan quantification is chromatographic separation with fluorescence detection [14]. The standard procedure includes enzymatic de-glycosylation of glycoproteins, labeling of released glycans with a fluorescent tag and cleanup procedures to remove excess reagents followed by separation on an HPLC instrument, recently replaced by UHPLC [9, 15]. Previously in our laboratory, a reproducible, quantitative, and robust method was developed for *N*-glycan analysis that involved glycoprotein immobilization in polyacrylamide [8, 16]. This method was successfully employed in many studies [17, 18] but immobilization in a gel block hampered the possibility for automation. Several sample preparation protocols have been published to date [19–23]. However, most of them have focused on analyzing antibodies, especially IgG.

Glycoprofiling of serum or plasma *N*-glycome is more challenging due to the higher complexity and overall heterogeneity of plasma glycoproteins. Although plasma glycoprofiling has substantial relevance and delivers a comprehensive view of released glycans from all plasma glycoproteins, it lacks information about the original glycan attachment sites and the identity of the carrier protein. Nevertheless, plasma glycoprofiling represents a recognized aim that can deliver interesting biological insights for further exploration.

While glycan analysis capabilities continue to improve and new approaches are developed, the sample preparation is still a labor-intensive and time consuming process. The manual workup is suitable for small sample sets but in order to meet requirements for large scale glycoprofiling, automated workflows are necessary. To that end, we have developed a high-throughput 96-well platform which is automated through the use of a robotic liquid handling system and applicable for IgG and plasma glycoprofiling (Fig. 1) [21, 24]. The method described here is an updated version and represents an advance in technology available for glycan analysis and should facilitate more routine analysis and large-scale studies.

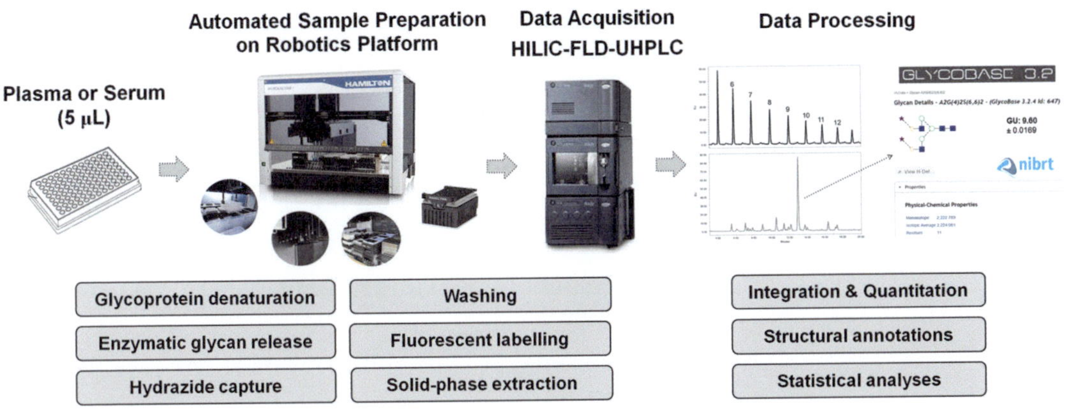

Fig. 1 Overview of the high-throughput workflow for sample preparation and analysis

2 Materials

2.1 Robotic Platform

Hamilton Robotics StarLet liquid handling workstation (Hamilton, Reno, NV) and safety cabinet with its key components: (1) eight independent pipetting channels with liquid-level detection and anti-droplet control, (2) robotic vacuum manifold, (3) plate transport tool, (4) temperature-controlled orbital shaker, (5) Hamilton Robotics Venus one software (*see* **Note 1**).

2.2 Sample Preparation

1. Denaturation solution—100 mM ammonium bicarbonate, 12 mM dithiothreitol (DTT).
2. 96-well V-bottom PP plate (cat.no.: 651201, Greiner Bio-One, Frickehausen, Germany).
3. Alkylating agent—120 mM iodoacetamide solution (IAA). This solution is light-sensitive and must be kept in the dark.
4. Trypsin (cat. no.: M150, Amresco, Solon, OH).
5. Greiner multiwell plate sealers (cat no.: A5596, Sigma-Aldrich).

2.3 PNGase F Digestion

1. Recombinant peptide-N-glycosidase F (PNGaseF; EC 3.5.1.52) expressed in *E. coli* (GKE-5006D, Prozyme, Glyco *N*-Glycanase).
2. PNGaseF solution—1.0 M ammonium bicarbonate.

2.4 Hydrazide-Assisted Glycan Cleanup

1. 96-well chemically inert filter plate with hydrophobic polytetrafluoroethylene (PTFE) membrane, 0.45 μm pore size (cat no.: MSRPN0450, Millipore, Billerica, MA).
2. Ultralink hydrazide resin (cat no.: 53149, ThermoFisher Scientific, Waltham, MA).

3. Methanol (MeOH), acetonitrile (MeCN), water (H_2O), and acetic acid (CH_3COOH) of the highest purity available.
4. 2.0 M Guanidine (CH_5N_3), triethylamine ($(C_2H_5)_3N$) (Sigma-Aldrich), acetic anhydride ($(CH_3CO)_2O$) (Sigma-Aldrich).

2.5 2AB Labeling

1. Dimethylsulfoxide (DMSO, cat no.: D8779, Sigma-Aldrich).
2. Acetic acid (CH_3COOH, cat no.: A6283, Sigma-Aldrich).
3. Sodium cyanoborohydride ($NaBH_3CN$, cat no.: 156159, Sigma-Aldrich).
4. 2-aminobenzamide (2-AB, anthranilamide, cat no.: A89804, Sigma-Aldrich).

2.6 Solid-Phase Extraction (SPE)

1. Polypropylene square well 2 mL/well plates (cat no.: AHO-7194, Phenomenex, Torrance, CA).
2. HyperSep™ Diol SPE cartridges (cat no.: 60108, ThermoFisher Scientific).
3. MeOH, MeCN, H_2O of the highest purity available.
4. AcroPrep™ Advance 96-Well filter plates with hydrophilic polypropylene membrane, 0.45 μm pore size (cat no.: 8084, Pall, Port Washington, NY).

2.7 HILIC-FLD- UHPLC

1. UHPLC instrument (Acquity UPLC H-class) consisting of a binary solvent manager, sample manager, and fluorescent detector.
2. Hydrophilic interaction liquid chromatography (HILIC) column for glycan separation, BEH Glycan column, 2.1 × 150 mm, 1.7 μm BEH particles (cat no.: 186004742, Waters, Milford, MA).
3. Empower 3 chromatography workstation software (Waters).
4. Dextran Calibration Ladder Standard (cat no.: 186004742, Waters).
5. 50 mM ammonium formate, pH 4.4 as solvent A and MeCN as solvent B.

3 Methods

3.1 Glycoprotein Denaturation

1. Plasma (or serum) aliquots of 5 μL per sample are dispensed into a 96-well V-bottom PP plate (*see* **Notes 2** and **3**).
2. Denaturation solution (55 μL per well, 100 mM ammonium bicarbonate, 12 mM DTT) is added and the plate is covered and placed on a robotic heater shaker and incubated at 65 °C with agitation at 700 rpm for 30 min.
3. The plate is left to cool down for 10 min and then iodoacetamide solution (10 μL per well, 120 mM) is added, the plate

is covered with an anti-evaporation lid and incubated at room temperature with agitation at 700 rpm for 30 min.

4. A trypsin solution (10 μL of a 40,000 U/mL solution per well) is added, the plate is covered with a foil seal and placed on a robotic heater shaker at 37 °C with agitation at 700 rpm for 120 min. Next, the temperature is increased to 105 °C and incubation is continued for 10 min.
5. The plate is left to cool to room temperature, briefly spun down and the seal removed.

3.2 N-glycan Release

1. To release glycans from glycoproteins prepare a PNGaseF solution (0.5 mU in 1.0 M ammonium bicarbonate).
2. PNGaseF solution (10 μL) is added and the plate is sealed with an anti-evaporation lid and incubated at 37 °C with agitation at 700 rpm for 120 min.

3.3 Hydrazide-Mediated Glycan Cleanup

1. A 96-well chemically inert filter plate is washed with 100 μL of MeOH.
2. Ultralink hydrazide resin (40 μL of a suspension in water) is dispensed to each well.
3. The resin is sequentially washed with MeOH (200 μL), H_2O (200 μL), and MeCN (200 μL).
4. To seal the membranes plate is placed on a heater (80 °C, 10 min).
5. MeCN–acetic acid (98:2) is added to the resin (180 μL), followed by 20 μL of the glycan solution (glycans released with PNGaseF, the remaining glycan sample is kept as a stock).
6. The filter plate is incubated with shaking at 700 rpm at 80 °C for 60 min.
7. MeCN–acetic acid (98:2) is added to the resin (50 μL) and shaking is continued at the same temperature (80 °C) for 10 min to disrupt resin aggregates.
8. The resin is washed sequentially with MeOH (2 × 300 μL per well), 2.0 M guanidine (2 × 200 μL), H_2O (2 × 200 μL), trimethylamine–MeOH (1:99, 2 × 200 μL), and MeOH (2 × 200 μL).
9. Fresh MeOH (180 μL) and acetic anhydride (20 μL) are added and the plate is incubated for 30 min with agitation at 700 rpm at ambient temperature.
10. Excess reagent is removed by filtration and the resin beads are washed sequentially with MeOH, H_2O, and MeCN (400 μL).
11. MeCN–acetic acid (98:2, 175 μL) and H_2O (25 μL) are sequentially added and the plate is incubated at 70 °C with agitation at 700 rpm for 90 min.

3.4 Fluorescent Labeling

1. 2-AB fluorescent labeling mix (50 μL, 350 mM 2-aminobenzamide, 1.0 M sodium cyanoborohydride in acetic acid–dimethyl sulfoxide (30:70) is dispensed into each well (*see* **Note 4**).
2. The plate is incubated at 65 °C with agitation at 800 rpm for 120 min. [AU1]
3. The labeling reaction is quenched by the addition of 400 μL MeCN–H_2O (95:5).

3.5 Glycan Solid-Phase Extraction

1. The suspension after 2-AB labeling is transferred to a 2 mL collection plate, the beads are left to settle and 200 μL of the supernatant is aspired and dispensed back into the filter plate.
2. The suspension is mixed several times and transferred back into the 2 mL collection plate. This cycle is repeated once more to ensure a quantitative transfer of the resins.
3. HyperSep Diol SPE cartridges are washed with 1 mL MeCN–H_2O (95:5), 1 mL H_2O, and 1 mL MeCN–H_2O (95:5).
4. The supernatants of the quenched reaction mixtures without the beads are transferred onto the SPE cartridges (900 μL).
5. Typically 5 min incubation leads to complete drainage of the solvent.
6. The SPE cartridges are washed three times with 750 μL MeCN–H_2O (95:5) to remove excess of 2-AB.
7. A collection plate is placed inside the robotic vacuum manifold and the SPE cartridges are washed under vacuum with 200 μL H_2O/MeCN (80:20), with an intermittent incubation period of 5 min to elute the retained glycans on the SPE cartridges.
8. Following glycan elution, samples are concentrated to dryness in a vacuum evaporator.

3.6 HILIC-FLD- UHPLC

1. Prepare fresh solvents for UHPLC separation: solvent A (50 mM ammonium formate solution, pH 4.4) and solvent B (MeCN) (*see* **Note 5**).
2. Set up the UHPLC system: prime the system, connect the column and turn on fluorescence detector.
3. Set up the instrument method in Empower software with the following parameters, flow: 0.56 mL/min, a linear gradient of 70–53% MeCN in 30 min separation run: $t=0$ min, 70% solvent B; $t=24.81$ min, 53% solvent B; $t=25.5$ min, 30% solvent B and $t=26.55$ min, 70% solvent B.
4. Adjust the settings of the fluorescence detector: excitation wavelength 330 nm and emission wavelength 420 nm, data units: emission, date rate: 2 pts/s and PTM gain 20.0.
5. Run a blank sample of H_2O.

6. Run 2-AB labeled dextran ladder (standard) that is be used for calibration, dissolve sample in 30 μL MeCN–H_2O (70:30) (*see* **Note 6**).
7. Run your samples of interest, bring the volume of your 2-AB labeled glycan samples to 30 μL with MeCN–H_2O (70:30) (*see* **Note 7**).

3.7 Data Analysis in Empower 3 Software

1. All samples are located in the sample sets; to process them you should view them as "Channels."
2. Highlight all the samples that you wish to process by holding "Ctrl" and clicking on each sample or to highlight all samples in the sample set click "E" in the top left corner and right click on highlighted samples and select "Alter sample."
3. Make sure to include a dextran standard, which is used to calibrate the runs. The GU value is calculated by fitting a fifth order polynomial distribution curve to the dextran ladder, then using this curve to allocate GU values from retention times.
4. Ensure that Sample Type is correct, narrow standard for dextran and broad unknown for samples.
5. Highlight the dextran line, go to "Edit" and select "Amount." In the Molecular Weight column enter 40,000 in line 1, all the way up to 150,000. Click OK and the window will close. Click "Save" in Alter Sample window.
6. To process samples, go back to "Channels" and right click on processed samples and select "Process."
7. Ensure that the "Clear Calibration" box is always ticked. Select "Use specified processing method" and choose appropriate method from the drop-down list. When you process a large set of samples, it is advisable to create your own processing method that will facilitate integration in a semiautomatic manner.
8. To analyze your processed samples go to "Channels" and right-click on highlighted samples and select "View as"-> "Results."
9. Open all samples in Review, first check the integration of dextran making sure that the peaks are integrated from peak 4 to 15 and delete any others.
10. To calibrate dextran, go to "Process" and select "Calibrate."
11. For each profile, once you have integrated as appropriate go to "Process" and select "Quantitate."
12. Always make sure to integrate the samples in the same manner, the same number of peaks to ensure proper comparison.
13. When finished, go to "File" and select "Save"-> "All."

14. All data can be exported, including peak list with % area, GU value, and other parameters that are of interest, to Excel file by creating Export method.
15. An excel spreadsheet can be then used for further statistical analysis of your samples.
16. For preliminary structural annotation, glycan structures are assigned by their GU value to the peaks and they can by matched to a database developed at NIBRT (http://glycobase.nibrt.ie). Glycobase is a tool that allows computer-assisted preliminary structural assignments [25] and to further elucidate the glycan structures, exoglycosidase digestions can be performed (*see* Chapter 18 for more details) (*see* **Note 8**).

4 Notes

1. Further increase in throughput can be achieved by converting the method to a 384-well plate format, it is possible with an optional 384 channel pipetting head available from Hamilton.
2. When performing glycoprofiling of many samples, make sure that all samples are of the same type, either plasma or serum to ensure high quality data and consistent results. There is a significant difference in the relative percentage area for several glycan structures, mainly biantennary digalactosylated monosialylated glycan (A2G2S1) between plasma and serum [26].

   The difference appears to be largely attributable to the presence of fibrinogen in plasma samples whereby a lack of this glycoprotein is observed in serum (Fig. 2).
3. To obtain more information about low abundant proteins, the depletion of high-abundant plasma proteins can be performed by affinity chromatography which significantly improves detection of low-level glycoproteins.
4. The protocol uses 2-AB as a fluorescent tag for glycans, however it might be also possible to take an advantage of AQC labeling agent (6-aminoquinolyl-N-hydroxysuccinimidyl carbamate) that has been recently applied for IgG analysis [27]. In this case, a hydrazide-mediated cleanup step is neither necessary nor possible. We have tested AQC label for analysis of

Fig. 2 (continued) available in Saldova et al. [29]. *Asterisk* denotes peaks glycan structures that in part may be originating from fibrinogen (**b**) Fibrinogen *N*-glycan profile with annotated structures as described in Adamczyk et al. [26]. (**c**) Serum sample from gastric cancer patient, some structures that were previously reported to be significantly altered with disease progression are shown, namely IgG G0 (FA2), IgG G1 (FA2G1), and sLe^x epitope (A3F1G3S3) [30, 31]. The structural symbols for the *N*-glycans and their linkages are shown in the boxes below the chromatograms

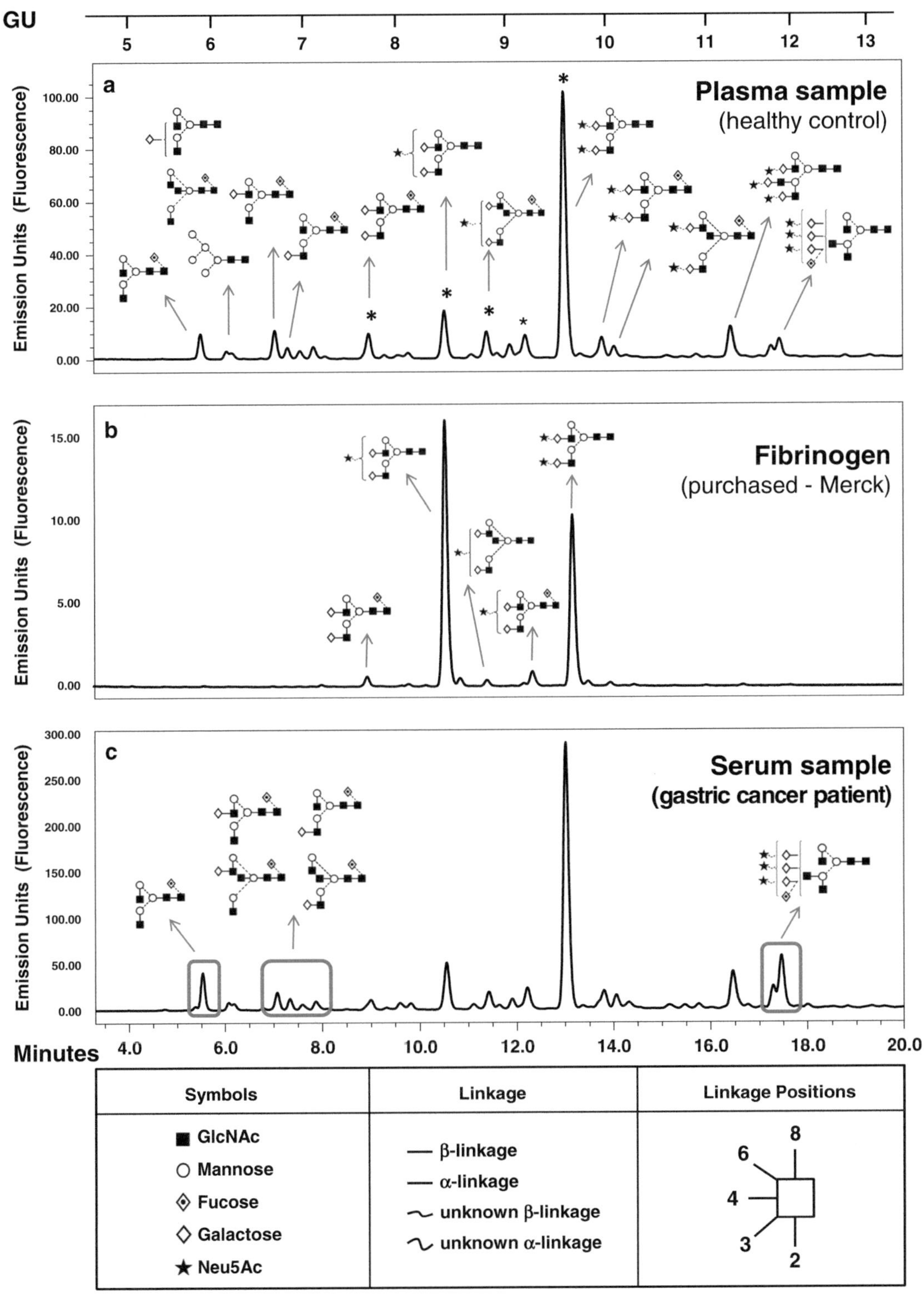

Fig. 2 HILIC-FLD-UHPLC chromatograms of 2AB labeled *N*-glycans (**a**) A typical plasma *N*-glycan profile; contains more than 160 glycans, some of the major structures are represented on the figure and abbreviated according to Harvey et al. [28]. The detailed description including monosaccharide sequences and linkage information is

a

Data Acquisition

on HILIC-FLD-UHPLC

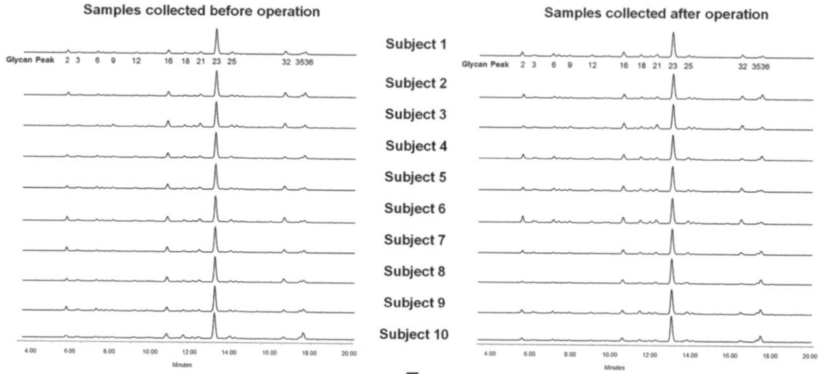

b

Sample Integration

to obtain % areas for all peaks

	1_Pre	2_Pre	3_Pre	4_Pre	5_Pre	6_Pre	7_Pre	8_Pre	9_Pre	10_Pre	1_Post	2_Post	3_Post	4_Post	5_Post	6_Post	7_Post	8_Post	9_Post	10_Post
GP1	0.11	0.14	0.08	0.19	0.27	0.16	0.22	0.26	0.22	0.15	0.14	0.12	0.14	0.22	0.38	0.23	0.13	0.13	0.14	0.17
GP2	3.73	3.88	1.74	2.71	3.28	5.03	5.19	3.26	5.33	2.96	5.03	4.34	2.35	5.25	3.55	6.53	3.49	1.99	3.91	3.47
GP3	1.54	1.52	1.32	1.84	2.2	1.68	2.05	2.18	2.33	1.73	2.24	1.5	1.84	1.92	2.55	2.9	1.72	1.21	2.18	1.31
GP4	0.05	0.06	0.06	0.11	0.16	0.06	0.06	0.16	0.07	0.05	0.08	0.05	0.08	0.1	0.19	0.09	0.05	0.05	0.04	0.06
GP5	0.1	0.24	0.14	0.17	0.24	0.15	0.1	0.24	0.28	0.19	0.14	0.22	0.21	0.17	0.27	0.21	0.09	0.1	0.19	0.2
GP6	2.07	1.71	1.26	2.11	1.99	2.08	1.61	1.98	2.62	1.6	3.39	2.09	1.9	2.99	2.7	3.18	1.34	1.07	2.3	1.49
GP7	0.99	0.79	0.51	0.75	0.92	1.02	0.69	0.92	1.06	0.92	1.9	1.28	0.82	1.12	1.36	1.56	0.64	0.48	1.15	0.92
GP8	0.62	0.69	0.72	0.83	0.87	0.68	0.6	0.87	1.21	1.05	0.94	0.77	0.94	0.93	1.16	1.16	0.76	0.46	1.28	0.68
GP9	1.4	1.13	2.7	1.56	1.62	1.46	1.46	1.61	1.47	1.49	1.6	1.19	2.35	1.53	1.76	1.92	1.38	1.12	1.49	1.09
GP10	0.18	0.18	0.19	0.36	0.32	0.2	0.22	0.32	0.22	0.12	0.29	0.15	0.28	0.29	0.44	0.29	0.17	0.16	0.14	0.12
GP11	0.08	0.13	0.12	0.15	0.16	0.1	0.09	0.16	0.16	0.1	0.14	0.12	0.17	0.13	0.22	0.15	0.08	0.08	0.1	0.09
GP12	1.38	1.19	1.3	1.9	1.65	1.33	1.08	1.64	1.62	1.66	2.58	1.59	1.77	2.11	2.35	2.17	1.07	0.79	1.77	1.14
GP13	0.26	0.27	0.43	0.4	0.41	0.3	0.26	0.41	0.47	0.23	0.39	0.29	0.49	0.42	0.53	0.48	0.31	0.15	0.57	0.21
GP14	0.76	1.05	1.11	0.85	0.99	0.91	0.83	0.97	1.07	1.21	1.08	1.04	1.4	0.89	1.17	1.14	0.75	0.68	0.89	0.88
GP15	0.73	0.65	0.81	0.66	0.77	0.86	0.64	0.78	0.82	0.95	1.07	0.92	0.81	0.9	0.97	1.11	0.59	0.46	0.99	0.78
GP16	5.91	6.69	8.68	7.78	7.73	7.44	7.86	7.68	6.59	6.65	7.27	6.45	7.7	7.14	7.36	7.47	7.15	6.69	5.01	5.17
GP17	0.51	0.75	0.7	0.59	1.03	0.47	0.44	1.03	0.81	0.81	0.68	0.78	1.11	0.58	1.32	0.7	0.52	0.57	0.77	0.63
GP18	2.23	1.72	2.94	3.03	2.43	2.57	2.2	2.43	2.22	4.93	3.3	2.53	3.17	3.84	3.37	4.12	3	1.71	4.31	2.47
GP19	0.65	0.57	1.16	0.69	0.7	0.57	0.45	0.69	0.45	0.74	0.69	0.59	0.94	0.66	0.86	0.55	0.48	0.29	0.69	0.49
GP20	1.33	1.16	2.23	1.29	1.75	1.32	1.06	1.73	1.71	1.7	1.69	1.41	2.33	1.63	2.28	2.13	1.04	0.75	2.17	1.24
GP21	3.83	3.89	6.38	3.93	3.69	4.87	4.93	3.67	2.89	2.77	4.79	3.5	6.95	3.5	3.81	4.9	3.64	3.32	2.58	2.72
GP22	0.18	0.14	0.19	0.17	0.21	0.18	0.25	0.21	0.15	0.22	0.17	0.14	0.19	0.19	0.24	0.18	0.19	0.2	0.23	0.17
GP23	42.03	37.9	37.18	42.54	40.55	38.64	41.69	40.39	40.44	35.74	37.95	36.68	34.97	35.76	36.58	31.66	41.91	45.97	35.83	39.86
GP24	0.97	0.88	0.91	0.97	1.07	0.89	1.17	1.07	0.95	0.65	0.91	0.75	0.89	0.88	1.15	0.79	0.9	1.1	0.98	0.94
GP25	4.08	2.93	3.84	3.95	2.68	3.81	3.25	2.67	3.15	4.58	3.49	3.35	3.53	4.54	2.87	3.96	3.91	4.63	4.11	3.75
GP26	1.28	1.8	2.76	1.17	1.76	1.27	0.87	1.73	1.82	1.86	1.55	2	2.59	1.56	2.12	1.8	0.9	0.89	1.92	1.7
GP27	1.2	1.65	2	1.35	1.41	1.61	1.56	1.39	1.32	1.18	1.28	1.53	1.72	1.47	1.5	1.59	1.37	1.11	1.22	0.96
GP28	0.86	1.05	0.77	0.83	1	0.91	0.79	0.98	1.1	0.65	0.84	0.94	0.83	0.79	0.97	0.81	0.76	0.59	0.73	0.58
GP29	0.47	0.63	0.4	0.64	0.41	0.39	0.47	0.4	0.6	0.61	0.43	0.65	0.4	0.71	0.37	0.35	0.56	0.76	0.6	0.61
GP30	0.67	0.92	1.26	0.51	0.66	1.12	0.82	0.65	0.43	0.43	0.68	0.76	1.33	0.55	0.69	1.04	0.56	0.32	0.39	0.33
GP31	0.28	0.24	0.23	0.23	0.18	0.29	0.29	0.17	0.14	0.17	0.33	0.21	0.3	0.2	0.2	0.27	0.23	0.2	0.16	0.25
GP32	6.3	7.6	6.9	3.71	6.68	7.92	5.86	6.77	4.51	3.36	4.84	6.02	6.59	3.36	6.19	6.39	4.84	2.27	3.93	3.04
GP33	0.39	0.57	0.56	0.46	0.3	0.41	0.5	0.3	0.3	0.65	0.33	0.63	0.58	0.57	0.29	0.38	0.53	0.64	0.46	0.67
GP34	0.52	0.44	0.59	0.29	0.29	0.56	0.39	0.29	0.28	0.22	0.34	0.37	0.53	0.28	0.29	0.41	0.34	0.21	0.28	0.25
GP35	2.71	2.82	1.19	1.96	2.63	2.08	2	2.7	2.74	2.65	1.39	2.93	1.07	2	1.96	1.38	2.7	2.41	3.02	3.15
GP36	5.15	7.32	4.51	5.94	4.27	3.8	4.79	4.35	5.46	9.55	3.43	8.08	4.2	7.01	3.53	3.34	7.34	9.82	8.11	10.6
GP37	0.43	0.38	0.24	0.3	0.23	0.33	0.3	0.23	0.33	0.34	0.25	0.37	0.24	0.34	0.21	0.27	0.36	0.49	0.43	0.42
GP38	0.5	0.57	0.44	0.52	0.24	0.33	0.4	0.24	0.42	0.66	0.3	0.63	0.43	0.66	0.23	0.27	0.59	0.9	0.66	0.76
GP39	0.53	0.57	0.25	0.45	0.42	0.35	0.41	0.43	0.5	0.55	0.38	0.52	0.32	0.53	0.43	0.37	0.46	0.54	0.64	0.61
GP40	0.51	0.56	0.33	0.25	0.32	0.49	0.42	0.34	0.21	0.29	0.37	0.35	0.47	0.22	0.32	0.47	0.36	0.24	0.32	0.34
GP41	0.77	0.78	0.3	0.48	0.56	0.54	0.56	0.6	0.47	0.6	0.45	0.57	0.38	0.5	0.49	0.48	0.61	0.64	0.7	0.71
GP42	0.5	0.54	0.23	0.31	0.29	0.33	0.37	0.32	0.24	0.55	0.29	0.4	0.29	0.33	0.25	0.3	0.48	0.42	0.47	0.78
GP43	0.73	0.77	0.22	0.57	0.41	0.33	0.49	0.46	0.45	1.23	0.35	0.66	0.25	0.62	0.3	0.27	0.9	1.23	1.07	1.91
GP44	0.33	0.36	0.09	0.33	0.18	0.12	0.22	0.21	0.25	0.86	0.16	0.36	0.14	0.38	0.15	0.14	0.57	1.39	0.77	1.64
GP45	0.14	0.17	0.04	0.16	0.08	0.04	0.09	0.09	0.1	0.38	0.08	0.17	0.07	0.22	0.07	0.07	0.21	0.78	0.29	0.65

c

Statistical analysis

to compare data sets and test significant differences

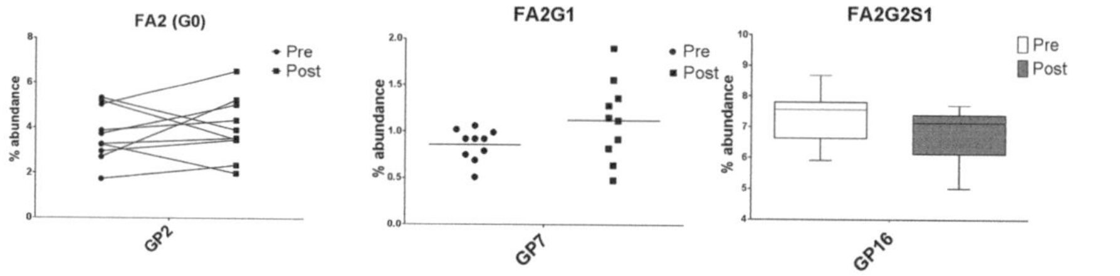

Fig. 3 Examples of the data output from a glycoprofiling experiment. (**a**) HILIC-FLD-UHPLC profiles of gastric cancer serum samples, *on the left* samples collected before operation and *on the right* samples collected after operation when patients are classified cancer-free (**b**) Analysis of the data in Empower results in the table with % area for all integrated peaks for both sets of patients (**c**) Statistical analysis to compare data sets using different approaches for data analysis, interpretation and presentation of the data (for example Boxplots, Heat Plots, Cluster Analysis, Principal Component Analysis)

plasma samples using manual in-gel-block method [8] and the obtained data were of good quality. However, before embarking on large-scale analysis more thorough tests and optimization should be performed.

5. Reagents used should be of the highest grade commercially available, especially for preparing buffers for HILIC-FLD-UHPLC analysis.
6. It is recommended to run a dextran ladder standard followed by the samples, for calibration and to check the performance of the instrument. The dextran should be run once every 24 h or every 48 samples.
7. The workflow presented in this book chapter has many applications; one of them is glycoprofiling of disease samples and comparing different stages of the disease. An example of glycoprofiling of serum samples, originating from gastric cancer patients, pre- and post-surgery is illustrated on Fig. 3.
8. The automated sample preparation workflow presented in this book chapter can be interfaced with other downstream analytical technologies, including mass spectrometry (MS) and capillary electrophoresis (CE), for rapid characterization and quantitation of oligosaccharides.

Acknowledgements

The authors acknowledge support from the European Union FP7 GastricGlycoExplorer ITN under grant agreement no. 316929. and HighGlycan under grant agreement no. 278535. We would like to acknowledge Dr. Karol Polom, Dr. Giovanni Corso, and Dr. Franco Roviello from University Hospital of Siena for providing gastric cancer serum samples. The UHPLC instrument was obtained with a grant from the IngaBritt and Arne Lundberg's Research Foundation.

References

1. Merry AH, Merry CL (2005) Glycoscience finally comes of age. EMBO Rep 6(10): 900–903
2. Varki A (1993) Biological roles of oligosaccharides: all of the theories are correct. Glycobiology 3(2):97–130
3. Lowe JB (2001) Glycosylation, immunity, and autoimmunity. Cell 104(6):809–812
4. Lauc G et al (2016) Mechanisms of disease: the human N-glycome. Biochim Biophys Acta 1860:1574–1582
5. Fuster MM, Esko JD (2005) The sweet and sour of cancer: glycans as novel therapeutic targets. Nat Rev Cancer 5(7):526–542
6. Freeze HH, Aebi M (2005) Altered glycan structures: the molecular basis of congenital disorders of glycosylation. Curr Opin Struct Biol 15(5):490–498
7. Szabo Z et al (2010) Improved sample preparation method for glycan analysis of glycoproteins by CE-LIF and CE-MS. Electrophoresis 31(8):1389–1395

8. Royle L et al (2008) HPLC-based analysis of serum N-glycans on a 96-well plate platform with dedicated database software. Anal Biochem 376(1):1–12
9. Ahn J et al (2010) Separation of 2-aminobenzamide labeled glycans using hydrophilic interaction chromatography columns packed with 1.7 um sorbent. J Chromatogr B 878(34):403–408
10. Marino K et al (2010) A systematic approach to protein glycosylation analysis: a path through the maze. Nat Chem Biol 6(10):713–723
11. North SJ et al (2009) Mass spectrometry in the analysis of N-linked and O-linked glycans. Curr Opin Struct Biol 19(5):498–506
12. Harvey DJ (2005) Proteomic analysis of glycosylation: structural determination of N-and O-linked glycans by mass spectrometry. Expert Rev Proteomics 2(1):87–101
13. Wuhrer M, Deelder AM, Hokke CH (2005) Protein glycosylation analysis by liquid chromatography–mass spectrometry. J Chromatogr B 825(2):124–133
14. Guile GR et al (1996) A rapid high-resolution high-performance liquid chromatographic method for separating glycan mixtures and analyzing oligosaccharide profiles. Anal Biochem 240(2):210–226
15. Tharmalingam T et al (2013) Strategies for the profiling, characterisation and detailed structural analysis of N-linked oligosaccharides. Glycoconj J 30(2):137–146
16. Royle L et al (2006) Detailed structural analysis of N-glycans released from glycoproteins in SDS-PAGE gel bands using HPLC combined with exoglycosidase array digestions. In: Brockhausen I (ed) Methods in molecular biology. Humana Press Inc, Totowa, NJ, pp 125–143
17. Knezevic A et al (2008) Variability, heritability and environmental determinants of human plasma N-glycome. J Proteome Res 8(2):694–701
18. Arnold JN et al (2011) Novel glycan biomarkers for the detection of lung cancer. J Proteome Res 10(4):1755–1764
19. Reusch D et al (2013) High-throughput work flow for IgG Fc-glycosylation analysis of biotechnological samples. Anal Biochem 432(2):82–89
20. Trbojević Akmačić I et al (2015) High-throughput glycomics: optimization of sample preparation. Biochemistry (Mosc) 80(7):934–942
21. Stöckmann H et al (2013) Automated, high-throughput IgG-antibody glycoprofiling platform. Anal Chem 85(18):8841–8849
22. Shubhakar A et al (2015) High-throughput analysis and automation for glycomics studies. Chromatographia 78(5–6):321–333
23. Ruhaak LR et al (2008) Hydrophilic interaction chromatography-based high-throughput sample preparation method for N-glycan analysis from total human plasma glycoproteins. Anal Chem 80(15):6119–6126
24. Stöckmann H et al (2015) Automated, high-throughput serum glycoprofiling platform. Integr Biol 7(9):1026–1032
25. Campbell MP, Royle L, Rudd PM (2015) GlycoBase and autoGU: resources for interpreting HPLC-Glycan data. Methods Mol Biol 1273:17–28
26. Adamczyk B et al (2012) Characterization of fibrinogen glycosylation and its importance for serum/plasma N-glycome analysis. J Proteome Res 12(1):444–454
27. Stöckmann H et al (2015) Ultrahigh throughput, ultrafiltration-based N-Glycomics platform for ultraperformance liquid chromatography (ULTRA3). Anal Chem 87(16):8316–8322
28. Harvey DJ et al (2009) Proposal for a standard system for drawing structural diagrams of N- and O-linked carbohydrates and related compounds. Proteomics 9(15):3796–3801
29. Saldova R et al (2014) Association of N-glycosylation with breast carcinoma and systemic features using high-resolution quantitative UPLC. J Proteome Res 13(5):2314–2327
30. Bones J et al (2010) Glycomic and glycoproteomic analysis of serum from patients with stomach cancer reveals potential markers arising from host defense response mechanisms. J Proteome Res 10(3):1246–1265
31. Bones J et al (2010) Ultra performance liquid chromatographic profiling of serum N-glycans for fast and efficient identification of cancer associated alterations in glycosylation. Anal Chem 82(24):10208–10215

Chapter 9

Site-Specific *N*- and *O*-Glycopeptide Analysis Using an Integrated C18-PGC-LC-ESI-QTOF-MS/MS Approach

Kathrin Stavenhagen, Hannes Hinneburg, Daniel Kolarich, and Manfred Wuhrer

Abstract

The vast heterogeneity of protein glycosylation, even of a single glycoprotein with only one glycosylation site, can give rise to a set of macromolecules with different physicochemical properties. Thus, the use of orthogonal approaches for comprehensive characterization of glycoproteins is a key requirement. This chapter describes a universal workflow for site-specific *N*- and *O*-glycopeptide analysis. In a first step glycoproteins are treated with Pronase to generate glycopeptides containing small peptide sequences for enhanced glycosylation site assignment and characterization. These glycopeptides are then separated and detected using an integrated C18-porous graphitized carbon-liquid chromatography (PGC-LC) setup online coupled to a high-resolution electrospray ionization (ESI)-quadrupole time-of-flight (QTOF)-mass spectrometer operated in a combined higher- and lower-energy CID (stepping-energy CID) mode. The LC-setup allows retention of more hydrophobic glycopeptides on C18 followed by subsequent capturing of C18-unbound (glyco)peptides by a downstream placed PGC stationary phase. Glycopeptides eluted from both columns are then analyzed within a single analysis in a combined data acquisition mode. Stepping-energy CID results in B- and Y-ion fragments originating from the glycan moiety as well as b- and y-ions derived from the peptide part. This allows simultaneous site-specific identification of the glycan and peptide sequence of a glycoprotein.

Key words Glycoproteomics, *N*-glycopeptide, *O*-glycopeptide, Mass spectrometry, Porous graphitized carbon, QTOF-MS, Stepping-energy CID, C18-PGC-LC

1 Introduction

A detailed characterization of protein glycosylation is essential to establish a basis for understanding and investigating its biological role and involvement in diseases. Glycopeptide-centric approaches play an important role in protein glycosylation analysis providing information on protein identity, glycosylation site and glycan structural features. Mass spectrometric analyses of *N*- and *O*-glycopeptides can be hindered by glycoprotein microheterogeneity (different glycans present at one glycosylation

Gordan Lauc and Manfred Wuhrer (eds.), *High-Throughput Glycomics and Glycoproteomics: Methods and Protocols*, Methods in Molecular Biology, vol. 1503, DOI 10.1007/978-1-4939-6493-2_9, © Springer Science+Business Media New York 2017

site), macroheterogeneity (site occupancy), as well as a lower ionization efficiency of glycopeptides compared to the equivalent unglycosylated peptide [1]. Despite these challenges recent analytical advances have facilitated and improved glycopeptide analyses [2–5].

Proteases exhibiting broad substrate specificity, such as Pronase or Proteinase K, are valuable tools for site-specific glycopeptide analysis by generating relatively short peptides (depending on the experimental conditions) [6–10]. This has the advantage to result in smaller glycopeptides with fewer glycosylation sites compared to other proteases such as trypsin. Subsequently, this feature enables a more unambiguous glycosylation site assignment and characterization.

In LC-mass spectrometry (MS) experiments C18-reversed-phase (RP) chromatography is one of the most commonly used stationary phase for glycopeptide separation [2], but shows only limited retention for highly hydrophilic glycopeptides that are often generated by Pronase treatment. The combination of operating C18-RP-LC with orthogonal stationary phases, such as PGC, has been shown to significantly enhance peptide sequence coverage [11]. Thus, we set up an integrated approach by coupling C18-RP-LC with PGC-LC for comprehensive glycopeptide analysis [12]. The Pronase-treated sample is loaded onto a C18 precolumn that retains more hydrophobic glycopeptides, and the hydrophilic (glyco)peptides present in the C18 flow through fraction are transferred to a downstream positioned PGC precolumn. The latter stationary phase is well known for glycan analysis [13–16] and has also shown to be of good use for glycopeptide analyses [7, 8, 17–20]. First C18- and then PGC-bound analytes are eluted from the respective precolums in a nested manner, followed by further separation of the (glyco)peptides using the corresponding analytical columns, and acquisition of the entire data within a single combined data file (Figs. 1 and 2).

Conventional collision-induced dissociation (CID) is a commonly applied technique for glycopeptide fragmentation when analyzed by MS. It mainly results in B- and Y-ions of the glycan moiety and therefore lacks valuable information to confirm the peptide sequence and thus protein identity [4]. This can be overcome by applying higher-energy CID which also generates b- and y-ions of the peptide backbone [21]. However, optimal results for glycopeptide assignment are provided when higher- and lower-energy CID (stepping-energy CID) are combined [22, 23], as recently systematically investigated [21]. This approach provides information complementary to electron-transfer dissociation (ETD) or electron-capture dissociation (ECD) based fragmentation techniques [4, 23–25].

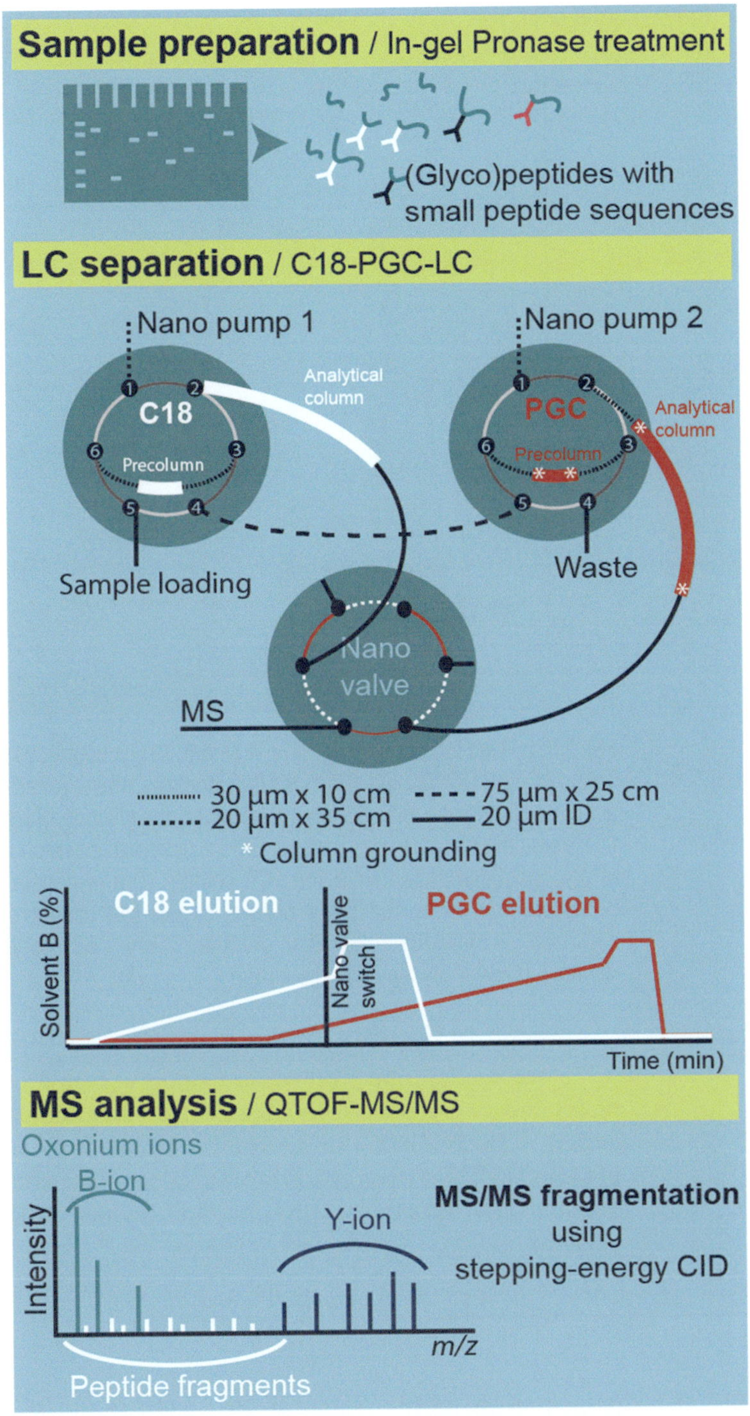

Fig. 1 C18-PGC-LC-QTOF-MS/MS workflow for glycopeptide analysis. In a first step the glycoprotein is treated in-gel with Pronase to generate glycopeptides with a small peptide backbone. These glycopeptides are then separated by an integrated C18-PGC-LC setup, where the sample is first loaded onto a C18 precolumn and the flow through is online directed towards a PGC precolumn. Glycopeptides with different physicochemical properties can be retained on both stationary phases and elution is performed sequentially over the respective analytical columns. The eluate from both analytical columns is sequentially directed to the mass spectrometer and data is acquired within a single run. The mass spectrometer is operated in stepping-energy CID mode to enable identification of the glycan composition as well as peptide sequence. Reprinted and modified with permission from [12]. Copyright 2016 American Chemical Society

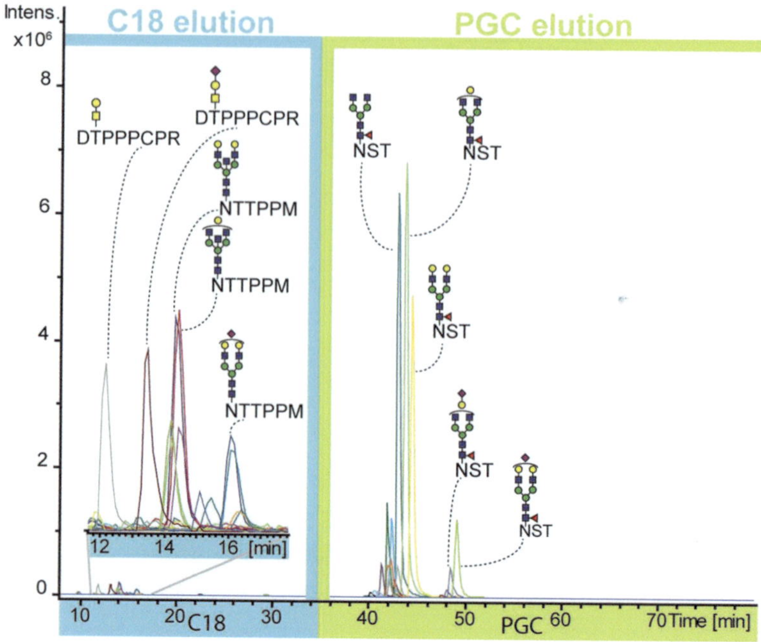

Fig. 2 Extracted ion chromatograms of *N*- and *O*-glycopeptides from human IgG3 obtained by C18-PGC-LC-QTOF-MS/MS analysis after Pronase treatment. Data acquired during the first 35 min is obtained for glycopeptides eluted from the C18 columns. After 35 min analysis time a nano valve switch enables detection of glycopeptides eluted from the PGC column. The glycopeptide retention and detection from both stationary phases shows the benefit of this integrated approach, as individual application of either C18 or PGC separation would have resulted in lower glycopeptide coverage

Here we describe a universal workflow that allows the analysis of *N*- and *O*-glycopeptides after Pronase treatment using an integrated C18-PGC-LC setup online coupled to high-resolution QTOF-MS. The mass spectrometer it set to operate in stepping-energy CID mode for glycopeptide tandem MS experiments (Fig. 1).

2 Materials

Prepare all solutions with ultrapure deionized water ($R > 18.2$ MΩ/cm) and MS grade solvents.

2.1 SDS-PAGE

1. Gel: 4–12% Bis-tris SDS-PAGE gel (e.g., NuPAGE; Life Technologies, Carlsbad, CA, USA) (*see* **Note 1**).
2. Sample buffer: 4× Laemmli sample buffer with 8% (w/v) SDS, 0.25 M Tris–HCl pH 6.8, 40% (w/v) glycerol, 5% 2-mercaptoethanol, bromophenol blue in water (stored at −20 °C) [26].

3. Running buffer: 50 mM MES, 50 mM Tris Base, 0.1% SDS, 1 mM EDTA, pH 7.3 (e.g., MES running buffer; NuPAGE; Life Technologies).
4. Staining solution: Coomassie-based staining solution (e.g., Simply blue safe stain, NuPAGE; Life Technologies).

2.2 In-Gel Reduction and Alkylation of Cysteine Bonds Followed by Pronase Treatment

1. AmBiC buffer: 25 mM ammonium bicarbonate.
2. Reduction buffer: 10 mM dithiothreitol in AmBiC buffer.
3. Alkylation buffer: 55 mM iodoacetamide in AmBiC buffer.
4. Digestion buffer: 30 μl of AmBiC buffer containing Pronase (*Streptomyces griseus*; e.g., Sigma-Aldrich; St. Louis, MO) in a ratio 1:3 (enzyme:protein) (*see* **Note 2**).

2.3 C18-PGC-LC Setup

1. NanoLC-system: Ultimate 3000 *x*2 Dual Analytical system (Dionex, part of Thermo Fisher, Breda, The Netherlands) featuring a dual binary pump system and an extra valve to separately direct the flow to the MS either from pump system one or two.
2. C18 precolumn: Acclaim C18 PepMap 100, 300 μm × 5 mm, 5 μm, 100 Å, placed in a μ-precolumn holder with 30 μm ID × 10 cm nanoViper™ connecting tubing (both Dionex/Thermo Fisher Scientific).
3. C18 analytical column: Acclaim PepMap RSLC, 75 μm × 15 cm, 2 μm, 100 Å, (Dionex/Thermo Scientific).
4. PGC precolumn setup: in-house made 100 μm ID × 1.5 cm, 3 μm Hypercarb material (Thermo Scientific), two 1/16″ stainless steel zero dead volume (ZDV) unions and two 30 μm ID × 10 cm nanoViper™ connecting tubing (*see* **Note 3**).
5. PGC analytical column setup: in-house made 50 μm ID × 15 cm, 3 μm Hypercarb material (Thermo Scientific), a 1/16″ stainless steel ZDV union, a conductive micro union (e.g., IDEX Europe GmbH, Erlangen, Germany), and 30 μm ID × 10 cm nanoViper™ connecting tubing (*see* **Note 3**).
6. Connection from the nano pumps to the valve with 20 μm ID × 35 cm nanoViper™ tubing.
7. Connection between the two valves (precolumns) with 75 μm ID × 25 cm nanoViper™ tubing.
8. Connection from the analytical column outlet to the MS with 20 μm ID fused silica tubing.

2.4 C18-PGC-ESI-LC-QTOF-MS/MS Analysis

1. Mass Spectrometer: maXis HD QTOF equipped with a CaptiveSpray ion source and nanoBooster™ (both Bruker, Bremen, Germany).
2. LC solvent A: 0.1% formic acid (FA) in water.

3. LC solvent B: 80% acetonitrile (ACN) and 0.1% FA in water.
4. Loading solvent: 0.1% FA in water (*see* **Note 4**).
5. NanoBooster solvent: ACN.

3 Methods

If not indicated otherwise all steps can be performed at room temperature. The workflow describes an in-gel Pronase treatment of glycoproteins [12]. Depending on the starting material purity and complexity, Pronase treatment can also be performed in solution [27].

3.1 SDS-PAGE

1. Prepare 5–10 μg of glycoprotein dissolved in water, mix it with sample buffer (ratio 3:1 (v/v)) and incubate the sample for 10 min at 70 °C for denaturation.
2. Load the sample onto the SDS-PAGE and separate proteins for 50 min at 200 V (*see* **Note 5**).
3. Wash the gel three times for 5 min with water on a shaker.
4. Stain the gel for 1 h on a shaker (*see* **Note 6**).
5. Destain the gel in water over night on a shaker.

3.2 In-Gel Reduction and Alkylation of Cysteine Bonds Followed by Pronase Treatment

1. Cut the gel bands of interest into small pieces of around 1 mm^3.
2. Transfer the gel pieces into a 1.5 ml reaction vial.
3. Wash the gel pieces with 100 μl AmBiC buffer for 5 min and then remove the liquid.
4. Shrink the gel pieces with 100 μl ACN for 2 min and discard the liquid, followed by another 100 μl of fresh ACN for 10 min and remove the liquid.
5. Swell the gel pieces with 50 μl reduction buffer and incubate it for 30 min at 60 °C to reduce cysteines and then remove the liquid (*see* **Note 7**).
6. Shrink the gel pieces with 50 μl ACN for 2 min and discard the liquid, followed by another 50 μl of fresh ACN for 5 min and remove the liquid.
7. Swell the gel pieces again with 50 μl alkylation buffer and incubate for 20 min in dark to alkylate cysteines and then remove the liquid.
8. Wash the gel pieces with 100 μl AmBiC buffer for 2 min and discard the liquid, followed by another 100 μl of fresh AmBiC buffer for 10 min and remove the liquid.
9. Shrink the gel pieces with 100 μl ACN for 5 min and remove the liquid.

10. Repeat the **steps 8** and **9** to destain the gel bands (*see* **Note 8**).
11. Dry down the gel pieces in a centrifugal vacuum concentrator for approximately 5 min.
12. Add 30 μl digestion buffer to the gel pieces and let it swell on ice for 45 min. Check after 45 min if the gel pieces are fully covered with liquid and add another 5–10 μl AmBiC buffer if necessary and let it incubate at 37 °C for 20 h.
13. Collect the liquid in a separate reaction vial.
14. Extract (glyco)peptides by adding 30 μl of AmBiC buffer for 1 h at maximum speed on a shaker.
15. Remove the liquid and combine it with the first one (**step 13**).
16. Keep the sample at −20 °C or dry down the liquid in a centrifugal vacuum concentrator if longer storage is needed.

3.3 C18-PGC-LC Setup

1. Connect the C18 columns and connecting tubing as indicated in Fig. 1.
2. Connect the PGC precolumn on both ends to 30 μm ID × 10 cm connecting tubing via 1/16″ stainless steel ZDV unions and the attach it to the valve.
3. Connect the PGC analytical column to the valve via a 1/16″ stainless steel ZDV union and 30 μm ID × 10 cm connecting tubing.
4. Connect the outlet of the PGC analytical column to the fused silica tubing by a conductive micro union.
5. Attach grounding cables to the stainless steel ZDV unions of the PGC precolumn and analytical column as well as to the conductive micro union (Fig. 1) (*see* **Note 9**).

3.4 C18-PGC-LC-ESI-QTOF-MS/MS Analysis of Glycopeptides

1. Calibrate the mass spectrometer according to the manufacturer's instructions.
2. Apply all QTOF-MS/MS settings according to Table 1.
3. Equilibrate both precolumn and analytical column with 1% solvent B.
4. Set the column oven temperature to 36 °C.
5. Dilute the sample in water to a protein concentration corresponding to approximately 20–50 ng/μl.
6. Inject 1 μl of sample (or any other volume if needed).
7. Set both valves to position 1_2 to load the sample for 6 min at 6 μl/min and then switch to 1_6 (Table 2).
8. Set the nano valve to guide the C18 elution to the MS.
9. Elute first analytes from the C18 columns with a gradient from 1% to 55% solvent B in 30 min (Table 2) (*see* **Note 10**).

Table 1
MS and MS/MS parameters [21]

Ion source parameters	
Capillary voltage	1200 V
nanoBooster pressure	0.2 bar
Dry gas	3 l/min
Dry gas temperature	150 °C
MS settings	
Mass range	50–2800 m/z
MS spectra rate	1 Hz
Funnel 1 RF	300 Vpp
Multipole RF	300 Vpp
Collision RF	500–1300 Vpp (each 50% of the time[a])
Quadrupole ion energy	3 eV
Collision energy	5 eV
Pre pulse storage	10 μs
Transfer time	90–130 μs (each 50% of the time[a])
MS/MS settings	
Spectra rate	0.5–2 Hz (depending on precursor intensity)
No. of precursors	3
Collision energy (stepping)	100% collision energy is applied at 80% of the time and half of the collision energy is applied 20% of the time[a]
Collision energy absolute	55 eV at m/z 700 (isolation width 8 m/z)
	124 eV at m/z 1800 (isolation width 10 m/z)

[a]In basic stepping mode

Table 2
LC parameters

Nano pump 1 (C18; 500 nl/min)		Nano pump 2 (PGC; 450 nl/min)		Valve switch		
Time [min]	**Solvent B [%]**	**Time [min]**	**Solvent B [%]**	**Time [min]**	**Valve 1 (C18)**	**Valve 2 (PGC)**
0	1	0	1	0	1_2	1_2
5	1	27	1	6	6_1	6_1
35	55	75	60	49	1_2	
36	90	77	90	85	6_1	1_2
46	90	83	90			
47	1	84	1	Nano valve switch at 35 min		
90	1	90	1	Loading pump: 6 μl/min		

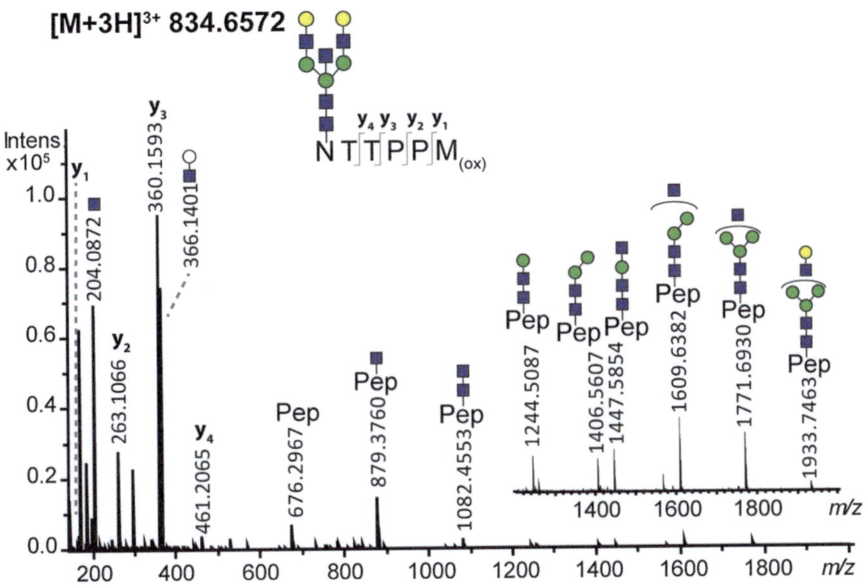

Fig. 3 MS/MS fragmentation spectrum of a *N*-glycopeptide from human IgG3 using stepping-energy CID, resulting in peptide as well as glycan-derived fragmentation

10. Switch the nano valve at 35 min to guide the elution of the PGC columns to the MS (Table 2) (*see* **Note 11**).
11. Elute analytes from the PGC column with a gradient from 1% to 60% solvent B in 48 min (Table 2) (*see* **Note 10**).

3.5 Data Analysis

1. Screen fragmentation spectra for oxonium ions 204.0867 [N-acetylhexosamine (HexNAc) + H]$^+$, 366.1394 [HexNAc + Hexose (Hex) + H]$^+$, 292.1027 [N-acetyl neuraminic acid (NeuAc) + H]$^+$, 657.2348 [HexNAc + Hex + NeuAc + H]$^+$ to classify spectra of glycopeptides (Fig. 3) (*see* **Note 12**).
2. Identify the intact peptide mass by following the glycan-derived B- and Y-ions (Fig. 3).
3. Search for potential peptide sequences using findpept (web.expasy.org/findpept) with a mass deviation of maximum 10 ppm.
4. Verify the peptide sequence manually by matching b- and y-ions (Fig. 3, **Note 12**).

4 Notes

1. The concentration of the gel can be adjusted depending on the nature and complexity of the sample.
2. The digestion buffer should be prepared fresh each time and Pronase can be stored as a stock solution of 5–20 μg/μl in water at −20 °C. The ratio of enzyme to substrate will have an

influence on the peptide length of the glycopeptides, resulting in shorter peptide sequences the more enzyme is added. Alternatively to Pronase other broad-specific proteases, such as Proteinase K, can be used.

3. Nanobore PGC columns are also commercially available (Thermo Scientific).
4. Trifluoroacetic acid is not recommended for the loading solvent as it may induce modification of sialic acid residues.
5. The conditions for SDS-PAGE can vary to yield desired protein separation, depending on the sample nature and complexity.
6. A variety of MS-compatible staining procedures are available that can be used for protein visualization. In case the glycoprotein of interest contains sialic acids, care should be taken to avoid any prolonged incubation times with high amounts of concentrated acid as this might lead to desialylation artifacts [28].
7. Reduction and alkylation could be also performed prior to SDS-PAGE.
8. This step needs to be repeated several times in case the bands are intensively stained.
9. An adequate grounding (clamps on ground wire) of the PGC columns on the in- and outlet of the precolumn and analytical column, respectively, is necessary. As PGC is a conductive stationary phase an electric current on the column, due to a backflow from the mass spectrometer (even if no voltage is applied to the spray tip), may result in modification of sialic acid residues or retention time shift [29].
10. The length and slope of the gradient can be adjusted depending on the sample.
11. An overlapping of both gradients during washing and start will help to optimize run times.
12. Glycopeptide identification can also be performed using automated software tools (e.g., Protein Scape (Bruker, Bremen, Germany) [21]).

References

1. Stavenhagen K, Hinneburg H, Thaysen-Andersen M et al (2013) Quantitative mapping of glycoprotein micro-heterogeneity and macro-heterogeneity: an evaluation of mass spectrometry signal strengths using synthetic peptides and glycopeptides. J Mass Spectrom 48:627–639
2. Thaysen-Andersen M, Packer NH (2014) Advances in LC-MS/MS-based glycoproteomics: getting closer to system-wide site-specific mapping of the *N*- and *O*-glycoproteomes. Biochim Biophys Acta 1844:1437–1452
3. Nilsson J, Halim A, Grahn A, Larson G (2013) Targeting the glycoproteome. Glycoconj J 30:119–136
4. Alley WR, Mann BF, Novotny MV (2013) High-sensitivity analytical approaches for the structural characterization of glycoproteins. Chem Rev 113:2668–2732

5. Desaire H (2013) Glycopeptide analysis, recent developments and applications. Mol Cell Proteomics 12:893–901
6. Zauner G, Koeleman CAM, Deelder AM, Wuhrer M (2010) Protein glycosylation analysis by HILIC-LC-MS of Proteinase K-generated *N*- and *O*-glycopeptides. J Sep Sci 33:903–910
7. Hua S, Nwosu CC, Strum JS et al (2012) Site-specific protein glycosylation analysis with glycan isomer differentiation. Anal Bioanal Chem 403:1291–1302
8. Nwosu CC, Huang J, Aldredge D et al (2013) In-gel nonspecific proteolysis for elucidating glycoproteins: a method for targeted protein-specific glycosylation analysis in complex protein mixtures. Anal Chem 85:956–963
9. Larsen MR, Højrup P, Roepstorff P (2005) Characterization of gel-separated glycoproteins using two-step proteolytic digestion combined with sequential microcolumns and mass spectrometry. Mol Cell Proteomics 4:107–119
10. Temporini C, Perani E, Calleri E et al (2007) Pronase-immobilized enzyme reactor: an approach for automation in glycoprotein analysis by LC/LC-ESI/MS pronase-immobilized enzyme reactor: an approach for automation in glycoprotein analysis by LC/LC-ESI/MSn. Anal Chem 79:355–363
11. Lewandrowski U, Sickmann A (2010) Online dual gradient reversed-phase/porous graphitized carbon nanoHPLC for proteomic applications. Anal Chem 82:5391–5396
12. Stavenhagen K, Plomp R, Wuhrer M (2015) Site-specific protein *N*- and *O*-glycosylation analysis by a C18-porous graphitized carbon-liquid chromatography-electrospray ionization mass spectrometry approach using pronase treated glycopeptides. Anal Chem 87: 11691–11699
13. Ruhaak LR, Deelder AM, Wuhrer M (2009) Oligosaccharide analysis by graphitized carbon liquid chromatography-mass spectrometry. Anal Bioanal Chem 394:163–174
14. Jensen PH, Karlsson NG, Kolarich D, Packer NH (2012) Structural analysis of *N*- and *O*-glycans released from glycoproteins. Nat Protoc 7:1299–1310
15. Stavenhagen K, Kolarich D, Wuhrer M (2014) Clinical glycomics employing graphitized carbon liquid chromatography–mass spectrometry. Chromatographia 78:307–320
16. Packer NH, Lawson MA, Jardine DR, Redmond JW (1998) A general approach to desalting oligosaccharides released from glycoproteins. Glycoconj J 15:737–747
17. Thaysen-Andersen M, Wilkinson BL, Payne RJ, Packer NH (2011) Site-specific characterisation of densely *O*-glycosylated mucin-type peptides using electron transfer dissociation ESI-MS/MS. Electrophoresis 32:3536–3545
18. Alley W, Mechref Y, Novotny MV (2009) Use of activated graphitized carbon chips for liquid chromatography/mass spectrometric and tandem mass spectrometric analysis of tryptic glycopeptides. Rapid Commun Mass Spectrom 23:495–505
19. Davies MJ, Smith KD, Harbin AM, Hounsell EF (1992) High-performance liquid chromatography of oligosaccharide alditols and glycopeptides on a graphitized carbon column. J Chromatogr 609:125–131
20. Wagner-Rousset E, Bednarczyk A, Bussat M-C et al (2008) The way forward, enhanced characterization of therapeutic antibody glycosylation: comparison of three level mass spectrometry-based strategies. J Chromatogr B 872:23–37
21. Hinneburg H, Stavenhagen K, Schweiger-Hufnagel U et al (2016) The art of destruction: optimizing collision energies in quadrupole-time of flight (Q-TOF) instruments for glycopeptide-based glycoproteomics. J Am Soc Mass Spectrom 27:507–519. doi:10.1007/s13361-015-1308-6
22. Kolli V, Dodds ED (2014) Energy-resolved collision-induced dissociation pathways of model *N*-linked glycopeptides: implications for capturing glycan connectivity and peptide sequence in a single experiment. Analyst 139:2144–2153
23. Dodds ED (2012) Gas-phase dissociation of glycosylated peptides ions. Mass Spectrom Rev 31:666–682
24. Mechref Y (2012) Use of CID/ETD mass spectrometry to analyze glycopeptides. Curr Protoc Protein Sci suppl 68:Unit 12.11
25. Wuhrer M, Catalina MI, Deelder AM, Hokke CH (2007) Glycoproteomics based on tandem mass spectrometry of glycopeptides. J Chromatogr B 849:115–128
26. Laemmli UK (1970) Cleavage of structural proteins during the assembly of the head of bacteriophage T4. Nature 227:680–685
27. Wuhrer M, Koeleman CAM, Hokke CH et al (2005) Protein glycosylation analyzed by normal-phase of glycopeptides. Anal Chem 77:886–894
28. Kolarich D, Weber A, Turecek PL et al (2006) Comprehensive glyco-proteomic analysis of human alpha1-antitrypsin and its charge isoforms. Proteomics 6:3369–3380
29. Pabst M, Altmann F (2008) Influence of electrosorption, solvent, temperature, and ion polarity on the performance of LC-ESI-MS using graphitic carbon for acidic oligosaccharides. Anal Chem 80:7534–7542

Chapter 10

Analysis of Milk Oligosaccharides by Mass Spectrometry

Lauren D. Wu, L. Renee Ruhaak, and Carlito B. Lebrilla

Abstract

Human milk oligosaccharides (HMOs) are a highly abundant constituent in human milk, and its protective and prebiotic properties have attracted considerable attention. HMOs have been shown to directly and indirectly benefit the overall health of the infant due to a number of functions including serving as a beneficial food for gut bacteria, block to pathogens, and aiding in brain development. Researchers are currently exploring whether these structures may act as possible disease and nutrition biomarkers. Because of this, rapid-throughput methods are desired to investigate biological activity in large patient sets. We have optimized a rapid-throughput protocol to analyze human milk oligosaccharides using micro-volumes of human breast milk for nutritional biomarkers. This method may additionally be applied to other biological fluid substrates such as plasma, urine, and feces. The protocol involves lipid separation via centrifugation, protein precipitation using ethanol, alditol reduction with sodium borohydride, and a final solid-phase extraction purification step using graphitized carbon cartridges. Samples are analyzed using HPLC-Chip/TOF-MS and data filtered on Agilent MassHunter using an in-house library. Individual structural identification is matched against a previously developed HMO library using accurate mass and retention time. Using this method will allow in-depth characterization and profiling of HMOs in large patient sets, and will ease the process of discovering significant nutritional biomarkers in human milk.

Key words Oligosaccharide, Human milk, Mass spectrometry, Glycan, Carbohydrate, Free oligosaccharide, HPLC, TOF

1 Introduction

HMOs are free sugars commonly found with a lactose (Galβ1-4Glc) core at the reducing end, and can further be elongated linearly or branched by N-acetyllactosamine units with fucose and sialic acid that can be present at the terminal regions [1, 2]. They are the third or fourth most abundant component in milk, which is dependent on the mother and lactation time point, but do not pose a direct nutritional benefit to the infant. Because of their biological importance in gut floral development and in pathogen blocking [3–5], their presence in the infant gut is vital. Recent experiments have shown that oligosaccharides are not only excreted into the urine, but also upregulated into the bloodstream of the

Gordan Lauc and Manfred Wuhrer (eds.), *High-Throughput Glycomics and Glycoproteomics: Methods and Protocols*, Methods in Molecular Biology, vol. 1503, DOI 10.1007/978-1-4939-6493-2_10, © Springer Science+Business Media New York 2017

infant [6, 7], supporting the previous findings that sialic acid is important in brain development and memory [8].

Further investigating these properties and future nutritional biomarker studies will require large patient sample sets, where high-throughput sample preparation and analysis are favored. Because of their heterogeneity and structural diversity, there has been difficulty in creating a high throughput method to investigate their biological activity in different patient sets. There are a number of current separation methods to analyze free oligosaccharides in milk, involving anionic exchange chromatography [9–11], capillary electrophoresis [12–14], reverse phase chromatography [15, 16], and hydrophilic interaction chromatography [17]. However, many of these methods require derivatization tags and have partial or limited isomer separation. Ruhaak et al. give a detailed comparison of the various protocols [18]. Using our method involving graphitized carbon coupled to nano-LC mass spec, we have been able to roughly monitor 250 free oligosaccharides, with 100 structurally identified isomers in human milk [19, 20].

The traditional human milk oligosaccharide protocol has been optimized into a rapid-throughput manner, taking into consideration each separation and extraction step to avoid sample loss, ensure reproducible MS signal, and to decrease sample handling time. Here we have developed a rapid-throughput 96-well plate method using separation and purification steps to isolate free oligosaccharides in milk. Method validation and reproducibility is explained in detail in Totten et al. [21].

This method involves an initial 1:1 dilution of sample and deionized water, proceeding with plate centrifugation at 3220 × *g* at 4 °C for 30 min, in order to separate lipids. The aqueous portion is carefully extracted using a multichannel pipette, and aliquoted onto a new plate. Two volumes of ethanol is then added to the samples and placed in −80 °C for 1.5 h to fully precipitate proteins. Samples are then thawed and centrifuged using the same parameters to separate aqueous from solids. The aqueous portion is collected and dried prior to alditol reduction. Sample is reconstituted using deionized water to its original volume, and an equivalent volume of 2 M sodium borohydride is added to make a final concentration of 1 M. Sample is incubated for 1.5 h in 65 °C. Sample is then applied onto a 96-well plate packed with graphitized carbon after column conditioning, and centrifuged at room temperature at 340 × *g* in order for compounds to have enough interaction time with the column. Deionized water is used to desalt, and 20% acetonitrile and 40% acetonitrile in 0.05% TFA in water is used to elute and collect oligosaccharides using the same centrifugal conditions. Eluate is dried to completion prior to a 50-fold dilution, using a 1 μL injection onto Agilent HPLC-Chip/TOF-MS. Because HMO standards are limited and costly, a pooled HMO sample collected from a number of mothers is analyzed in parallel in order to closely match retention times with the already built library and

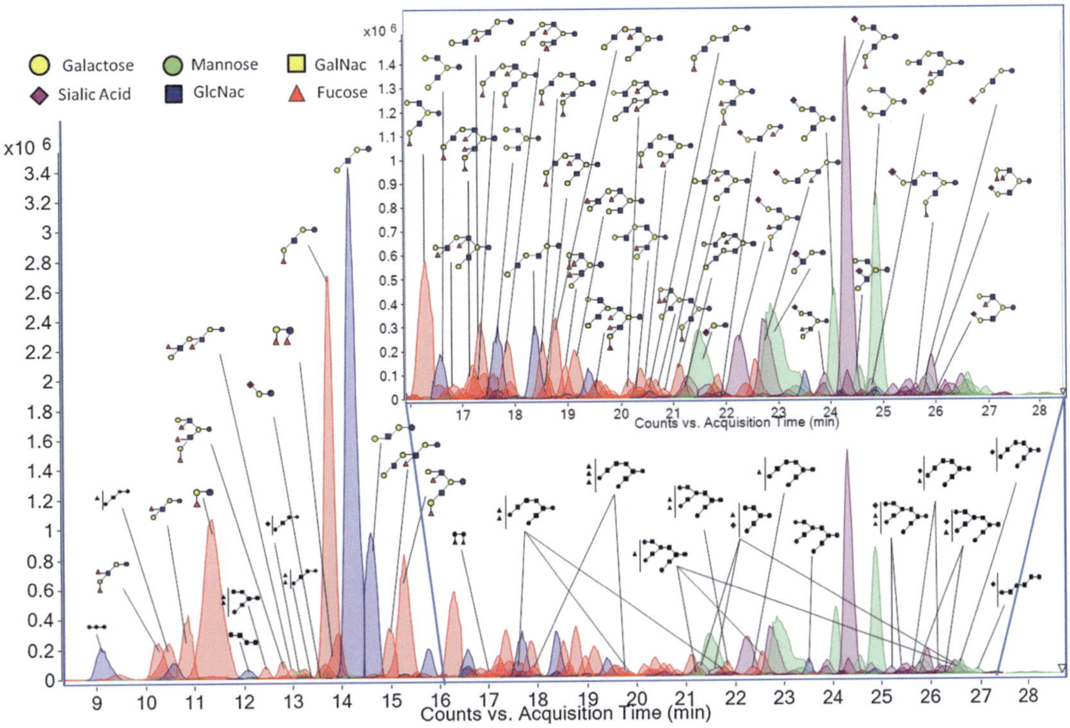

Fig. 1 An HMO pool sample was analyzed using HPLC Chip/TOF-MS to determine glycan signal. Sample representing a 50-fold dilution of milk was injected into the nano-LC instrument. All known compounds were extracted, overlaid, and were structurally identified and annotated using an in-house library. *Each color* indicates a different glycan class—*blue* (non-fucosylated neutrals), *red* (fucosylated), *green* (sialylated), and *purple* (fucosylated and sialylated)

also to monitor retention time shift and signal intensity throughout a large run. Figure 1 is an example of an extracted compound chromatogram of an HMO pool sample.

This rapid-throughput approach has already been applied to global lactation and nutritional studies as large as 2000 samples, and has already given promising results between the different groups. Not only can this process be applied to human milk, but can be applied to other mammalian milks as well, and other biological matrices such as feces, serum, and urine [6, 22], with minimal alternations to the protocol. With this method we can compare the overall HMO profiles in different patients and additionally monitor the individual isomers in human milk at the same time.

2 Materials

2.1 Separation Reagents

1. Nanopure water.
2. 100% ethanol (200 Proof).
3. 2 M sodium borohydride ($NaBH_4$) in deionized H_2O.
4. Folch solution (2:1 chloroform–methanol (v/v))—optional.

2.2 Solid-Phase Extraction Solvents

1. 20% ACN in H_2O (v/v).
2. 40% ACN in H_2O (0.05% TFA) (v/v).
3. 80% ACN in H_2O (0.1% TFA) (v/v).

2.3 Specialized Supplies and Equipment

1. 96-well carbon plate for solid-phase extraction (Glygen, Cat #FNSCAR).
2. 0.2 mL 96-well plate (GeneMate, Cat #T-3183-1) with compatible 96-well plate lid (Phenix Research Products, Cat #SMX-PCR96).
3. 0.8 mL 96-well polypropylene plate (Thermo Scientific, Cat #AB0765) with compatible 96-well plate lid (Thermo Scientific, Cat #AB0566).
4. 1.3 mL 96-well polypropylene plate (Thermo Scientific, Cat #260251) with compatible 96-well plate lid (Thermo Scientific Nunc, Cat #12565560).
5. Refrigerated bucket centrifuge for plate centrifugation and for solid-phase extraction.

2.4 MALDI-MS Analysis of Oligosaccharides

1. Matrix: 2,5-dihydroxybenzoic acid (DHB) (5 mg/100 μL in 50% ACN in H_2O).
2. Positive mode dopant: 0.1 M NaCl in 50:50 H_2O–ACN.

2.5 Nano LC/MS Analysis of Oligosaccharides

1. 3% ACN in 0.1% FA (v/v) (A1 Solvent).
2. 90% ACN in 0.1% FA (v/v) (B1 Solvent).
3. Compatible 96-well autosampler plate (Eppendorf, Cat #951020401).
4. Compatible pierceable 96-well plate lid (Phenix Research Products, Cat #SMX-PCR96).
5. PGC Chip (40 nL with 43 × 0.075 mm analytical column, Agilent, Cat#G4240-64010).

3 Methods

3.1 Separation and Purification of Free Oligosaccharides

1. Aliquot 25 μL of milk onto a 0.2 mL 96-well plate (*See* **Note 1**).
2. Add 25 μL water using a multichannel pipette and vortex lightly with a sealed plate lid (*See* **Note 2**).
3. Centrifuge samples at 4 °C for 30 min at maximum speed (3220 × *g*).
4. Extract aqueous layer between the top lipid layer and the bottom protein/cell detritus layer and transfer to 0.8 mL plate for reduction (*See* **Note 3**).

5. Protein precipitation (*See* **Note 4**).
   (a) Add 2 sample volumes of ethanol, and vortex completely. Spin down briefly.
   (b) Place samples in −80 °C for a minimum of 1.5 h. Samples may be stored overnight.
6. Thaw and centrifuge at 4 °C for 30 min at 3220 × *g*.
7. Carefully collect supernatant without disturbing the protein pellet, and dry to completion on low heat using a centrifugal evaporator. Samples may be stored in −80 °C overnight.
8. Reduction from aldehyde to alditol.
   (a) Add 25 μL of water to dried sample, and place on plate vortex for 30 min with a sealed plate on highest speed setting (*See* **Note 5**). Add an equivalent volume of 2 M $NaBH_4$, vortex, and briefly centrifuge. Final concentration of $NaBH_4$ buffer is 1 M.
   (b) Incubate for 1.5 h at 65 °C in a dry incubator (*See* **Note 6**).

3.2 Solid-Phase Extraction Using 96-Well Carbon Packed Plates

1. Add 2 volumes of nanopure water to Glygen carbon plate (200 μL well capacity) to wet the column. Waste collection should be placed below cartridge plate and should be discarded for each run.
2. Centrifuge at 340 × *g* for 1 min at room temperature. Ensure water has passed through the column.
3. Add 2 volumes of 80 % acetonitrile in 0.1 % TFA to the SPE plate to remove contaminants. Centrifuge using same parameters.
4. Add 3 volumes of nanopure water to the SPE plate to equilibrate the column and centrifuge. (*See* **Note 7**)
5. Load sample to the SPE plate. Centrifuge at 340 × *g* for 2–3 min until sample has been completely loaded. Ensure that sample has been fully loaded onto column before continuing.
6. Desalt a total of 5 times with nanopure water (*see* **Note 8**), centrifuging and discarding waste between each water addition.
7. Replace waste plate with 1.3 mL collection plate. Be sure not to discard any of the collection eluate, as these fractions contain the analytes of interest.
8. Collect 0.4 mL of 20 % ACN and 0.4 mL of 40 % ACN in 0.05 % TFA (2 volumes each) in the same 1 mL Thermo/Nunc plate with centrifugation steps in between. Final elution volume will be 0.8 mL. (*See* **Note 9**).
9. Dry to completion using a centrifugal evaporator on low to medium heat for 6–8 h (*See* **Note 10**).
10. Store between −80 °C to −20 °C with a sealed lid until ready for analysis. Wrap 96-well plate in foil to prevent ice forming around the lid seal. Recommended storage is up to 1 year.

3.3 Sample Reconstitution

1. To the dried samples, add the amount of water equivalent to original volume of milk aliquotted (in this case, 25 μL).
2. Vortex on high speed setting for 30 min with secure lid.
3. Spin down briefly.

3.4 MALDI-MS Analysis

1. Pipette mix glycan solution (1 μL), matrix solution (0.7 μL), and NaCl dopant (0.3 μL) onto the MALDI probe for positive mode analysis. In negative mode, NaCl dopant is excluded.
2. Dry sample in vacuum chamber prior to analysis.
3. Data was generated using an IonSpec HiResMALDI equipped with external MALDI source, a 7.0 T superconducting magnet, and a pulsed Nd:YAG laser (355 nm).

3.5 Nano LC/MS Analysis

1. To a 0.2 mL 96-well autosampler compatible plate, add a 50× dilution of sample using nanopure water. Vortex with a secured lid and spin down briefly. Secure an autosampler compatible lid to prevent sample evaporation.
2. Apply a 45 min gradient to each method (2.5–20 min, 0–16% B1; 20–30 min, 16–44% B1; 30–35 min, 100% B1; 20 min equilibration, 0% B1) with a flow rate of 0.3 μL/min.
3. Load 1 μL of sample onto a 40 nL enrichment column and analytical column (porous graphitized carbon) using a 4 μL/min flow rate (Solvent A).
4. Data was generated using an Agilent 1200 series HPLC coupled to an Agilent Technologies 6210 TOF mass spectrometer and Agilent 1200 series HPLC coupled to Agilent 6520 Q-TOF mass spectrometer, with both detectors having a chip-cube interface.

3.6 Human Milk Oligosaccharide Identification for LC/MS

1. Using an in-house created HMO library, we are able to filter all the detected compounds using accurate mass, and at this point can separate compounds by glycan class and composition.
2. Once all compositions have been identified, we then can accurately assign specific linkage structures to these compositions using the published HMO library (*See* **Note 11**).
3. We have further developed an in-house peak alignment software to correct for retention time shift. The description of this software is explained in Totten et al. [21].

4 Notes

1. 10–50 μL of original milk may be used as starting volume. Using more than 50 μL of milk may saturate the SPE plate cartridge [21]. Figure 2 depicts the complete sample preparation workflow for HMO extraction.

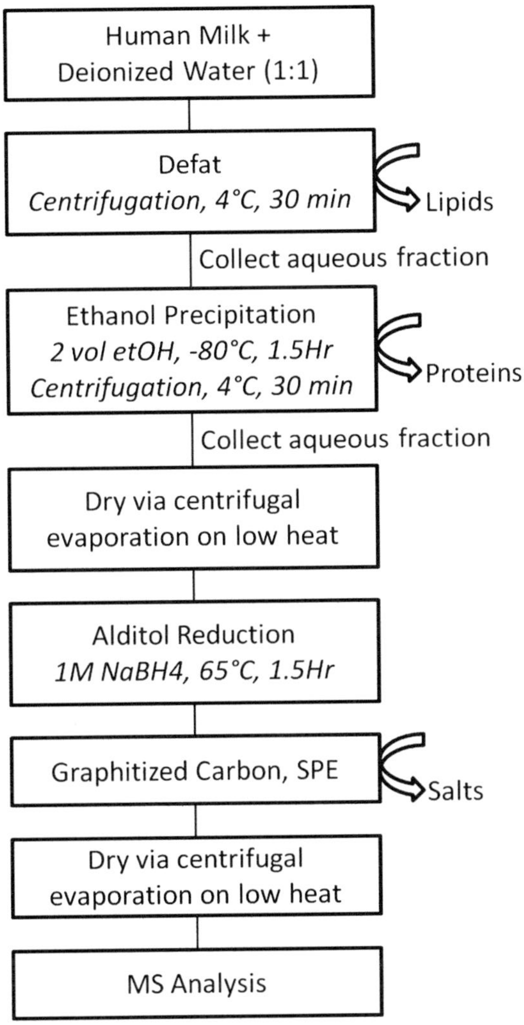

Fig. 2 Chart depicts the complete sample preparation workflow for rapid-throughput HMO analysis. The complete protocol for 96 samples prior to MS analysis can be completed in 2–3 work days with drying time incorporated

2. A 1:1 water to sample addition is typical. But because this method uses such small volumes, separation may be difficult. If this step is repeatedly becoming an issue, bring total volume to 100 μL using nanopure water prior to centrifugation. For large nutritional studies, our methods have optimized 25 μL of milk as the standard.
3. It is important to avoid collecting the upper layer of lipids. Trace amounts are unavoidable during the puncturing of the lipid layer. To ensure an ultraclean final mixture, a Folch step may be included in-between the initial defatting step and the protein removal step. Add 4 volumes of Folch solution (2:1 chloroform–methanol [v/v]) with glass pipette tips, vortex, and centrifuge at maximum speed at 4 °C. Carefully

collect the upper layer, and proceed with **step 5** for protein precipitation in Subheading 3. This step is currently omitted in the rapid-throughput method due to insignificance of its exclusion using starting volumes of 10–50 μL of milk [21].

4. An alternative method to ethanol precipitation is protein removal using C8 cartridges. Ethanol precipitation is published in the high-throughput protocol since it is a much cheaper and simpler alternative. The following steps are provided if protein removal using C8 is desired. Load sample onto C8 SPE plate and collect flow-through containing HMOs and salts. Add a one cartridge volume of nanopure water and combine with previous flow-through. Collection mixture would then be reduced and directly added to carbon cartridge in solid-phase extraction section under methods.
5. It is important to vortex microplate with nanopure water to ensure sample is homogenous prior to sodium borohydride addition. Vortex on a microplate vortexer for 30 min at maximum speed setting with a secured lid.
6. Before incubation, it is important to properly secure lid to microplate to avoid sample evaporation due to the chemical reaction. Apply lab tape to all sides of the lid creating a tight seal to completely prevent gases from escaping the wells. Incubating in a water bath causes condensation formation on the lid and wells and may introduce cross-contamination. For reduction purposes, our study has utilized a dry incubator to avoid this issue. SPE must be administered immediately after this step. Temporary sample storage should be avoided.
7. Because acetonitrile is an elution solvent, small analytes may wash off during the loading and desalting step if there are leftover traces of ACN prior to sample loading. Make sure ACN is completely washed off of the column by adding an extra column volume of water during the equilibration step.
8. Small compounds may also wash off during the desalting step if excess water has passed through the column. 7 column volumes of water should be the maximum amount to wash off salts and small peptides when using 50 μL of sample. The amount of desalting volume also depends on the original sample volume. 5 column volumes for desalting is enough for a starting volume of 25 μL of sample.
9. If using MALDI-MS, keep fractions separate in order to prevent ion suppression. If using HPLC, these elutions may be combined.
10. On higher heat settings, sialic acid on the terminal end may be unstable and will detach from the oligosaccharide during the drying process. It is important to dry on low to medium heat.
11. Refer to Wu et al. [19, 20] for the complete method on oligosaccharide identification.

References

1. Kunz C, Rudloff S, Baier W, Klein N, Strobel S (2000) Oligosaccharides in human milk: structural, functional, and metabolic aspects. Annu Rev Nutr 20:699–722
2. Ninonuevo MR, Perkins PD, Francis J, Lamotte LM, LoCascio RG, Freeman SL, Mills DA, German JB, Grimm R, Lebrilla CB (2008) Daily variations in oligosaccharides of human milk determined by microfluidic chips and mass spectrometry. J Agric Food Chem 56(2):618–626
3. Smilowitz JT, Lebrilla CB, Mills DA, German JB, Freeman SL (2014) Breast milk oligosaccharides: structure-function relationships in the neonate. Annu Rev Nutr 34:143–169
4. Newburg DS, Ruiz-Palacios GM, Morrow AL (2005) Human milk glycans protect infants against enteric pathogens. Annu Rev Nutr 25:37–58
5. Zivkovic AM, German JB, Lebrilla CB, Mills DA (2011) Human milk glycobiome and its impact on the infant gastrointestinal microbiota. Proc Natl Acad Sci U S A 108(Suppl 1): 4653–4658
6. Ruhaak LR, Stroble C, Underwood MA, Lebrilla CB (2014) Detection of milk oligosaccharides in plasma of infants. Anal Bioanal Chem 406(24):5775–5784
7. Goehring KC, Kennedy AD, Prieto PA, Buck RH (2014) Direct evidence for the presence of human milk oligosaccharides in the circulation of breastfed infants. PLoS One 9(7), e101692
8. Wang B (2009) Sialic acid is an essential nutrient for brain development and cognition. Annu Rev Nutr 29:177–222
9. Finke B, Mank M, Daniel H, Stahl B (2000) Offline coupling of low-pressure anion-exchange chromatography with MALDI-MS to determine the elution order of human milk oligosaccharides. Anal Biochem 284(2): 256–265
10. Grey C, Edebrink P, Krook M, Jacobsson SP (2009) Development of a high performance anion exchange chromatography analysis for mapping of oligosaccharides. J Chromatogr B Analyt Technol Biomed Life Sci 877(20-21): 1827–1832
11. Coppa GV, Pierani P, Zampini L, Bruni S, Carloni I, Gabrielli O (2001) Characterization of oligosaccharides in milk and feces of breast-fed infants by high-performance anion-exchange chromatography. Adv Exp Med Biol 501:307–314
12. Monti L, Cattaneo TM, Orlandi M, Curadi MC (2015) Capillary electrophoresis of sialylated oligosaccharides in milk from different species. J Chromatogr A 1409:288–291
13. Bao Y, Newburg DS (2008) Capillary electrophoresis of acidic oligosaccharides from human milk. Electrophoresis 29(12):2508–2515
14. Shen Z, Warren CD, Newburg DS (2000) High-performance capillary electrophoresis of sialylated oligosaccharides of human milk. Anal Biochem 279(1):37–45
15. Chaturvedi P, Warren CD, Ruiz-Palacios GM, Pickering LK, Newburg DS (1997) Milk oligosaccharide profiles by reversed-phase HPLC of their perbenzoylated derivatives. Anal Biochem 251(1):89–97
16. Leo F, Asakuma S, Fukuda K, Senda A, Urashima T (2010) Determination of sialyl and neutral oligosaccharide levels in transition and mature milks of Samoan women, using anthranilic derivatization followed by reverse phase high performance liquid chromatography. Biosci Biotechnol Biochem 74(2):298–303
17. Marino K, Lane JA, Abrahams JL, Struwe WB, Harvey DJ, Marotta M, Hickey RM, Rudd PM (2011) Method for milk oligosaccharide profiling by 2-aminobenzamide labeling and hydrophilic interaction chromatography. Glycobiology 21(10):1317–1330
18. Ruhaak LR, Lebrilla CB (2012) Advances in analysis of human milk oligosaccharides. Adv Nutr 3(3):406S–414S
19. Wu S, Tao N, German JB, Grimm R, Lebrilla CB (2010) Development of an annotated library of neutral human milk oligosaccharides. J Proteome Res 9(8):4138–4151
20. Wu S, Grimm R, German JB, Lebrilla CB (2011) Annotation and structural analysis of sialylated human milk oligosaccharides. J Proteome Res 10(2):856–868
21. Totten SM, Wu LD, Parker EA, Davis JC, Hua S, Stroble C, Ruhaak LR, Smilowitz JT, German JB, Lebrilla CB (2014) Rapid-throughput glycomics applied to human milk oligosaccharide profiling for large human studies. Anal Bioanal Chem 406(30):7925–7935
22. De Leoz ML, Gaerlan SC, Strum JS, Dimapasoc LM, Mirmiran M, Tancredi DJ, Smilowitz JT, Kalanetra KM, Mills DA, German JB, Lebrilla CB, Underwood MA (2012) Lacto-N-tetraose, fucosylation, and secretor status are highly variable in human milk oligosaccharides from women delivering preterm. J Proteome Res 11(9):4662–4672

Chapter 11

N- and *O*-Glycomics from Minor Amounts of Formalin-Fixed, Paraffin-Embedded Tissue Samples

Hannes Hinneburg, Falko Schirmeister, Petra Korać, and Daniel Kolarich

Abstract

The availability of well-defined samples in sufficient numbers represents a major bottleneck for any biomarker related research. The utilization of preserved, archived and clinically well-described samples therefore holds a great potential to bridge this gap. This chapter describes a universal workflow for the comprehensive characterization of *N*- and *O*-glycans released from whole formalin-fixed, paraffin-embedded tissue sections, including an option for further partitioning using laser microdissection of specific tissue areas/cell populations. Glycoproteins are extracted and subsequently immobilized onto a PVDF membrane prior enzymatic release of *N*-glycans. Following *N*-glycan retrieval *O*-glycans are released using reductive β-elimination from the same sample spot, significantly reducing the required amount of starting material. Released and reduced glycan structures are characterized using porous graphitized carbon liquid chromatography online coupled to an electrospray ionization-ion trap mass spectrometer. This technique provides information on the relative abundances of individual glycans along with detailed structural information, including isomer differentiation and functional epitope characterization of *N*- and *O*-glycans obtained from minimal amounts of tissue down to a few thousand cells.

Key words Glycoprotein, *N*-glycan, *O*-glycan, Glycomics, Formalin-fixed, paraffin-embedded (FFPE), Liquid chromatography, Mass spectrometry, Porous graphitized carbon, Laser capture microdissection

1 Introduction

Protein glycosylation is a key aspect in numerous important biological processes such as immune modulation, cell–cell interaction, and signal transduction [1–3]. Changes in protein glycosylation are also a hallmark of severe diseases such as cancer and chronic inflammatory diseases and its elucidation bears a great potential to improve initial diagnosis and prognosis. Furthermore, monitoring these changes may facilitate and advance personalized medicine applications by adjusting treatment protocols according to actual patient conditions [4–6]. Among the many glycan types known to date, *N*- and mucin-type *O*-glycans belong to the most abundant and best-studied forms of post-translational modifications.

Gordan Lauc and Manfred Wuhrer (eds.), *High-Throughput Glycomics and Glycoproteomics: Methods and Protocols*, Methods in Molecular Biology, vol. 1503, DOI 10.1007/978-1-4939-6493-2_11, © Springer Science+Business Media New York 2017

These glycans have also been shown to be crucial for the function of their respective protein carriers [2, 7–9]. Thus, detailed knowledge on the particular structures associated with specific cell types, body fluids or specific proteins is a crucial prerequisite for studying and understanding their functional role in health and disease.

Glycosylation pattern analyses of cell subsets [10] or specific tissue parts [11, 12] have demonstrated that glycomic information on localized alterations represents an important step towards this aim. For the analysis of tissues, cryo-sections or formalin-fixed, paraffin-embedded (FFPE) sections are highly interesting and valuable sample sources. FFPE tissue sections are of particular interest as they can be stored for decades, comprise various patient cohorts, and are available from pathology archives around the world, along with clinical data. Additionally, the protocols to prepare and preserve these tissue samples are highly standardized across all histology labs, increasing the comparability of samples from different places. Therefore FFPE tissue samples have recently created great interest in the field of biomarker discovery studies to analyze proteins [13, 14], DNA/RNA [15, 16], but also glycans [12, 17, 18].

Laser capture microdissection (LCM) provides a superior technique for isolating specific tissue parts or cell populations from tissue sections [19]. It significantly decreases the risk of contaminations by other tissue parts that, if present, might cover disease specific signatures. Thus, LCM based tissue fractionation is of particular benefit for the analysis of any tissues of interest, such as tumor tissue, which often invades surrounding areas or gets infiltrated with populations of non-tumor cells. LCM has also been used extensively to isolate specific tissue parts from sections for following analysis of DNA/RNA, proteins, and metabolites [20].

In this work we describe a well-established protocol for the extraction of *N*- and *O*-glycans from entire FFPE tissue sections as well as from cells isolated by LCM including their subsequent characterization using a porous graphitized carbon nano liquid chromatography electrospray ionization (PGC-nanoLC-ESI)-tandem mass spectrometric (MS/MS) based glycomics approach [21, 22].

Using the protocol described here around 80 *N*-glycan structures could be identified in samples from FFPE tissue sections prepared following standard preservation protocols. The nature of the PGC separation makes it possible to visualize isomers, which are otherwise difficult to differentiate by mass spectrometry only (Fig. 1).

Besides *N*-glycan profiles it is also possible to obtain *O*-glycan profiles from the same spot of immobilized protein, minimizing the sample amount needed for analysis (Fig. 2).

2 Materials

If not stated otherwise, all materials are of the highest grade possible.

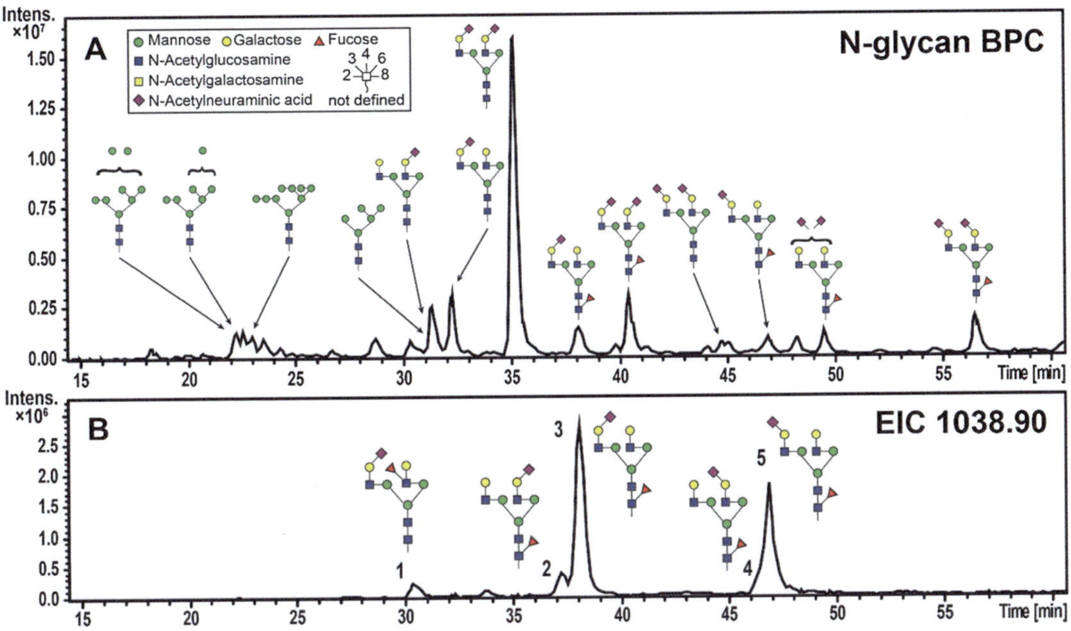

Fig. 1 (**a**) PGC-LC base peak chromatogram (BPC) acquired for the *N*-glycans released from a single H&E stained FFPE tissue section (~100 mm^2) obtained from a hepatocellular carcinoma patient. (**b**) Extracted ion chromatogram (EIC) showing isomer separation of a representative *N*-glycan (Hex5HexNAc4NeuAc1Fuc1, $[M-2H]^{2-} = 1038.9$ Da). *N*-glycans eluted after carbon purification were dried, dissolved in 10 μL of water and 3 μL were injected for a single analysis. The 13 most abundant *N*-glycans out of 80 identified structures are shown in the BPC

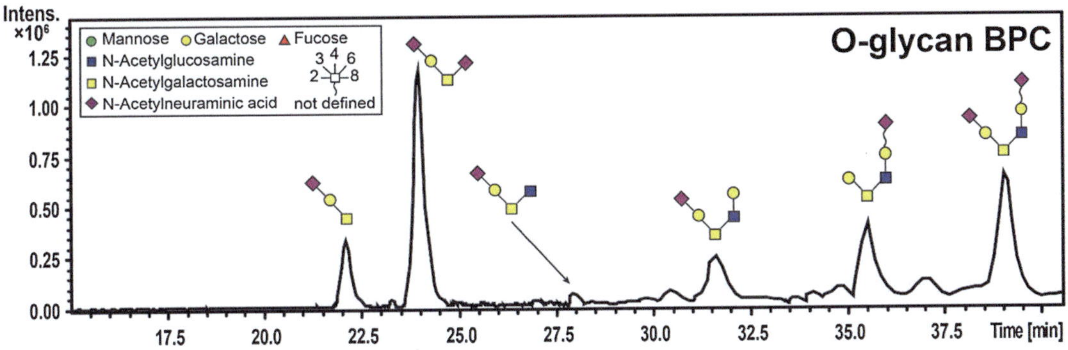

Fig. 2 PGC-LC base peak chromatogram (BPC) acquired for the *O*-glycans released from a single FFPE tissue section (~100 mm^2). *O*-glycans eluted after carbon purification were dried, dissolved in 10 μL of water and 3 μL were injected for a single analysis

2.1 Tissue Slide Preparation, Microdissection, and HE Staining

1. Ethanol, water, xylene, hematoxylin, and eosin.
2. Tissue samples in paraffin blocks (e.g. from pathology archives).
3. Membrane slides (MembraneSlide 1.0 PEN [polyethylene naphthalate], Carl Zeiss Microscopy).
4. Microtome (Accu-Cut SRM 200 Rotary Microtome, Sakura Finetek Europe, Leiden, Netherlands).

5. Drying oven.
6. Tubes for sample collection (Sample AdhesiveCap 500 opaque (D) PCR Tube, Carl Zeiss Microscopy).
7. Laser microdissection system (Palm Zeiss MicroBeam, Axiovert 200 M with PALM RoboSoftware, Carl Zeiss Microscopy).

2.2 Retrieval of Proteins

1. Water, xylene, ethanol, methanol, chloroform.
2. Retrieval buffer: 4% SDS (w/v, sodium dodecyl sulfate) in 0.1 M Tris–HCl pH 8 at 25°C, 0.1 M DTT (dithiothreitol) (*see* **Note 1**).
3. 8 M urea (*see* **Note 2**).
4. Razor blades.
5. Branson sonifier B-12 sonicator (Branson Sonic Power) or comparable device.
6. Sample tube centrifuge (up to 14,000 rcf).
7. Vortex mixer.
8. Sample tube heating block with agitation (up to 99°C).
9. Sonicator bath.

2.3 Dot-Blotting of Proteins onto PVDF Membrane

1. Water, ethanol.
2. Glass petri dish.
3. Lint free tissues (e.g., Kimtech wipes).
4. PVDF membrane (0.22 µm or 0.45 µm, Millipore).
5. Clean pair of scissors.
6. Direct Blue stock solution: 0.1% (w/v) Direct Blue 71 (Sigma-Aldrich) in water.
7. Direct Blue wash solution: 40% (v/v) ethanol, 10% (v/v) acetic acid. Solution can be stored for 3 months at RT.
8. Shaker.

2.4 Glycan Release and Reduction

1. Water, methanol, acetic acid.
2. Flat-bottom polypropylene 96-well plate (*see* **Note 3**).
3. Parafilm or lid to seal 96-well plate.
4. Clean scalpel.
5. Polyvinylpyrrolidone (PVP40, Sigma-Aldrich) solution: 1% (w/v) solution in 50% (v/v) methanol. Solution can be stored for 3 months at RT.
6. Peptide:N-glycosidase F (PNGase F) from *Flavobacterium meningosepticum*, recombinant from *E. coli*, glycerol-free (Roche or NEB).
7. Potassium hydroxide: 50 mM in water. Solution can be stored for 1 month at RT.

8. Sodium borohydride in 50 mM potassium hydroxide solution. Needs to be prepared fresh every time. 1 M for *N*-glycans; 0.5 M for *O*-glycans.
9. Incubator at 37°C/50°C.
10. Centrifuge for 96-well plates.
11. Ultrasonic bath.
12. Centrifugal evaporator (e.g., Savant SpeedVac).

2.5 Desalting and Purification of Reduced Glycans

1. Water, methanol, hydrochloric acid, acetonitrile, trifluoroacetic acid (TFA).
2. Cation exchange resin AG 50 W X8; 50–100 dry mesh size (Bio-Rad Laboratories).
3. ZipTip C18, 0.6 μL (Millipore), adapters to hold ZipTips in place within tubes.
4. Carbon material from "Extract-Clean" carbon SPE cartridge (Grace).
5. 80% (v/v) acetonitrile and 0.1% (v/v) TFA.
6. 0.1% (v/v) TFA in water.

2.6 PGC-nanoLC-ESI-MS/MS Analysis

1. 100 mM ammonium bicarbonate (filtered, 0.2 μm pore size).
2. LC solvents (*see* **Note 4**):
   (a) LC solvent A: 10 mM ammonium bicarbonate.
   (b) LC solvent B: 60% (v/v) acetonitrile in 10 mM ammonium bicarbonate.
   (c) Loading pump LC solvent C: Same as LC solvent A.
   (d) Loading pump LC solvent D: 90% (v/v) acetonitrile in 10 mM ammonium bicarbonate.
3. Hypercarb PGC columns (precolumn: Hypercarb KAPPA 30 × 0.32 mm, 5 μm particle size; analytical column: 100 mm × 75 μm, 3 μm particle size, both Thermo Scientific).
4. Tubings and appropriate connections [e.g., nanoViper finger-tight fittings and peeksil tubing (Thermo Fisher)].
5. Mass spectrometer (e.g., amaZon speed ETD ion trap, Bruker Daltonics) or comparable instrument equipped with appropriate source (e.g., CaptiveSpray™, Bruker Daltonics).

3 Methods

Glycans can be extracted from whole FFPE tissue sections or after further tissue partitioning using LCMs. The extraction protocols differ slightly depending on the starting material. Once the proteins are extracted they are spotted onto a PVDF membrane and glycans are subsequently released by PNGase F (*N*-glycans) and reductive β-elimination (*O*-glycans) [21]. Herein room temperature (RT) is defined as 20–25°C.

3.1 Preparation of FFPE Tissue Sections, Laser Capture Microdissection and HE Staining

1. 2–5 μm thick tissue sections are cut from paraffin blocks and mounted on histology glass slides for direct extraction or mounted on membrane slides for microdissection (*see* **Note 8**).
2. Histology glass slides or membrane slides are dried overnight at 56°C.
3. Slides are deparaffinized using xylene or similar solvent for 2 min (2×).
4. Tissue sections are washed with absolute ethanol for 2 min (2×)
5. Sections are air-dried and histology glass slides can now be stained or used for direct extraction, while membrane slides can be stained or used for LCM.
6. Staining (optional): tissue sections mounted on histology glass slides or membrane slides can be stained with hematoxylin and eosin for better recognition of different tissue areas/ cell populations.
   (a) Deparaffinization with xylene (as above) and rehydration through a series of ethanol (absolute, 85%, and 70% ethanol, each 2 min) followed by distilled water.
   (b) Hematoxylin is applied for 2 min.
   (c) Slides are washed with distilled water for 2 min.
   (d) Eosin is applied for 10 s.
   (e) Slides are dehydrated with increasing ethanol series (70%, 85%, and absolute ethanol, each 2 min).
   (f) Sections are briefly air-dried.
   (g) Sections on histology glass slides can be used for direct extraction, while sections on membrane slides can be used for LCM.
7. Laser capture microdissection (optional). Typical tissue sections used for LCMs are shown in Fig. 3. The tissue section is mounted on a membrane slide and different parts, areas of interest, are selected for microdissection, laser captured, and prepared for downstream analysis.
   (a) Stained or unstained slides are analyzed under the microscope and specific tissue parts or cells are selected using PALM RoboSoftware according to manufacturer's manual
   (b) Non-contact LCM is performed using the protocol described in the manufacturer's manual and specific tissue parts are collected in adhesive caps of 500 μL collection tubes
   (c) Microdissected samples are stored at −80°C

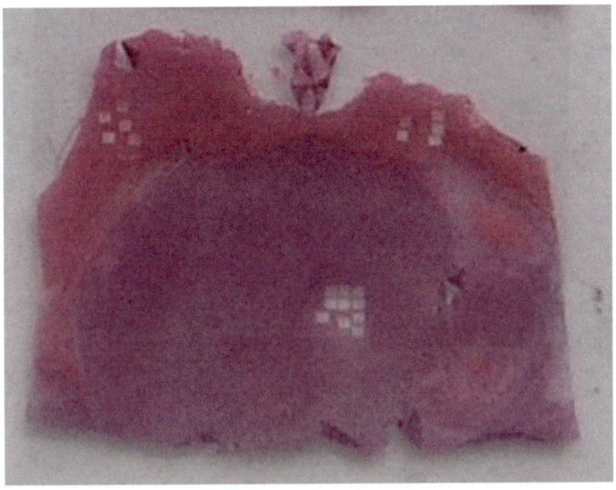

Fig. 3 Laser capture microdissection of tumor (*central area*) and non-tumor parts (*left and right upper areas*) of hepatocellular carcinoma tissue

3.2 Retrieval of Glycoproteins from FFPE Tissue Sections and Laser Capture Microdissected Material

For microdissections start with 3B.

1. Sections (2–5 μm thick, unstained or H&E stained) are washed 3 times with xylene followed by washing 3 times with absolute ethanol (~1 min per washing step) in a staining jar. If only a few slides (~3–5) are used, the washing steps might be performed directly on the slides to reduce solvent consumption (*see* **Note 5**).
2. The dry tissue is scraped off the slides using a razor blade and transferred into a sample tube (*see* **Note 6**).
   (a) 100 μL of retrieval buffer are added to the entire tissue sections (tissue buffer ratio: ~1:20) and the suspension is sonicated for 10–15 s (output power control is set to 1.5 out of 10; *see* **Notes 1**, **7**, **8**).
   (b) 100 μL of retrieval buffer are added to the microdissected tissues (>2,000 cells), which are subsequently incubated with the tubes upside down at room temperature for 10 min prior to sonication as described in **step 3**(**a**) (*see* **Notes 1**, **7**, **8**).
4. Incubate at 99°C for 60 min with mild agitation (~600 rpm, *see* **Note** 7).
5. Remove tubes carefully from the heating block and allow them to slowly cool down to room temperature. Centrifuge tubes at 2,000 rcf for 20 min at room temperature until the solution is clear. Supernatant S1 (~100 μL) is then transferred into a new tube. Centrifugation is usually not necessary for microdissections since tissue amounts are low.
6. Precipitate the proteins dissolved in S1 using the chloroform-methanol precipitation according to Wessel and Flügge [23]. Adjust used volumes according to sample volume. In short,

400 μL of methanol are added to supernatant S1 (~100 μL), followed by 100 μL chloroform and 300 μL water followed by vigorous vortexing. Centrifuge for 5 min at 14,000 rcf. The precipitated proteins are localized at the interphase between the chloroform (bottom) and methanol/water layer (top). The supernatant on top (S2) is carefully removed. Add 300 μL of methanol to the lower chloroform phase and vortex vigorously. Centrifuge the sample (10 min; 14,000 rcf) to pellet the protein. The entire supernatant (S3) is removed. The (glyco) proteins of interest are found in the pellet (*see* **Note 9**).

7. Dissolve the pellet in 8 M urea with the smallest volume possible (~20–50 μL for whole sections; ~10 μL for microdissections, *see* **Notes 2** and **10**).

3.3 Dot-Blotting of Proteins onto PVDF Membranes

The dot-blotting is performed according to Jensen et al. [21].

1. Cut a piece of PVDF membrane according to the number of protein spots to be applied (approx. 1.5 cm^2 for each spot is needed, which includes ~1 cm of space between spots). Draw desired spotting pattern on a piece of paper and put this template underneath the petri dish to facilitate spotting, especially if samples have to be applied onto the same spot multiple times.
2. Position the membrane on top of an ethanol-wetted lint free tissue in a petri dish. The ethanol will soak into the membrane and thereby activate it, which facilitates protein capture. Ensure that no excess ethanol covers the membrane surface before applying the samples. Keep the tissue underneath thoroughly wetted throughout the spotting procedure by applying drops of ethanol next to the rim of the tissue in regular intervals (2–4 min).
3. Apply the protein solution in discrete spots of max. 2 μL at a time. In total 5 μg of protein per spot is a suitable amount; this also applies for any standard glycoprotein included. Keep the tissue wet with ethanol until all spots have been applied and they dried up (usually < 5 min). If volumes larger than 2 μL need to be applied, repeat the spotting procedure and place additional 2 μL aliquots on top of the previously applied and dried spot until the desired protein amount is reached.
4. Dry the PVDF membrane at RT. Drying overnight is advised to ensure proper binding of proteins to the membrane.
5. Rewet the membrane for 15 min with ethanol while shaking.
6. Wash the membrane for 15 min with water to remove salts. Its hydrophobic nature makes it necessary to turn it over and over again for several times.

7. The spots are stained using a mixture of 20 mL of wash solution and 1.6 mL of Direct Blue 71 stock solution. Gently shake the membrane until the spots become visible.
8. Destain briefly in wash solution and then remove acid by washing with water. Leave the membrane at RT until dry. The dried membrane can be stored for months at RT.

3.4 N-Glycan Release and Reduction

1. Fill the required number of wells in a 96-well plate with 100 μL of the PVP40 solution.
2. Cut the stained protein spots with a scalpel and put them (protein facing up!) into the wells. Do not turn the cut spots during the following steps.
3. Shake the plate gently for 5 min before removing the PVP40 solution.
4. Wash spots three times for ~5 min with 100 μL water.
5. Remove the water and subsequently add 5 μL of PNGase F solution (*see* **Note 11**) and incubate for 10 min at 37°C.
6. Add 10 μL of water to each well and incubate overnight at 37°C. To limit evaporation add ~100 μL to the empty space surrounding a sample well. Seal the plate with a lid and/or Parafilm.
7. Following incubation sonicate the plate for 5 min and collect the samples at the bottom of the wells by centrifugation (500 rcf, 1 min). Transfer the liquid of each well into different tubes (the collected supernatant contains the released *N*-glycans).
8. Wash the wells twice with 20 μL of water (dispense and aspirate several times) and pool the washes with the *N*-glycan supernatant (*see* **step** 7 above).
9. Dry the samples in a centrifugal evaporator without heating.
10. Add 20 μL of 1 M $NaBH_4$ in 50 mM KOH and incubate for 3 h at 50°C to reduce the reducing end of the *N*-glycans (*see* **Note 12**).
11. Briefly (~30 s) spin the samples to collect them at the bottom of the sample tube.
12. Add 2 μL of glacial acetic acid to neutralize the reaction and mix the samples vigorously. Some effervescence might be seen upon the addition of the acid. Spin samples again and desalt them subsequently (*see* Subheading 3.6).

3.5 O-Glycan Release

1. After the *N*-glycans have been successfully cleaved from the glycoproteins rewet the PVDF membrane spots left in the 96-well plate with 2.5 μL of methanol.
2. Add 20 μL of 0.5 M $NaBH_4$ in 50 mM KOH to each well and incubate at 50°C for 16 h. Seal plates with a lid or Parafilm.

Additionally, placing the plate into a bag with wetted tissue is beneficial to limit solvent evaporation.

3. Collect the samples into fresh sample tubes and wash wells twice with 20 μL water. Combine the washing solution with the samples.
4. Neutralize the reaction by adding 2 μL of glacial acetic acid and spin sample down. Desalt the samples subsequently.

3.6 Desalting and Purification of Reduced Glycans

1. Prepare one cation exchange column per sample by depositing cation exchange resin (~30 μL) onto a ZipTip C18 tip (*see* **Note 13**).
2. Wash the column by centrifugation with:

   40 μL of 1 M HCl (3×)

   40 μL of methanol (3×)

   40 μL water (3×), avoid that the columns run dry during the last centrifugation step if columns are not immediately used. For the washing procedure the use of 2 mL sample tubes is recommended.
3. Place the column into a new tube. Add the reduced glycan sample (~20–60 μL) to the column and spin in a centrifuge.
4. Wash the original sample tube with 20 μL of water. Add this to the column and spin again.
5. Wash the column twice with 50 μL of water to elute potential bound glycans directly into same tube.
6. Discard the column and dry the glycans (completely!) in a centrifugal evaporator.
7. Wash the glycans with 150 μL of methanol and dry again (~10–15 min). Repeat this step three to five times until no borate is visible (*see* **Note 14**). Dried samples can be stored at −20°C for months.

Optional procedure: Purification using carbon material

8. Preparation of carbon clean up columns: Load approx. 25 μL of the carbon material onto a C18 ZipTip (*see* **Note 15**).
9. Place the column into a 2 mL sample tube, spin at 4,000 rcf at RT for about 10 s to pack column and remove any fines (small carbon particles) that may pass through the C18 material.
10. Wash the column three times with 40 μL of 80% (v/v) acetonitrile containing 0.1% (v/v) TFA by centrifugation at 2,000 rcf at RT for about 10 s.
11. Wash the column three times with 40 μL of 0.1% (v/v) TFA.
12. Dissolve the sample in 20 μL of 0.1% (v/v) TFA and load it onto the column. Wash the original sample tube again with 20 μL of 0.1% (v/v) TFA, add this to the column and centrifuge. The glycans bind to the carbon material.

13. Wash the column twice with 40 µL 0.1% (v/v) TFA for *N*-glycans. For *O*-glycans wash the sample only once. Discard the flow through and place the column into a new 1.5 mL sample tube (final sample tube).
14. Elute the bound glycans with 40 µL of 80% (v/v) acetonitrile containing 0.1% (v/v) TFA. Repeat.
15. Dry the eluted sample in the centrifugal evaporator at RT. Dried samples can be stored at −20°C for months.

3.7 Analysis of N- and O-Glycans Using PGC-nanoLC-ESI-MS/MS

The PGC-LC setup is based on the procedures described by Jensen et al. and Kolarich et al. [21, 22].

The setup uses a trapping (precolumn) and analytical PGC column. Equilibration of the precolumn is done in solvent A. Purified and reduced glycans are loaded for 5 min at 6 µL/min prior to valve switching. The starting conditions for the analytical column are 3% solvent B.

After trapping a linear gradient of solvent B is used as follows: reduced *N*-glycans are separated using a gradient from 3% solvent B to 15.8% in 1 min before increasing to 40.3% solvent B over 49 min. *O*-glycans are analyzed using a linear gradient from 2% to 35% solvent B over 33 min. The flow rate in both cases is set to 0.8–1 µL/min (depending on the individual column backpressure) and column temperatures are set to 40°C.

The LC is directly coupled to an amaZon speed ETD ion trap mass spectrometer equipped with CaptiveSpray™ source for online detection of glycans in negative ion mode. UltraScan mode (MS) within an m/z range of 380–1,800 is used; SPS tune is set to m/z 900, ICC to 40,000, and max. acquisition time to 200 ms.

The three most abundant precursors are used to generate fragment spectra in a range of m/z 100–2,500 with an isolation width of m/z 3 and fragmentation cutoff of 27% with 100% fragmentation amplitude using the Enhanced SmartFrag option from 30 to 120% in 32 ms. ICC is set to 150,000 and ion scan mode to Enhanced.

4 Notes

1. Prepare a stock solution of 0.1 M Tris–HCl pH 8 (at 25°C), containing 4% SDS [(w/v) in Tris–HCl] and add fresh DTT each time. Do not use the complete retrieval buffer for more than 1 day. Stock solutions should not be used longer than 2 weeks. Tris–HCl buffers should be stored at 4°C to prevent microbial growth. Note that the buffer needs to warm up to room temperature.

2. Urea solutions should always be prepared freshly, as they develop a significant concentration of reactive cyanate ions when kept at room temperature for a longer time.
3. This microplate material is used because of its suitable solvent stability. It has to withstand the *O*-glycan release conditions.
4. LC solvents need to be degassed prior to use. Ammonium bicarbonate buffers change their pH over time and therefore it should be avoided to use buffers longer than 2 weeks to reduce influence on retention times. LC buffers suggested can be used for the initial glycan analysis and might be adapted to the indented purpose and/or the properties of the samples.
5. For washing tissue sections directly on the slides without using a jar, put them onto a bench top with the tissue section facing up. Add enough liquid (~1 mL) to cover the section and proceed as follows: Wash with xylene by pipetting the liquid slowly up and down and discard used solution (3×). Then wash 3 times with absolute ethanol (~1 min per washing step). Stained tissue sections are usually deparaffinized, but when it is unclear whether that step has been performed previously, perform a xylene washing. Even if stained sections have been treated for staining, wash as described with ethanol before starting with scraping procedure.
6. It is recommended to apply small amounts of retrieval buffer (~1–2 drops) to small tissue sections (~15 mm^2) in order not to lose them. Working without gloves at this point might also be beneficial to avoid static repulsion.
7. Do not put more than 100 μL of retrieval buffer in a 1.5 mL sample tube. For larger amounts use 2 mL tubes, however, the volume of retrieval buffer should not exceed 200 μL as the use of larger volumes can result in unintended opening of the tubes during the heating procedure. Furthermore, the capacity of the tube will not be sufficient to hold the volumes needed during the chloroform-methanol precipitation. In case problems with excessive foaming occur during the sonification process, an alternative buffer of 0.1 M Tris–HCl pH 8 containing 0.1 M DTT can be used. SDS is then added to a final concentration of 4% just before the heating procedure.
8. The use of polyethylene naphthalate (PEN) membrane glass slides is not an issue for the analysis of laser microdissected material. If entire tissue sections with the membrane are intended to be used for analysis the protein yield can be decreased as the cell disruption is disturbed because the membrane tends to hold large pieces of tissue together. It is recommended to use an entire tissue section without the membrane.
9. The protein precipitation can be skipped if large amounts of protein (e.g., when combining several sections) are obtained

and S1 can be directly spotted onto the PVDF membrane. However, the reproducibility of relative glycan quantification is more reliable when the precipitation step is included. It is also recommended when using H&E stained slides, since the dye will also be removed during the procedure.

10. In order to dissolve the pellet pipetting small sample volumes (~10 μL) up and down several times should be avoided as foam is created that will result in sample losses. For these instances the use a sonicator bath for resolubilization has provided better results.

11. The required amount of PNGase F is usually specified in units of enzyme activity (U). The unit definition depends highly on the respective enzyme activity test employed by the manufacturer. For example, 1–2.5 U of PNGase F from Roche are required per sample. Alternatively, if using the product from New England Biolabs (NEB) it is recommended to use 125 U of PNGase F per sample. In both cases the final working solution needs to be adjusted to a total volume of 5 μL per sample by the addition of water to the enzyme solution. For these small volumes the preparation of a master mix is advised.

12. The reduction of the reducing end of *N*-glycans converts the α- and β-anomers to sugar alditols. This results in single chromatographic peaks during the PGC-LC for each particular glycan structure.

13. The AG 50W X8 cation exchange resin should be washed at least three times with excess of methanol prior to use. A larger volume of resin can be prewashed and stored in fresh methanol. It is possible to use other types of tips for packing the cation exchange columns. The ZipTip C18 tips are used for their convenient size. As the released glycans are not retained by the reversed phase, the C18 packing only serves as a means of support to hold back the cation exchange resin in the tip. The volume stated for the resin to be used refers to the absolute volume of the material needed in the end. Make sure not only solvent is sucked into the pipette. Pipette tips can be cut or wide tips can be used to ensure that the resin can be pipetted properly. For convenience it is recommended to have suitable adapters to hold columns in the right position within tube.

14. The addition of methanol is used to remove residual borate by evaporation of the volatile methyl borate. Ensure that the dried samples are completely dissolved prior to each drying step.

15. The volume stated for the carbon material to be used refers to the absolute volume of the material needed in the end. Make sure not only solvent is sucked into the pipette. Pipette tips can be cut or wide tips can be used to ensure that the resin can be pipetted properly.

Acknowledgements

We acknowledge support by the Max Planck Society and European Union (Seventh Framework Programme "Glycoproteomics" project, grant number PCIG09-GA-2011-293847, IBD-BIOM project, grant number 305479).

References

1. Ohtsubo K, Marth JD (2006) Glycosylation in cellular mechanisms of health and disease. Cell 126:855–867
2. Varki A, Cummings RD, Esko JD, Freeze HH, Stanley P, Bertozzi CR et al (2009) 2 edn. Cold Spring Harbor Laboratory Press, New York
3. Kolarich D, Lepenies B, Seeberger PH (2012) Glycomics, glycoproteomics and the immune system. Curr Opin Chem Biol 16:214–220
4. Almeida A, Kolarich D (2016) The promise of protein glycosylation for personalized medicine. Biochim Biophys Acta 1860:1583–1595
5. Christiansen MN, Chik J, Lee L, Anugraham M, Abrahams JL, Packer NH (2014) Cell surface protein glycosylation in cancer. Proteomics 14:525–546
6. Vasconcelos-Dos-Santos A, Oliveira IA, Lucena MC, Mantuano NR, Whelan SA, Dias WB et al (2015) Biosynthetic machinery involved in aberrant glycosylation: promising targets for developing of drugs against cancer. Front Oncol 5:138
7. Spiro RG (2002) Protein glycosylation: nature, distribution, enzymatic formation, and disease implications of glycopeptide bonds. Glycobiology 12:43R–56R
8. Shade KT, Platzer B, Washburn N, Mani V, Bartsch YC, Conroy M et al (2015) A single glycan on IgE is indispensable for initiation of anaphylaxis. J Exp Med 212:457–467
9. Carvalho S, Catarino TA, Dias AM, Kato M, Almeida A, Hessling B et al (2015) Preventing E-cadherin aberrant *N*-glycosylation at Asn-554 improves its critical function in gastric cancer. Oncogene 35:1619–1631
10. Thaysen-Andersen M, Venkatakrishnan V, Loke I, Laurini C, Diestel S, Parker BL et al (2015) Human neutrophils secrete bioactive paucimannosidic proteins from azurophilic granules into pathogen-infected sputum. J Biol Chem 290:8789–8802
11. Toghi Eshghi S, Yang S, Wang X, Shah P, Li X, Zhang H (2014) Imaging of *N*-linked glycans from formalin-fixed paraffin-embedded tissue sections using MALDI mass spectrometry. ACS Chem Biol 9:2149–2156
12. Turiak L, Shao C, Meng L, Khatri K, Leymarie N, Wang Q et al (2014) Workflow for combined proteomics and glycomics profiling from histological tissues. Anal Chem 86:9670–9678
13. Ostasiewicz P, Zielinska DF, Mann M, Wisniewski JR (2010) Proteome, phosphoproteome, and *N*-glycoproteome are quantitatively preserved in formalin-fixed paraffin-embedded tissue and analyzable by high-resolution mass spectrometry. J Proteome Res 9:3688–3700
14. Casadonte R, Caprioli RM (2011) Proteomic analysis of formalin-fixed paraffin-embedded tissue by MALDI imaging mass spectrometry. Nat Protoc 6:1695–1709
15. Klopfleisch R, Weiss AT, Gruber AD (2011) Excavation of a buried treasure—DNA, mRNA, miRNA and protein analysis in formalin fixed, paraffin embedded tissues. Histol Histopathol 26:797–810
16. Kotorashvili A, Ramnauth A, Liu C, Lin J, Ye K, Kim R et al (2012) Effective DNA/RNA co-extraction for analysis of microRNAs, mRNAs, and genomic DNA from formalin-fixed paraffin-embedded specimens. PLoS One 7, e34683
17. Gustafsson OJ, Briggs MT, Condina MR, Winderbaum LJ, Pelzing M, McColl SR et al (2015) MALDI imaging mass spectrometry of *N*-linked glycans on formalin-fixed paraffin-embedded murine kidney. Anal Bioanal Chem 407:2127–2139
18. Dwek MV, Brooks SA, Streets AJ, Harvey DJ, Leathem AJ (1996) Oligosaccharide release from frozen and paraffin-wax-embedded archival tissues. Anal Biochem 242:8–14

19. Espina V, Wulfkuhle JD, Calvert VS, VanMeter A, Zhou WD, Coukos G et al (2006) Laser-capture microdissection. Nat Protoc 1:586–603
20. Cheng L, Zhang S, MacLennan GT, Williamson SR, Davidson DD, Wang M et al (2013) Laser-assisted microdissection in translational research: theory, technical considerations, and future applications. Appl Immunohistochem Mol Morphol 21:31–47
21. Jensen PH, Karlsson NG, Kolarich D, Packer NH (2012) Structural analysis of *N*- and *O*-glycans released from glycoproteins. Nat Protoc 7:1299–1310
22. Kolarich D, Windwarder M, Alagesan K, Altmann F (2015) Isomer-specific analysis of released *N*-glycans by LC-ESI MS/MS with porous graphitized carbon. Methods Mol Biol 1321:427–435
23. Wessel D, Flügge UI (1984) A method for the quantitative recovery of protein in dilute solution in the presence of detergents and lipids. Anal Biochem 138:141–143

Chapter 12

Carbohydrate Microarray Technology Applied to High-Throughput Mapping of Plant Cell Wall Glycans Using Comprehensive Microarray Polymer Profiling (CoMPP)

Stjepan Krešimir Kračun, Jonatan Ulrik Fangel, Maja Gro Rydahl, Henriette Lodberg Pedersen, Silvia Vidal-Melgosa, and William George Tycho Willats

Abstract

Cell walls are an important feature of plant cells and a major component of the plant glycome. They have both structural and physiological functions and are critical for plant growth and development. The diversity and complexity of these structures demand advanced high-throughput techniques to answer questions about their structure, functions and roles in both fundamental and applied scientific fields. Microarray technology provides both the high-throughput and the feasibility aspects required to meet that demand. In this chapter, some of the most recent microarray-based techniques relating to plant cell walls are described together with an overview of related contemporary techniques applied to carbohydrate microarrays and their general potential in glycoscience. A detailed experimental procedure for high-throughput mapping of plant cell wall glycans using the comprehensive microarray polymer profiling (CoMPP) technique is included in the chapter and provides a good example of both the robust and high-throughput nature of microarrays as well as their applicability to plant glycomics.

Key words Carbohydrate microarrays, Plant cell walls, Plant glycans, Plant glycomics, Glycan profiling

1 Introduction

1.1 Introduction to Microarray Technology

Microarray technology has revolutionized DNA and RNA research [1, 2] and has lived up to the slogan—lab-on-a-chip. The high-throughput and multiplexity taken together with decreasing prices of equipment needed for production of microarrays have brought this technology close to every research laboratory, making it an accessible resource for scientists worldwide. Today, in a repertoire of a modern research facility, a microarray printer can be considered as commonplace as a mass spectrometer or a high-performance liquid chromatography system.

Gordan Lauc and Manfred Wuhrer (eds.), *High-Throughput Glycomics and Glycoproteomics: Methods and Protocols*, Methods in Molecular Biology, vol. 1503, DOI 10.1007/978-1-4939-6493-2_12, © Springer Science+Business Media New York 2017

Carbohydrates play important roles in cell functions such as signaling and trafficking as well as structural roles. Many of these functions are mediated through interactions of carbohydrates with other biomolecules such as proteins, peptides and even other carbohydrates. The binding partners can be completely natural molecules (e.g., lectins) and some may be engineered (e.g., antibodies artificially raised against a carbohydrate epitope), and the presence or absence of binding can reveal important physiological and structural information.

In order to evaluate and determine the binding specificity, binding affinity, and other parameters of such interactions, a high-throughput solution is needed, a solution that enables screening a multitude of carbohydrates against a multitude of binding partners in the most time- and cost-effective fashion and that is what microarray technology provides.

One of the hallmarks of microarray technology is that the amounts of material needed for printing microarrays are very small compared to other analytical methods, in some cases as little as 1 fmol of material needed per spot with a volume range in picoliters [3]. Considering how precious carbohydrate material is, synthetic material in particular, this is a very important advantage. Because of their small size, the amount of probing material required is also small. A microarray the size of a microscope slide enabling 48 or more probing experiments at once by being divided into so-called subarrays which are basically copies of the sample set on the same microarray. Lithographic methods are a worthy mention as they are making the microarray world even smaller, effectively entering the realm of "nanoarrays" [4].

1.2 Carbohydrate Microarrays

1.2.1 Oligosaccharides

A microarray experiment comprises several components: the microarray printer, the actual microarray surface, the material that is to be printed onto the surface, the probing material whose binding to the printed material is to be evaluated, and a means of detecting and evaluating that binding. While all other components are easy to come by as they are usually commercial, the material to be printed is usually the most difficult and it is important to mention that well-established methods for oligonucleotide and peptide synthesis are in place and automated most of the time. That enables easy access to materials needed for oligonucleotide/DNA and peptide/protein microarrays. However, chemical synthesis of oligosaccharides hitherto remains the greatest challenge for carbohydrate microarrays. Carbohydrate synthesis remains a complicated issue because of both the complexity of the synthetic targets that is inferred from the diversity of carbohydrate structures found in nature and because of synthetic challenges that carbohydrate chemistry faces. Generation of different anomeric configurations, linkages between monosaccharide building blocks and specialized decorations such as sulfonate and phosphate groups is an issue that arises from nature alone; but complicated multistep chemical transformations, protecting group strategies, the need for purification after each step and high variability in coupling conditions, to name a few, add on additional difficulty to an already difficult problem [5]. Even though significant advances towards automated synthesis of carbohydrates have been made [6], traditional synthetic approaches are still the most common route [5].

Oligosaccharides obtained by controlled degradation of polysaccharides followed by purification are an easier route to oligosaccharides for polysaccharides which are structurally uniform or repetitive along the chain. Their breakdown into oligosaccharides can be achieved by enzymatic, acidic, or physical hydrolysis [7].

In order to print them, oligosaccharides need to be able to strongly bind to the microarray surface, either through a covalent bond or through a strong non-covalent interaction. While strong non-covalent interactions mostly involve conjugation to large and/or hydrophobic molecules and immobilization on membranes such as nitrocellulose, covalent attachment methods are mediated by functional groups on the oligosaccharides and on the microarray surface and can be divided into (1) direct functionalization where the functional group on the oligosaccharide is incorporated during oligosaccharide synthesis into the protected oligosaccharide and (2) indirect functionalization where the functional group is introduced post-synthetically into a free (unprotected) oligosaccharide. Based on specificity, the methods of immobilization can also be (1) site-specific, reacting with a specific group on the oligosaccharide, or oligosaccharide conjugate or (2) site-nonspecific where their site of attachment on the saccharide cannot be accurately predicted.

Site-specific immobilization with direct functionalization of oligosaccharides can only be achieved through chemical synthesis and that is the major bottleneck of this method. This method involves chemical synthesis of oligosaccharides bearing a useful functional group such as an amine, azide, thiol, cyclopentadiene, maleimide, alkyne, and others [6, 8–12]. Each functional group needs a compatible reactive group or other method of immobilization on the microarray surface. Some of the available surfaces that have a defined surface chemistry are coated with aryltrifluoromethyldiazirine, 4-azido-2,3,5,6-tetrafluorophenyl, phthalimide, phenylboronic acid, maleimide, thiols, 2-pyridyl disulfide, benzoquinone, tetrazine, cyanuric chloride, NHS ester, aldehyde, amine, epoxide, alkyne, azide, and phosphane [12].

Site-specific immobilization with indirect functionalization is normally facilitated through a linker or tag molecule that facilitates functionalization of free oligosaccharides through reductive amination [8, 13, 14] or reactions with hydroxylamine and hydrazine derivatives [15–18], which then enables either covalent or non-covalent immobilization on microarrays. The advantage of the linker approach is that the linkers can often be used on free oligosaccharides, making it possible to avoid complicated synthetic manipulations involved in site-specific immobilization with direct functionalization.

Site nonspecific methods involve functionalized surfaces such as aryltrifluoromethyldiazirine, the 4-azido-2,3,5,6-tetrafluorophenyl group, phthalimide-derivatized surfaces and phenylboronic acid-coated surfaces [12, 19] and are performed with free oligosaccharides or polysaccharides.

Fluorous-, biotin-, and oligonucleotide-tagged oligosaccharides can be non-covalently immobilized through surfaces that are covered with molecules that bind to the tags, such as streptavidin in the case of biotin. In some cases, the tags or linkers can be large enough to enable direct immobilization of tagged oligosaccharides onto a surface like nitrocellulose. Examples of such large conjugates are neoglycolipids [8], BSA conjugates [14], and fluorous tags if they are large enough [12].

It should be added that consortia such as the Consortium for Functional Glycomics (CFG) greatly contributed to the library of synthetic oligosaccharides, microarray screening services, and online resources that are still publicly available [20, 21].

1.2.2 Polysaccharides

Non-covalent immobilization of polysaccharides is relatively straightforward as they naturally adhere to nitrocellulose or other membranes such as positively charged nylon because of their relatively high molecular weight [22, 24]. Non-covalent immobilization was also the basis for development of the first glycan array using polysaccharide immobilization on nitrocellulose and oxidized polystyrene [23, 24]. Covalent immobilization of polysaccharides can be achieved by using aryltrifluoromethyldiazirine and phthalimido functionalized surfaces mentioned earlier [12, 19].

1.3 Microarray Production

1.3.1 Microarray Printers

Based on the method of spot formation, microarray printers can be divided into contact and non-contact printers. While contact printers using technologies such as pin printing and microstamping are more widely used, new developments in non-contact printing such as inkjet piezoelectric printing, photochemistry-based techniques, laser writing, and electrospray deposition are faster and geared towards more high-throughput microarray production [25].

1.3.2 Detection Methods

As with most analytical assays, microarray detection relies on measuring the disappearance or more often—the appearance of some sort of signal, and that is indicative of a change that has occurred on the surface of the array. The most commonly used detection method of events such as binding on microarrays is fluorescence detection which is very sensitive [11, 12, 17, 26]. Other methods, particularly on nitrocellulose membranes, include colorimetric techniques with alkaline phosphatase or horseradish peroxidase conjugates [27]. While fluorescence detection requires a fluorescence scanner, which is commonplace in modern laboratories as it often also detects radioactivity and chemiluminescence therefore being useful for other work apart from microarray analysis, colorimetric detection can be successfully achieved with a regular office-grade high-resolution scanner. Alternatives include detection of radioactivity, oblique incidence reflectivity difference, microscopy, electrochemoluminescence, complementary metal oxide semiconductor-based electric signal readout and cantilever-based detection [12].

1.3.3 Quantification and Analysis

Analyzing and interpreting the data is an essential part of working with microarrays. There are several commercial software packages available with implemented spot-finding, background subtraction, normalization, and noise-reduction algorithms. An important part of microarray analysis is artifact-handling. Artifacts can result from carryover during printing, proximity artifacts resulting from proximity of spots on the array, doughnut-shaped spots and many others [28, 29], and these artifacts need to be addressed before automated microarray quantification.

Once a microarray has been quantified, the data can be represented in a variety of ways. The format is often determined by personal preference but is also based on the experimental setup. Some of the data representation methods include bar-graphs, box-plots, heat maps, MA-plots, *p*-value histograms, and volcano plots [30]. Awareness of the experimental parameters that are part of microarray printing, probing, detection, and analysis determines how the data should be interpreted. To enable unambiguous interpretation and reproducibility of results from microarrays, a form of consensus has been reached on what rules should be implemented in the whole process [30, 31]. These rules were generally developed for DNA-microarrays but are widely applicable to carbohydrate microarrays as well.

1.3.4 On-Slide Reactions

On-slide modification assays add a whole new dimension to carbohydrate microarrays and can be used in either a constructive [26] or deconstructive [32] manner. As with just regular microarrays, the amount of material needed for an enzymatic or chemical reaction performed on the printed material is minute and such an assay has many advantages in terms of yield and purity added to the already mentioned high-throughput and multiplexity, just like solid-phase synthesis is advantageous compared to solution-phase synthesis when it is a viable option.

1.4 Plant Cell Walls and Microarray-Based Analytical Techniques

1.4.1 Introduction to Plant Cell Walls

Plant cell walls are the most significant and distinguishing feature of plant cells. They are a glycan rich combination of load-bearing structures and matrix polymers and are collectively the largest source of biomass on Earth. And while studies of glycosylation of plant glycoproteins are to some degree a separate field that has yielded some very interesting results such as the discovery of N-acetyl- and N-glycolylneuraminic acids on plant glycoproteins

[33] and extensive studies of plant N-glycans and their relation to plant physiology [34] as well as studies of O-glycans in relation to evaluation of plants as expression systems [35], plant cell walls are the most abundant body of complex carbohydrate structures in a plant and they are far more dynamic than previously believed [36]. They dynamically respond to external stimuli such as pathogens and environmental factors [36] as well as internal processes such as growth and cell division and differentiation [37].

During plant growth and cell expansion, the plant cell is surrounded by a primary wall with cellulose and hemicelluloses as its load-bearing structure embedded in a pectin matrix. The primary wall gives structural support and determines the plant cell shape while resisting turgor pressure from within the cell. After cell expansion is complete some cell types deposit a secondary cell wall, which often has a unique function, composition, and organization relative to cell differentiation [38]. The secondary cell wall is often rich in lignin which increases structural support [39]. The three major groups of polysaccharides: cellulose, hemicelluloses, and pectins are within themselves also very diverse. Cellulose, although always a β-1,4-glucan, has regions of crystalline (microfibrils) and amorphous forms in the plant cell wall. Hemicelluloses are a large group of polysaccharides such as xyloglucan, xylans, and mannans which support the cellulose microfibrils. Pectins contain galacturonic acid and members of this group such as homogalacturonan and rhamnogalacturonans I and II provide additional support to the cellulose and hemicellulose matrix [40]. The ratio between the different polymers and their substructures varies greatly between the primary and secondary wall and even within these, the structural composition diverges between species and different cell types [41]. This complexity originates from a vast cooperative assortment of biosynthetic enzymes. Plant genomes have revealed hundreds of glycosyltransferases (GTs) that catalyze the linkage-formation between sugars and their numbers greatly exceeds the number of GTs found in genomes of animals [42]. The great quantity of GTs and the possibility for each to synthesize a distinct linkage provides endless possibilities for cell wall construction. The polysaccharides can further be modified with methyl- and acetyl groups and subjected to modifications by various glycosyl hydrolases (GHs) after deposition in the wall [41, 43].

This machinery has created plant cell walls with enormous diversity both in wall and polymer composition and the many different functions of plant cell wall polysaccharides have made them a central renewable resource with various important industrial applications in society. Timber has always been used for construction and the pulp papermaking process has been used thousands of years before the discovery of plant cell walls. With the increasing understanding of cell walls many new applications have emerged such as functional food ingredients and nutraceuticals.

Most recently, a major focus has been on the cell wall as a renewable energy source, an area where increased understanding of the cell wall and its polysaccharides can provide more energy-efficient solutions as opposed to simple combustion [44].

The need for a better understanding of the plant cell wall and its underlying mechanisms are evident; however, the analysis of cell wall polysaccharides and their biosynthetic pathways is a challenging process. Since polysaccharides are not easily synthesized or sequenced like DNA and proteins, conventional molecular biology methods are often less readily applicable and the characterization of the genes has also proven very difficult as no optimal system for protein expression has been identified yet [45]. Over the last several decades new and more advanced methods have been developed and as a result the field of plant cell wall research has moved considerably forward. Three recent techniques are based on plant cell wall carbohydrate microarrays and are described in more detail in the text below [14, 27, 46].

1.4.2 Comprehensive Microarray Polymer Profiling (CoMPP)

The methodology referred to as comprehensive microarray polymer profiling (CoMPP) was originally published in 2007 [27] with a follow-up publication of the video describing the procedure in 2012 [47]. A combination of a sequential extraction of the major classes of cell wall glycans was described together with construction of carbohydrate microarrays which are subsequently probed with monoclonal antibodies (mAbs) with specificities for plant cell wall components. The extracted glycan arrays generated in this high-throughput process are populated with a heterogeneous mixture of polysaccharides from a given sample or number of samples. By probing the microarrays with a panel of plant cell wall specific probes glycomic profiles across selected sample sets can quickly be generated. The CoMPP technique has been used successfully in many studies including the areas of cell wall evolution and food crops [48–51] and can now be regarded as a well-established instrument in the plant cell wall tool box.

The plant samples are homogenized and polysaccharides are extracted using sequential treatment with solvents that release the major cell wall polymer classes. For example, when applied to cell walls, a calcium chelator is used to remove pectins that are normally crosslinked with Ca^{2+} ions, a strong base such as aqueous NaOH removes hemicelluloses while formulations such as cadoxen are used to extract cellulose. The extracted material is then printed on multiple microarrays, each of which is probed with appropriate probes (mAbs or carbohydrate-binding modules (CBMs)) with specificity for cell wall components. The resulting spot signals provide information about the relative abundance of the epitopes recognized across the sample set. In one example, the fates of cell wall polysaccharides were assessed during processing of leaf material in the fungal gardens of leaf cutting ants [52]. In another, a survey of cell walls in phylogenetically diverse species across the plant kingdom was conducted with the aim of gaining insight into plant cell wall evolution and diversity [53]. Recent applications of the technique include a study of dietary polysaccharides on the swine microbiome [54], cell wall composition profiling of the parasitic giant dodder *Cuscuta reflexa* and its hosts [55], and determining grape cell wall changes during winemaking [56].

Importantly, the extracted glycan arrays are only as powerful as the panel of cell wall specific probes available for their analysis. To this date probes have been produced with specificities towards a considerable selection of the plant cell wall polysaccharides, especially for the structures commonly found in the primary cell walls of higher plants. Nevertheless, the available collection is still only scratching the surface when compared to the abundance of complex glycan structures found in plant cell walls and more probes are needed to decipher the structures and substructures of the polymers in greater detail. Unfortunately, the production of antibodies and especially the subsequent characterization of their epitopes can be a slow and difficult process. Therefore, the future usefulness of the extracted glycan arrays and the following glycome mapping depends heavily on the development of a high-throughput and high-resolution technique for characterizing novel cell wall specific probes.

1.4.3 Defined Plant Oligosaccharide Microarrays

The defined microarray is complementary to the CoMPP analysis as it is populated with structurally defined glycan samples. The purpose of these arrays is to screen molecular probes with unknown specificities towards plant cell wall derived glycans. Polysaccharides can easily be immobilized non-covalently onto various microarray surfaces in a process referred to as passive adsorption [57]. Structural complexity of polysaccharides often entails that they contain multiple distinct carbohydrate substructures, effectively making them heterogeneously multivalent antigens, which in turn can lead to probes with different binding patterns recognizing the same antigen, albeit different substructures of it. Hence polysaccharide arrays lack the analytical resolution to provide detailed information about a probe's specificity on epitope level.

Defined oligosaccharide microarrays have the potential to provide the high resolution necessary for analyzing the specificities of plant cell wall directed probes on epitope levels. Their development has however been difficult due to the aforementioned challenges with acquiring adequate oligosaccharide structures and finding a reliable way of immobilizing them in a microarray format. The latter challenge was overcome by coupling the oligosaccharides to a protein scaffold (BSA) via reductive amination making it possible to immobilize the neoglycoconjugates onto a selection of microarray surfaces suitable for either non-covalent passive adsorption or for covalent immobilization of proteins [14]. The assembly of a large plant cell wall derived oligosaccharide library has been aided by the increasing knowledge of the plant cell wall polymers and related hydrolases, which has made it easier to generate plat cell wall oligosaccharides through enzymatic degradation of plant cell wall polymers. This combined with the increasing number of commercially available oligosaccharides made it possible to collect a comprehensive library covering a large part of the substructures found in the plant cell wall polymers. The microarrays display high resolution in screening probes with undefined specificity as highlighted by the characterization of two novel xyloglucan binding mAbs (LM24 and LM25), the only difference between the two antibodies being their tolerance for galactosyl substitution in the side chains of the xyloglucan. The defined oligosaccharide microarrays also supported the binding profiles of already analyzed cell wall mAb. Another important usage of the defined oligosaccharide arrays was their ability to assist in GT characterization. In these experiments the immobilized oligosaccharides are possible acceptors and using an array populated with a large range of defined oligosaccharides, makes it is possible to quickly screen for enzyme activity towards all the different acceptors in one single experiment. Recently, this technology has been used for characterizing novel oligosaccharide-derived probes designed for detection of homogalacturonan and chitin [58].

1.4.4 Microarray-Based Screening of Enzymatic Degradation of Plant Glycans

Effective exploitation of plant biomass is an essential step towards a bio-sustainable, modern society. The basis for effective exploitation lies in optimal usage of plant-biomass-degrading enzymes. Significant advances in genome and transcriptome sequencing have identified a very large number of carbohydrate-active enzymes with putative activities as identified in the carbohydrate-active enzyme database (CAZy) [59]. However, there is a severe lack of high-throughput biochemical methods to actually pinpoint the exact activity of these enzymes and that presents a major bottleneck in efforts exerted towards efficient biomass degradation.

Amongst some other recently developed and comparably efficient techniques [60], a microarray-based technique developed recently holds promise that the gap between the predicted and empirical activity of these enzymes will narrow [46].

The method is based on mixing polysaccharides or polysaccharide mixtures with enzymes in 384-well plates and then, after incubation, printing the reaction mixtures onto microarrays followed by probing with antibodies and CBMs. The disappearance or diminution of a probe signal indicates enzyme digestion. The method is very sensitive, in some cases capable of detecting enzyme activity from solutions as dilute as 1 μU/mL. It is also very high-throughput, enabling analysis of at least 200 enzymes or broths against 15 different polysaccharides in 48 h [46].

This technique has recently successfully been used for discovery of enzymatic degradation of hemicelluloses by lytic polysaccharide monooxygenases (LPMOs) [61], analysis of enzymes from the parasitic plant *Cuscuta reflexa* mentioned earlier [55], and characterization of GHs from the gut microbiome of *Pseudacanthotermes militaris*, a fungus-growing termite [46].

1.5 CoMPP Practical Procedure

There are two critical steps that have a very significant impact on quality of the results from a CoMPP analysis. The first one is plant material sample preparation where the raw plant material is transformed into an alcohol-insoluble residue (AIR). This step has been found to be crucial in terms of reducing background and artifacts in analysis. The second step is the actual extraction step where different groups of cell wall polysaccharides, roughly divided into cellulose, pectins, and hemicelluloses, are sequentially extracted using specialized solvents. The protocol is quite simple after these two steps, as it only involves printing of the extractions onto microarrays, probing with antibodies or CBMs, detection, scanning, and analysis. The focus of the protocol included in this chapter is optimized for plants, but it is important to mention that some similar samples such as algae may require different solvents for both AIR preparation and the extraction steps.

1.6 Conclusions

In conclusion, it is important to recognize the high-throughput potential of microarray technology as applied to plant glycomics as well as to glycomics of other organisms as many steps of the microarray fabrication process overlap. Carbohydrate synthesis remains the greatest bottleneck in analysis of oligosaccharide interactions with other biomolecules; however, many oligosaccharides derived from plant polysaccharides are commercially available and an abundance of easy and cheap methods of immobilizing plant glycans regardless of their size makes it feasible to produce carbohydrate microarrays. In addition, one of the most important aspects of plant glycomics: plant cell wall analysis, is complemented with a selection of contemporary and high-throughput techniques based on carbohydrate microarrays: analysis of plant cell wall composition using techniques such as CoMPP for which a protocol is provided in this chapter, epitope mapping of plant-polysaccharide related antibodies, CBMs or other probes, and biochemical specificity characterization of carbohydrate-active enzymes. These techniques have a significant impact on fundamental plant science dealing with plant evolution and diversity, plant physiology and cellular machinery as well as on industrial aspects such as optimization of plant biomass degradation and utilization.

2 Materials

1. Liquid nitrogen.
2. Long metal forceps.

3. Freeze-dryer.
4. Ball mill or tissue homogenizer.
5. Solvents for plant material processing.
   (a) AIR preparation solvents.
      - 70 % ethanol in water.
      - Methanol–chloroform = 1:1.

   (b) Extraction solvents.
      - 50 mM 1,2-diaminocyclohexanetetraacetic acid (CDTA) pH 7.5 in water.
      - 4 M NaOH with 0.1 % $NaBH_4$ in water.
      - Cadoxen (31 % V/V 1,2-diaminoethane with 0.78 M CdO).
6. Phosphate-buffered saline (PBS).
7. Blocking buffer: 5 % skimmed milk powder in PBS.
8. Primary antibodies or CBMs against plant cell wall glycans that need to be probed. The concentration should be adjusted according to manufacturer's instructions, but the concentration specified for use with Western blots can be used as a starting point.
9. Secondary antibody/probe, determined by the host animal where the primary antibody was produced (such as mouse or rat), antibody class (such as IgG or IgM) or by a specific tag on the primary antibody or CBM (such as a His-tag, FLAG-tag, biotin, or other). Again, the concentration used for Western blots specified by the manufacturer is a good starting point. In this particular example, the microarrays will be analyzed colorimetrically, so the secondary antibody will be an alkaline phosphatase conjugate. Horseradish peroxidase conjugates can be used as well albeit requiring changing the development substrate. Fluorescently labeled antibodies are also an option but they require a microarray fluorescence scanner, and although the procedure is very simple, the details of that option are not discussed in this particular protocol.
10. Alkaline phosphatase buffer: 100 mM Tris buffer with 100 mM NaCl and 5 mM $MgCl_2$ with pH adjusted to 9.5.
11. Alkaline phosphatase substrate: 33 mL of a 50 mg/mL solution of 5-bromo-4-chloro-3-indolyl phosphate disodium salt (BCIP) in water and 330 mL of a 10 mg/mL solution of nitroblue tetrazolium (NBT) mixed with 10 mL of alkaline phosphatase buffer, should be prepared fresh.
12. Centrifuge capable of spinning high-volume 96-well plates at 2500 × *g* and small tubes at 10,000 × *g*.
13. 8-tube strips with caps similar to PCR-tube strips with a volume capacity of around 1 mL are recommended.

14. Tabletop shaker.
15. ArrayJet Sprint (ArrayJet, Roslin, UK) microarray printer, print buffer, and system liquid used is 55.2% glycerol with 44% deionized water and 0.8% Triton X-100 equipped with atmosphere control and a vacuum pump (*see* **Note 1**). The software controlling the microarray printer is the Command Centre software from ArrayJet.
16. 384-well plates (non-stick).
17. DKP 49 black regular inkjet printer ink. Other types of ink can be used as long as they are compatible with the microarray printer, and do not produce artifacts or smear during microarray probing and analysis.
18. Nitrocellulose membrane (0.45 μm pore size, cut into sheets that fit the microarray printer printing surface, in this case 26.85 × 7.3 cm).
19. 12-well plastic plates (non-stick).
20. High-resolution flatbed scanner capable of scanning images at a minimum resolution of 2400 dpi.
21. Office paper glue stick and thick paper sheets.

3 Methods

3.1 Plant Material (See Note 2)

1. Snap-freeze the sample plant material in liquid nitrogen and freeze-dry it. Snap freezing, in this case, is achieved by submerging an appropriate plastic vessel (depends on the size of the sample) into liquid nitrogen using a long metal forceps for around 5 s.
2. Homogenize the plant sample in an ample amount (*see* **Note 3**) in a tissue homogenizer/ball mill to a fine powder while being careful that the temperature does not rise significantly as a result of the friction involved. A rise in temperature can alter the composition of the analytes. Use appropriate protection for your ears as homogenization at 30 Hz, especially if using a ball mill and steel balls, produces a lot of noise that can damage your hearing.
3. To produce alcohol-insoluble residue (AIR), transfer the material to an appropriate vessel using a spatula (*see* **Note 4**), add 5 volumes (V/w) compared to sample amount of 70% ethanol in water, vortex thoroughly and spin down at 10,000 × g for 10 min and then decant the solvent.
4. Add 5 volumes (V/w) compared to sample amount of a 1:1 mixture of methanol and chloroform and vortex thoroughly. After that, spin down at 10,000 × g for 10 min and then decant the solvent.

5. Add 5 volumes (V/w) compared to sample amount of acetone, vortex thoroughly and spin down at 10,000 × *g* for 10 min—decant the solvent and leave to dry until the scent of acetone cannot be detected anymore. Depending on sample amount this can be between a few hours to days. After drying, the AIR sample is ready for the next step.

3.2 Sequential Extraction of Cell Wall Glycans

1. Weigh off 10 mg of each AIR sample into each tube of the 8-tube strip and add a glass ball that can freely move inside the tube.
2. Add 300 μL of CDTA solution to each tube and cap the tubes.
3. Agitate the tubes at a frequency of 30 Hz for 2 min. Agitation is very important here as the shaking of the liquid suspension together with the glass ball ensures successful extraction of target analytes.
4. Check that all the pellets have been suspended properly in the tubes. If not, repeat the 2-min shake from the previous step. In special cases it can be necessary to open the tubes and use a small inoculation pin to successfully suspend the pellet. When all samples have been suspended, begin shaking again at 30 Hz and after 10 s reduce the speed slowly to 8–10 Hz. Leave the samples at this speed for 2 h.
5. Spin down samples at 2500 × *g* for 10 min and collect the supernatant.
6. Add 4 M NaOH with 0.1% $NaBH_4$ to the pellets at a concentration of 30 μL/mg and repeat the extraction steps (**steps 3** and **4**).
7. Spin down samples at 2500 × *g* for 10 min and collect the supernatant.
8. Spin all extractions for 10 min at 10,000 × *g* to get rid of any residual particulate material.
9. The extracted material is now ready for printing onto microarrays (*see* **Note 5**).

3.3 Printing of Microarrays

1. All extracted samples are deposited into the 384-well plate according to the printing system of choice. Each sample is diluted 3 times (the first dilution is a 1:1 dilution in printing buffer, and the following dilutions are fivefold dilutions of the previous one) to a total of four spots per extract. Ink should be added in the top and bottom to outline the array. Ink is a coloring compound that will, when printed, produce a visible grid on the microarray that will enable unambiguous fitting of the quantification grid later in the experiment.
2. The 384-well source plate is fixed into its designated position in the microarray printer, the nitrocellulose membrane is fitted

to the membrane holder part of the microarray printer, with the vacuum pump on to keep the membranes in place, the microarray printer system liquid contained should be full and the waste bottle empty. Finally, the microarray printer lid is closed and the atmosphere/humidity control is turned on.

3. Microarrays are printed at optimized conditions: six droplets per spot, printing at 19 °C with 55% relative humidity with at least one technical replicate onto the nitrocellulose membrane,
4. After the print run is finished, the system is turned off and the membrane sheets removed from the printer. The sheets can be stored at room temperature and in the protective paper that they are normally packed in by the manufacturer.

3.4 Probing of Microarrays

1. Cut nitrocellulose membrane into arrays according to the ink outline. Cut a couple of millimeters away from the border to enable easy manipulation of the microarray with a forceps. The vessels that the subsequent steps will be conducted in depend mainly on the size of the microarray. The amount of liquid in the vessel should be generous, but also cost-effective and practical. Regular plastic 12-well plates can be used for arrays with about 144 spots (12 spot rows and 12 spot columns).
2. Block arrays in blocking buffer for 1 h while agitating gently (30 rpm) using the tabletop shaker.
3. Discard the blocking buffer and add primary antibody diluted appropriately in blocking buffer. Incubate for 2 h at room temperature while agitating gently (30 rpm) using the tabletop shaker. Remember to use microarrays for secondary antibody controls—these are incubated with blocking buffer only in this step.
4. Discard the primary antibody solutions and wash the microarrays beginning with 2× with PBS for a short period (5 min) and a third time with PBS for 10 min while agitating gently (30 rpm)—discard the PBS.
5. Add alkaline phosphatase (AP) conjugated secondary antibodies diluted appropriately in blocking buffer for 2 h at room temperature while agitating gently (30 rpm).
6. Discard the secondary antibody solutions from the arrays and wash according to **step 5**. After discarding of PBS, add water to the arrays.
7. Discard the water and develop the arrays by adding the alkaline phosphatase substrate. Stop the development by washing the microarrays with water. A rule of thumb for knowing when development should be stopped is when the background staining of the array begins to rise.
8. Allow the microarrays to dry and then glue them carefully to a thick paper sheet using a paper glue stick (*see* **Note 6**).

3.5 Quantification of Microarrays

1. Scan the microarrays in a commercial flatbed scanner at 2400 dpi grey scale and save as 16-bit Tagged Image File Format (TIFF) or other format as required by the next step.
2. Upload the images to a microarray analysis software program of choice and quantify the signal strength for each individual spot. Most commercially available software packages for microarray analysis do this almost automatically.
3. Depending on the choice of data representation format, data should be statistically analyzed appropriately. Data presented in literature is often represented as signal mean values with standard deviations or standard errors depicted as heat maps or bar graphs.

4 Notes

1. The choice of a microarray printer governs the printing conditions, the source plate layout, and the microarray surface format, and this particular protocol is optimized for use with the ArrayJet Sprint microarray printer. If you plan to use a different microarray printer, which is a viable option, the system settings, printing conditions and layout may change. The target result in printing microarrays is always achieving optimal consistency in spot shape and size and minimizing artifacts such microsatellites (splatter caused by inadequate printing conditions), spot shape defects such as "doughnut" shapes, spot merging caused by inadequate spot spacing and many others. Most of these problems can be solved by modifying the solvents used for printing where adding compounds such as glycerol or polyethyleneglycol can adjust the viscosity of the printing solvent and the use of mild detergents can help with adjusting the surface tension. Additionally, contact and non-contact printers have different requirements. With all this taken into account, it is always a good start to ask the manufacturer of the printer about their experience and recommendations.
2. During snap freezing, precautions should be taken to use protection such as protective glasses and temperature-resistant gloves. In most cases, only specific organs are dissected from the plant and included in the analysis (such as leaves or stems); however, sometimes several organs can be homogenized together as part of the same analysis.
3. 20 mg is enough for one analysis without replicates, but you may want to have some spare material in case you need to increase the number of replicates or repeat the analysis.
4. Because of the volatile and toxic nature of AIR solvents and extraction solvents, perform most of these procedures in the

fume hood or a well-ventilated area. When working with the methanol–chloroform mixture, use the fume hood without exception as both are very toxic and harmful solvents.

Organic solvents need to be disposed of properly, so make sure you follow the safety guidelines when discarding them.

Also, the amount of the sample determines the volume of solvents needed which in turn governs the vessel that the experiment is conducted in. It can range from 1.5 mL Eppendorf tubes to 50 mL Falcon tubes. The size of the vessel should be such that there is roughly 5 times more solvent than solid in volume—that ensures successful solvent removal and sample recovery during centrifugation/decanting steps.

5. It should be noted that CDTA and NaOH do not have to be used per se *per se* but can be replaced or supplemented with any desired extraction solvent deemed better suited for material in question. An example is to follow the NaOH extraction with cadoxen (31% V/V 1,2-diaminoethane with 0.78 M CdO) which is a harsher extraction solvent. However, it is important to check the compatibility of any given extraction solvent with the experimental setup down the line, such as the microarray printer machine and the microarray surface. Extracts can be kept for later by freezing them at −20 °C, but it is not recommended to do so for longer than 48 h.
6. It is convenient to have arrays glued to a sheet of thick paper because otherwise small nitrocellulose arrays can easily be blown away by draft and that can lead to blemishes and artifacts on their surface. Having all of the arrays from one experiment on the same sheet of paper also helps with organizing multiple experiments.

References

1. Singh A, Kumar N (2013) A review on DNA microarray technology. Int J Cur Res Rev 5(22):01–05
2. Trevino V, Falciani F, Barrera-Saldana HA (2007) DNA microarrays: a powerful genomic tool for biomedical and clinical research. Mol Med 13(9–10):527–541. doi:10.2119/2006-00107.Trevino
3. Wang CC, Huang YL, Ren CT, Lin CW, Hung JT, Yu JC, Yu AL, Wu CY, Wong CH (2008) Glycan microarray of Globo H and related structures for quantitative analysis of breast cancer. Proc Natl Acad Sci U S A 105(33):11661–11666. doi:10.1073/pnas.0804923105
4. Scheible MB, Pardatscher G, Kuzyk A, Simmel FC (2014) Single molecule characterization of DNA binding and strand displacement reactions on lithographic DNA origami microarrays. Nano Lett 14(3):1627–1633. doi:10.1021/nl500092j
5. Galan MC, Benito-Alifonso D, Watt GM (2011) Carbohydrate chemistry in drug discovery. Org Biomol Chem 9(10):3598–3610. doi:10.1039/c0ob01017k
6. Schmidt D, Schuhmacher F, Geissner A, Seeberger PH, Pfrengle F (2015) Automated synthesis of arabinoxylan-oligosaccharides enables characterization of antibodies that recognize plant cell wall glycans. Chemistry 21(15):5709–5713. doi:10.1002/chem.201500065
7. de Moura FA, Macagnan FT, da Silva LP (2015) Oligosaccharide production by hydrolysis of polysaccharides: a review. Int J Food Sci Tech 50(2):275–281. doi:10.1111/ijfs.12681
8. Liu Y (2010) Neoglycolipid (NGL)-based oligosaccharide microarrays and highlights of

their recent applications in studies of the molecular basis of pathogen-host interactions. Biochem Soc Trans 38:1361–1367. doi:10.1042/Bst0381361

9. Pachamuthu K, Schmidt RR (2006) Synthetic routes to thiooligosaccharides and thioglycopeptides. Chem Rev 106(1):160–187. doi:10.1021/Cr040660c
10. Blixt O, Head S, Mondala T, Scanlan C, Huflejt ME, Alvarez R, Bryan MC, Fazio F, Calarese D, Stevens J, Razi N, Stevens DJ, Skehel JJ, van Die I, Burton DR, Wilson IA, Cummings R, Bovin N, Wong CH, Paulson JC (2004) Printed covalent glycan array for ligand profiling of diverse glycan binding proteins. Proc Natl Acad Sci U S A 101(49):17033–17038, doi:0407902101 [pii] 10.1073/pnas.0407902101
11. Shilova N, Navakouski M, Khasbiullina N, Blixt O, Bovin N (2012) Printed glycan array: antibodies as probed in undiluted serum and effects of dilution. Glycoconj J 29(2–3):87–91. doi:10.1007/s10719-011-9368-8
12. Park S, Gildersleeve JC, Blixt O, Shin I (2013) Carbohydrate microarrays. Chem Soc Rev 42(10):4310–4326. doi:10.1039/C2cs35401b
13. Feizi T, Chai WG (2004) Oligosaccharide microarrays to decipher the glyco code. Nat Rev Mol Cell Biol 5(7):582–588. doi:10.1038/Nrm1428
14. Pedersen HL, Fangel JU, McCleary B, Ruzanski C, Rydahl MG, Ralet MC, Farkas V, von Schantz L, Marcus SE, Andersen MCF, Field R, Ohlin M, Knox JP, Clausen MH, Willats WGT (2012) Versatile high resolution oligosaccharide microarrays for plant glycobiology and cell wall research. J Biol Chem 287(47):39429–39438. doi:10.1074/jbc.M112.396598
15. Maolanon NN, Blaise M, Sorensen KK, Thygesen MB, Clo E, Sullivan JT, Ronson CW, Stougaard J, Blixt O, Jensen KJ (2014) Lipochitin oligosaccharides immobilized through oximes in glycan microarrays bind LysM proteins. Chembiochem 15(3):425–434. doi:10.1002/cbic.201300520
16. Bohorov O, Andersson-Sand H, Hoffmann J, Blixt O (2006) Arraying glycomics: a novel bi-functional spacer for one-step microscale derivatization of free reducing glycans. Glycobiology 16(12):21c–27c. doi:10.1093/glycob/cwl044
17. Clo E, Blixt O, Jensen KJ (2010) Chemoselective reagents for covalent capture and display of glycans in microarrays. Eur J Org Chem 3:540–554. doi:10.1002/ejoc.200901234
18. Park S, Lee MR, Shin I (2007) Fabrication of carbohydrate chips and their use to probe protein-carbohydrate interactions. Nat Protoc 2(11):2747–2758. doi:10.1038/nprot.2007.373
19. Rillahan CD, Paulson JC (2011) Glycan microarrays for decoding the glycome. Annu Rev Biochem 80(80):797–823. doi:10.1146/annurev-biochem-061809-152236
20. Raman R, Venkataraman M, Ramakrishnan S, Lang W, Raguram S, Sasisekharan R (2006) Advancing glycomics: implementation strategies at the consortium for functional glycomics. Glycobiology 16(5):82r–90r. doi:10.1093/glycob/cwj080
21. Venkataraman M, Sasisekharan R, Raman R (2015) Glycan array data management at Consortium for Functional Glycomics. Methods Mol Biol 1273:181–190. doi:10.1007/978-1-4939-2343-4_13
22. Jeffree CE (1993) Immobilization of pectin, xyloglucan and other soluble plant polysaccharides on blotting membranes. New Phytologist 125 (4), 695–706, doi: 10.1111/j.1469-8137.1993.tb03918.x
23. Wang D, Liu S, Trummer BJ, Deng C, Wang A (2002) Carbohydrate microarrays for the recognition of cross-reactive molecular markers of microbes and host cells. Nat Biotechnol 20(3):275–281. doi:10.1038/nbt0302-275
24. Willats WG, Rasmussen SE, Kristensen T, Mikkelsen JD, Knox JP (2002) Sugar-coated microarrays: a novel slide surface for the high-throughput analysis of glycans. Proteomics 2(12):1666–1671. doi:10.1002/1615-9861(200212)2:12<1666::AID-PROT1666>3.0.CO;2-E
25. Barbulovic-Nad I, Lucente M, Sun Y, Zhang MJ, Wheeler AR, Bussmann M (2006) Bio-microarray fabrication techniques—a review. Crit Rev Biotechnol 26(4):237–259. doi:10.1080/07388550600978358
26. Clo E, Kracun SK, Nudelman AS, Jensen KJ, Liljeqvist JA, Olofsson S, Bergstrom T, Blixt O (2012) Characterization of the viral O-glycopeptidome: a novel tool of relevance for vaccine design and serodiagnosis. J Virol 86(11):6268–6278. doi:10.1128/Jvi.00392-12
27. Moller I, Sorensen I, Bernal AJ, Blaukopf C, Lee K, Obro J, Pettolino F, Roberts A, Mikkelsen JD, Knox JP, Bacic A, Willats WGT (2007) High-throughput mapping of cell-wall polymers within and between plants using novel microarrays. Plant J 50(6):1118–1128. doi:10.1111/j.1365-313X.2007.03114.x
28. Yu HY, Nguyen K, Royce T, Qian J, Nelson K, Snyder M, Gerstein M (2007) Positional artifacts in microarrays: experimental verification and construction of COP, an automated detection tool. Nucleic Acids Res 35(2):8, doi:ARTN E8, 10.1093/nar/gkl871

29. Pappaert K, Ottevaere H, Thienpont H, Van Hummelen P, Desmet G (2006) Diffusion limitation: a possible source for the occurrence of doughnut patterns on DNA microarrays. Biotechniques 41(5):609–616. doi:10.2144/000112293
30. Allison DB, Cui XQ, Page CP, Sabripour M (2006) Microarray data analysis: from disarray to consolidation and consensus (vol 7, pg 55, 2006). Nat Rev Genet 7(5):406. doi:10.1038/nrg1869
31. Glaab E, Garibaldi JM, Krasnogor N (2009) ArrayMining: a modular web-application for microarray analysis combining ensemble and consensus methods with cross-study normalization. BMC Bioinformatics 10:358, doi: Artn 358, 10.1186/1471-2105-10-358
32. Uttamchandani M, Huang X, Chen GYJ, Yao SQ (2005) Nanodroplet profiling of enzymatic activities in a microarray. Bioorg Med Chem Lett 15(8):2135–2139. doi:10.1016/j.bmcl.2005.02.019
33. Shah MM, Fujiyama K, Flynn CR, Joshi L (2003) Sialylated endogenous glycoconjugates in plant cells. Nat Biotechnol 21(12):1470–1471. doi:10.1038/nbt912
34. Strasser R (2014) Biological significance of complex N-glycans in plants and their impact on plant physiology. Front Plant Sci 5:363. doi:10.3389/fpls.2014.00363.ARTN363
35. Strasser R (2012) Challenges in O-glycan engineering of plants. Front Plant Sci 3:218. doi:10.3389/fpls.2012.00218.ARTN218
36. Bellincampi D, Cervone F, Lionetti V (2014) Plant cell wall dynamics and wall-related susceptibility in plant-pathogen interactions. Front Plant Sci 5:228. doi:10.3389/fpls.2014.00228
37. Perrot-Rechenmann C (2010) Cellular responses to auxin: division versus expansion. Csh Perspect Biol 2(5):a001446. doi:10.1101/cshperspect.a001446
38. Carpita NC, Gibeaut DM (1993) Structural models of primary-cell walls in flowering plants—consistency of molecular-structure with the physical-properties of the walls during growth. Plant J 3(1):1–30. doi:10.1111/j.1365-313X.1993.tb00007.x
39. Boerjan W, Ralph J, Baucher M (2003) Lignin biosynthesis. Annu Rev Plant Biol 54:519–546. doi:10.1146/annurev.arplant.54.031902.134938
40. Rytioja J, Hilden K, Yuzon J, Hatakka A, de Vries RP, Makela MR (2014) Plant-polysaccharide-degrading enzymes from basidiomycetes. Microbiol Mol Biol Rev 78(4):614–649. doi:10.1128/MMBR.00035-14
41. Burton RA, Gidley MJ, Fincher GB (2010) Heterogeneity in the chemistry, structure and function of plant cell walls. Nat Chem Biol 6(10):724–732. doi:10.1038/nchembio.439
42. Coutinho PM, Starn M, Blanc E, Henrissat B (2003) Why are there so many carbohydrate-active enzyme-related genes in plants? Trends Plant Sci 8(12):563–565. doi:10.1016/j.tplants.2003.10.002
43. Popper ZA, Fry SC (2004) Primary cell wall composition of pteridophytes and spermatophytes. New Phytol 164(1):165–174. doi:10.1111/j.1469-8137.2004.01146.x
44. Pauly M, Keegstra K (2010) Plant cell wall polymers as precursors for biofuels. Curr Opin Plant Biol 13(3):305–312. doi:10.1016/j.pbi.2009.12.009
45. Petersen BL, Egelund J, Damager I, Faber K, Jensen JK, Yang Z, Bennett EP, Scheller HV, Ulvskov P (2009) Assay and heterologous expression in Pichia pastoris of plant cell wall type-II membrane anchored glycosyltransferases. Glycoconj J 26(9):1235–1246. doi:10.1007/s10719-009-9242-0
46. Vidal-Melgosa S, Pedersen HL, Schuckel J, Arnal G, Dumon C, Amby DB, Monrad RN, Westereng B, Willats WGT (2015) A new versatile microarray-based method for high throughput screening of carbohydrate-active enzymes. J Biol Chem 290(14):9020–9036. doi:10.1074/jbc.M114.630673
47. Moller IE, Pettolino FA, Hart C, Lampugnani ER, Willats WG, Bacic A (2012) Glycan profiling of plant cell wall polymers using microarrays. J Vis Exp 70, e4238. doi:10.3791/4238
48. Sorensen I, Pettolino FA, Bacic A, Ralph J, Lu FC, O'Neill MA, Fei ZZ, Rose JKC, Domozych DS, Willats WGT (2011) The charophycean green algae provide insights into the early origins of plant cell walls. Plant J 68(2):201–211. doi:10.1111/j.1365-313X.2011.04686.x
49. Zhang H, Fangel JU, Willats WGT, Selig MJ, Lindedam J, Jorgensen H, Felby C (2014) Assessment of leaf/stem ratio in wheat straw feedstock and impact on enzymatic conversion. GCB Bioenergy 6(1):90–96. doi:10.1111/gcbb.12060
50. Moore JP, Nguema-Ona E, Fangel JU, Willats WGT, Hugo A, Vivier MA (2014) Profiling the main cell wall polysaccharides of grapevine leaves using high-throughput and fractionation methods. Carbohydr Polym 99:190–198. doi:10.1016/j.carbpol.2013.08.013
51. Fangel JU, Ulvskov P, Knox JP, Mikkelsen MD, Harholt J, Popper ZA, Willats WGT (2012) Cell wall evolution and diversity. Front Plant Sci 3:152. doi:10.3389/fpls.2012.00152

52. Moller IE, De Fine Licht HH, Harholt J, Willats WG, Boomsma JJ (2011) The dynamics of plant cell-wall polysaccharide decomposition in leaf-cutting ant fungus gardens. PLoS One 6(3), e17506. doi:10.1371/journal.pone.0017506
53. Sorensen I, Pettolino FA, Wilson SM, Doblin MS, Johansen B, Bacic A, Willats WGT (2008) Mixed-linkage (1 -> 3), (1 -> 4)-beta-D-glucan is not unique to the poales and is an abundant component of Equisetum arvense cell walls. Plant J 54(3):510–521. doi:10.1111/j.1365-313X.2008.03453.x
54. Umu OC, Frank JA, Fangel JU, Oostindjer M, da Silva CS, Bolhuis EJ, Bosch G, Willats WG, Pope PB, Diep DB (2015) Resistant starch diet induces change in the swine microbiome and a predominance of beneficial bacterial populations. Microbiome 3:16. doi:10.1186/s40168-015-0078-5
55. Johnsen HR, Striberny B, Olsen S, Vidal-Melgosa S, Fangel JU, Willats WG, Rose JK, Krause K (2015) Cell wall composition profiling of parasitic giant dodder (Cuscuta reflexa) and its hosts: a priori differences and induced changes. New Phytol 207(3):805–816. doi:10.1111/nph.13378
56. Gao Y, Fangel JU, Willats WGT, Vivier MA, Moore JP (2015) Dissecting the polysaccharide-rich grape cell wall changes during winemaking using combined high-throughput and fractionation methods. Carbohydr Polym 133:567–577
57. Wang D, Collins N (2013) The 10 anniversary of carbohydrate microarrays (2002–2012). J Proteomics Bioinform 6(2), e22
58. Mravec J, Kracun SK, Rydahl MG, Westereng B, Miart F, Clausen MH, Fangel JU, Daugaard M, Van Cutsem P, Licht HHD, Hofte H, Malinovsky FG, Domozych DS, Willats WGT (2014) Tracking developmentally regulated post-synthetic processing of homogalacturonan and chitin using reciprocal oligosaccharide probes. Development 141(24):4841–4850. doi:10.1242/Dev.113365
59. Lombard V, Ramulu HG, Drula E, Coutinho PM, Henrissat B (2014) The carbohydrate-active enzymes database (CAZy) in 2013. Nucleic Acids Res 42(D1):D490–D495. doi:10.1093/Nar/Gkt1178
60. Kracun SK, Schuckel J, Westereng B, Thygesen LG, Monrad RN, Eijsink VG, Willats WG (2015) A new generation of versatile chromogenic substrates for high-throughput analysis of biomass-degrading enzymes. Biotechnol Biofuels 8:70. doi:10.1186/s13068-015-0250-y
61. Agger JW, Isaksen T, Varnai A, Vidal-Melgosa S, Willats WGT, Ludwig R, Horn SJ, Eijsink VGH, Westereng B (2014) Discovery of LPMO activity on hemicelluloses shows the importance of oxidative processes in plant cell wall degradation. Proc Natl Acad Sci U S A 111(17):6287–6292. doi:10.1073/pnas.1323629111

Chapter 13

Analysis of Invertebrate and Protist *N*-Glycans

Alba Hykollari, Katharina Paschinger, Barbara Eckmair, and Iain B.H. Wilson

Abstract

N-glycans from invertebrates and protists have often unusual structures which present analytical challenges. Both core and antennal modifications can be quite different from the more familiar vertebrate glycan motifs; thereby, contrary to the concept that "simple" organisms have "simple" N-glycans, rather complex oligosaccharides structures, including zwitterionic and anionic ones, have been found in a range of species. Thus, to facilitate the optimized elucidation of the maximal possible range of structures, the analytical workflow for glycomics of these organisms should include sequential release and fractionation steps. Peptide:N-glycosidase F is sufficient to isolate N-glycans from fungi and some protists, but in most invertebrates core α1,3-fucose is present, so release of the glycans from glycopeptides with peptide:N-glycosidases A is required. Subsequent solid-phase extraction with graphitized carbon and reversed phase resins enables different classes of N-glycans to be separated prior to high-pressure liquid chromatography (HPLC) and matrix-assisted laser desorption/ionization time-of-flight mass spectrometry (MALDI-TOF MS). Depending on the types and numbers of glycans present, either reversed- or normal-phase HPLC (or both in series) enable even single isomeric or isobaric structures to be separated prior to MALDI-TOF MS and MS/MS. The use of enzymatic or chemical treatments allows further insights to be gained, although some glycan modifications (especially methylation) are resistant. Using a battery of methods, sometimes up to 100 structures from a single organism can be assigned, a complexity which raises evolutionary questions regarding the function of these glycans.

Key words Glycosylation, Glycome, Mass spectrometry, Phosphorylated glycans, Sulphated glycans

1 Introduction

The modification of asparagine residues of proteins by glycans (N-glycosylation) occurs in almost all eukaryotes as well as in some prokaryotes. While the N-glycans of higher organisms often have defined functions in terms of organismal viability or protein function, the biology of those of lower organisms is less well understood, partly due to restricted glycomic knowledge or limitations in the possibilities for genetic manipulation of many species. Nevertheless, it is known that in flies the formation of the neural system depends on proper N-glycan processing, whereas

Gordan Lauc and Manfred Wuhrer (eds.), *High-Throughput Glycomics and Glycoproteomics: Methods and Protocols*, Methods in Molecular Biology, vol. 1503, DOI 10.1007/978-1-4939-6493-2_13, © Springer Science+Business Media New York 2017

for nematodes some glyco-modifications play a role in inter-organismal interactions [1, 2]. For parasites, whether unicellular or multicellular, glycans are involved in binding to host cells or modulation of the host immune system [3]. Certainly, even now, a detailed glycomic exploration is required before we can even begin to understand phylogenetic distribution of glycomotifs and their function.

Although the basic principles of N-glycan processing in eukaryotes (formation of a lipid-linked precursor, followed by transfer and further processing) are similar, the exact structures differ greatly between species. While prokaryotic N-glycans are quite unusual, only a few parasitic eukaryotes do not N-glycosylate their proteins at all. One source of variation in eukaryotes is that a number of unicellular species do not synthesize the $Glc_3Man_9GlcNAc_2$ dolichol-linked precursor familiar from for example yeast and mammalian systems [4]; instead *Giardia* and *Plasmodium* only transfer $GlcNAc_{1-2}$ to proteins, *Trichomonas vaginalis* $Man_5GlcNAc_2$, *Tetrahymena* $Glc_3Man_5GlcNAc_2$, and trypanosomatids $Man_{5-9}GlcNAc_2$. This source of variation is due to the absence of different *alg* genes encoding glycosyltransferases of the endoplasmic reticulum; also, the oligosaccharyltransferase complexes necessary for transfer of the glycan precursor to asparagine residues often have a lower number of subunits in some protists.

After removal/transfer of terminal glucose residues to the protein-bound glycan (dependent on the species), transport of the folded glycoproteins from the endoplasmic reticulum, a large variety of genus-, species-, and cell-specific processing events take place in the Golgi apparatus. Residues are removed and others added; not just monosaccharides such as xylose, fucose, glucuronic acid, *N*-acetylglucosamine, galactopyranose, and galactofuranose are transferred, but also other moieties such as phosphate, sulfate, phosphorylcholine, and phosphoethanolamine. On the other hand, sialic acids tend to be rare outside the vertebrates. While our knowledge of vertebrate and plant N-glycosylation is extensive, the wide variability, accompanied by a lack of predictability, of invertebrate, fungal and protist N-glycans [5] means that the release, fractionation and analysis methods must be "open" to unexpected outcomes. Here we summarize methods successfully applied to N-glycans from a number of quite different organisms, including protist, fungal, molluscan, insect and nematode species [6–17]. A particular focus is on sample preparation, solid-phase extraction, HPLC, MALDI-TOF MS, and chemical or enzymatic treatments.

2 Materials

2.1 Reagents, Buffers, and Columns (See Note 1)

2.1.1 Cell/Tissue Disruption, Proteolytic Digestion, and Glycopeptide Purification

1. Tight fitting glass homogenizer (e.g., glass Dounce-type), porcelain mortar and pestle, or sonifier.
2. Pepsin A from porcine gastric mucosa (Sigma).
3. Thermolysin (Promega).
4. Sequencing grade modified trypsin dissolved to 1 mg/ml in 1 mM HCl (Promega or Roche).
5. Dowex 50 W×8, 200–400 mesh, H^+ form (Sigma-Aldrich), pre-equilibrated with 2 % acetic acid.
6. 0.5 %, 2 %, and 10 % (v/v) acetic acid.
7. 0.5 M ammonium acetate (pH 6.0; prepared from 0.5 M acetic acid adjusted to pH 6 with ammonia).
8. 20-ml polypropylene columns (Bio-Rad, EconoPac).
9. Sephadex™ G25 (medium; GE Healthcare).
10. Glass columns (Bio-Rad) of 1.5 cm diameter and 50 cm length.
11. Orcinol monohydrate (Sigma), 200 mg dissolved in 100 ml of 20 % (v/v) H_2SO_4, suitable for spraying or pretreatment of TLC plates (Silica Gel 60 F254 plates (Merck)).

2.1.2 Glycan Release

1. N-glycosidase F (PNGase F) from *Flavobacterium meningosepticum* (Roche, now sold through Sigma-Aldrich; recombinant).
2. N-glycosidase A (PNGase A) from almond meal (Roche, now sold through Sigma-Aldrich). Recombinant PNGase A (prepared in-house) can also be used.
3. 50 mM ammonium hydrogen carbonate (pH 8) or ammonium acetate (pH 5).
4. Chromatography media as in Subheading 2.1.2.

2.1.3 Solid-Phase Extraction for Glycan Purification and Subfractionation

1. Acetonitrile (MeCN; e.g., VWR, LC-MS grade).
2. Methanol (MeOH; e.g., Roth, HPLC gradient grade).
3. C_{18}-SepPak (100 mg) cartridge or C18 material (Lichroprep, Merck) placed into a 1 ml solid-phase extraction (SPE) column (e.g., Supelco). Pre-equilibrate by sequential application of 100 % MeOH and then water.
4. Non-porous graphitized carbon (PGC) column: 250 mg ENVI™ Carb bulk material (Sigma-Aldrich) per 1-ml SPE tube. Pre-equilibrate by sequential application of 100 % MeCN and then water.
5. Syringe mounted on an adaptor for expelling solutions as well as clamp for holding column.

2.1.4 Glycan Derivatization or Modification

1. 2-aminopyridine (2AP or PA; ≥99%, Sigma-Aldrich); prepare fresh PA-solution of 100 mg of PA in a mixture of 76 μL of concentrated HCl and 152 μL of water.
2. Sodium cyanoborohydride (95%, Sigma-Aldrich); prepare fresh cyanoborohydride/PA solution with 4.5 mg of sodium cyanoborohydride in a mixture of 9 μL of the PA solution and 13.5 μL of water (consider bubble formation).
3. Concentrated hydrochloric acid (37% HCl; Roth).
4. 96-well F black plates (Nunc).
5. PA-labeled partial dextran hydrolysate, 2–20 glucose units (bought in non-pyridylaminated form from Sigma-Aldrich and labeled as described below).
6. Sephadex™ G15 (GE Healthcare).
7. Glass columns (Bio-Rad) of 1 cm diameter and 50 cm length.

2.1.5 HPLC

1. Tosoh Amide-80 column (4.6×250 mm, 5 μm; stored in 95% acetonitrile) with a guard column, for normal phase (NP)-HPLC. Pre-equilibrate in a 1:3 mixture of 10 mM ammonium formate (pH 7.0; i.e., 0.1 M formic acid adjusted to pH 7 with ammonia) and 95% acetonitrile in water. Alternative columns are available from other suppliers (e.g., Waters or Phenomenex).
2. Agilent Hypersil ODS (4 mm×250 mm, 5 μm; stored in 30% MeOH) with a guard column, for reverse phase (RP)-HPLC. Pre-equilibrate in 0.1 M ammonium acetate, pH 4.0 (i.e., 0.1 M acetic acid adjusted to pH 4 with ammonia). Alternatives are fused core RP-HPLC columns with a superior, but also subtly different, resolution: e.g., Ascentis® Express RP-Amide (150×4.6 mm, 2.7 μm; Supelco) or Kinetex™ 5 μm XB-C18 (250×4.6 mm; Phenomenex).
3. Thermo Dionex IonPac AS11 (4×250 mm) with guard column, for hydrophilic interaction/anion exchange (HIAX); stored in 95% acetonitrile; pre-equilibrate with 80% acetonitrile in ammonium acetate, pH 3.85 (i.e., 800 mM ammonia adjusted with acetic acid to pH 3.85).
4. Acetonitrile and methanol should be of LC-MS and HPLC gradient grade respectively.

2.1.6 Mass Spectrometry

1. MALDI matrices.
   (a) 3 mg/ml 6-aza-2-thiothymine (ATT; Sigma-Aldrich) in 50% EtOH.
   (b) 20 mg/ml (i.e., 2% (w/v)) 2,5-dihydroxybenzoic acid (DHB; Sigma-Aldrich) in 30% (v/v) MeCN/water.

(c) 10 mg/ml (i.e., 1% (w/v)) α-cyano-4-hydroxycinnamic acid (ACH; Sigma-Aldrich) in 0.1% trifluoroacetic acid, 50% MeCN/water.

2. Seven-component peptide M_r standard mixture (Bruker Daltonics).
3. Appropriate ground or polished steel MALDI plate.

2.1.7 Enzymatic and Chemical Treatments

1. Mannosidases: jack bean α-mannosidase from Sigma, *Aspergillus* (α1,2-specific) from Prozyme, *Xanthomonas* (α1,2/3-specific or α1,6-specific) from New England Biolabs (NEB), *Helix pomatia* β-mannosidase from Sigma or bacterial endo-α-mannosidase expressed in-house [9].
2. Fucosidases: bovine kidney (α1,2/6-preferring) from Sigma, almond (α1,3-specific) from Calbiochem, *Xanthomonas* (α1,2-specific) from NEB, *Corynebacterium* (α1,2-specific) from Takara or microbial (α1,2-specific) from Megazyme.
3. Galactosidases: recombinant *Aspergillus niger* or *oryzae* β-galactosidases (prepared in-house [18]), *Xanthomonas* (β1,3-specific) from NEB, *Bacillus fragilis* (β1,4-specific) from NEB or green coffee bean α-galactosidase from Sigma.
4. Hexosaminidases: recombinant *Apis mellifera* FDL linkage-specific β-*N*-acetylglucosaminidase or *C. elegans* HEX-4 β-*N*-acetylgalactosaminidase (both prepared in-house [19]), jack bean β-*N*-acetylhexosaminidase from Sigma, *Streptomyces* chitinase (β1,3/4/6-*N*-acetylhexosaminidase from NEB) or chicken liver (α1,3/4-*N*-acetylgalactosaminidase from Sigma).
5. Glucuronidase: *E. coli* β-glucuronidase from Megazyme (should be subject to desalting, e.g., with a 10000 MWCO ultrafiltration device and dilution/re-concentration).
6. Hydrofluoric acid (48%; e.g., from Sigma-Aldrich).

2.1.8 Western Blotting

1. Standard SDS-PAGE electrophoresis apparatus (e.g., Bio-Rad) and standard electrophoresis buffers.
2. Trans blot SD-semidry transfer cell (e.g., Bio-Rad), standard nitrocellulose membrane, and transfer buffers.
3. Antibody dilution and blocking solution (also used for antibody dilution): 0.5% (w/v) crystalline bovine serum albumin (Roth) in 100 mM Tris–HCl (pH 7.4), 150 mM NaCl, 0.05% (v/v) Tween-20.
4. Anti-horseradish peroxidase (anti-HRP; Sigma-Aldrich) from rabbit; use at 1 μg/ml (1:20,000) in blocking solution.
5. Goat anti-rabbit IgG, conjugated with alkaline phosphatase (Vector Labs); use at 1:2000 in blocking solution.

6. Anti-phosphorylcholine (TEPC-15; Sigma-Aldrich) from mouse; use at 1:200 in blocking solution.
7. Goat anti-mouse IgA, conjugated with horseradish peroxidase or alkaline phosphatase (Sigma-Aldrich); use at 1:1000 in blocking solution.
8. Biotinylated lectins (Vector Labs); typically use at 1:2000 in blocking solution [20].
9. Anti-Biotin from goat conjugated with alkaline phosphatase (Sigma); use at 1:10,000 in blocking solution.

2.1.9 Glycan Data Analysis

1. Glycoworkbench 2.1 (free download at www.glycoworkbench.org/).
2. flexAnalysis (Bruker Daltonics).
3. Calculator for manual interpretation.

2.2 Equipment

1. Vacuum centrifuge (e.g., SpeedVac, Thermo).
2. MALDI-TOF-TOF-MS: Autoflex Speed or UltrafleXtreme MALDI-TOF-TOF (Bruker Daltonics, Billerica, MA). Alternatives are available commercially from Shimadzu or Applied Biosystems.
3. Liquid chromatograph with fluorescence detector; e.g., LC-30 AD with RF 20 AXS (Shimadzu).
4. Fraction collector, such as Bio-Rad model 2110.
5. Multifunctional microplate reader (such as Infinite M200 monochromator based instrument; Tecan).
6. Microcentrifuge, such as Heraeus (Thermo).
7. Probe sonifier, e.g., model 250 (Branson).
8. Freeze-drier, e.g, FreeZone 4.5 (Labconco).

3 Methods (*See* flowchart in Fig. 1)

3.1 Cell/Tissue Disruption, Proteolytic Digestion, and Glycopeptide Purification

Biological material (*see* **Note 2**) can be whole organisms, tissues, cells, (semi-)purified proteins, media/buffer containing secreted proteins, etc., stored at −80 °C after collection; large volumes of proteinaceous sample should be reduced by precipitation using 5 volumes of methanol or lyophilised.

1. Suspend cells or other biological material (1–6 g wet weight) in 10 ml of boiling water for 5–10 min.
2. After cooling, disperse cells or tissue using a tight fitting glass homogenizer (for cellular samples, a probe sonifier may be suitable to generate subcellular particles as determined using

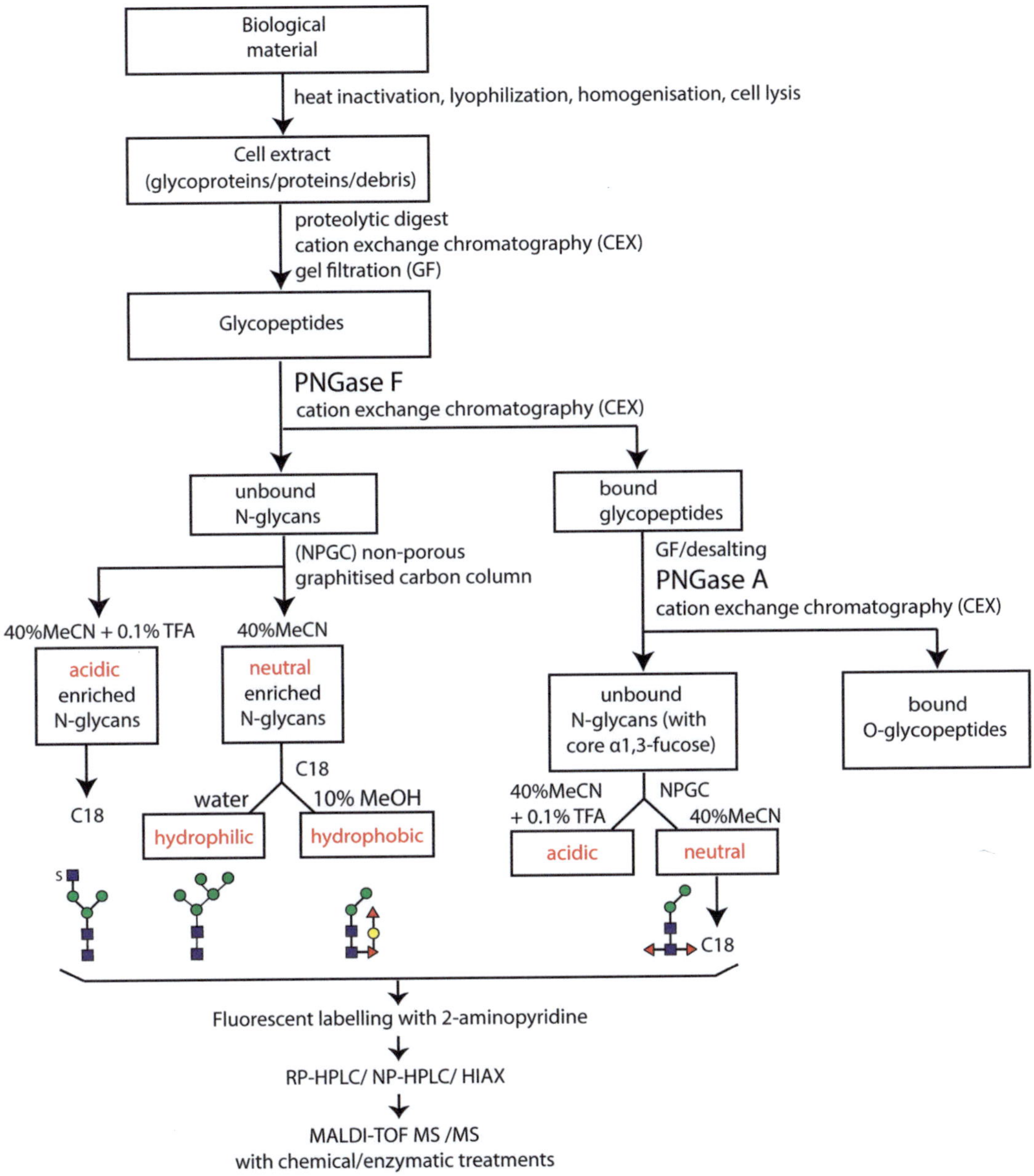

Fig. 1 Glycomic workflow for the analysis of N-glycans. Initially, samples are proteolyzed, the glycopeptides enriched by cation exchange (CEX) and gel filtration (GF) and the glycans released enzymatically, whereby PNGase A (and not PNGase F) is capable of releasing the core α1,3-fucosylated N-glycans. Subsequent subfractionation by non-porous graphitized carbon (NPGC) and/or reversed-phase (C18) resins results in pools differing in terms of anionic and zwitterionic modifications. Finally, all N-glycans are analyzed by different types of HPLC (reversed or normal phase or hydrophilic interaction/anion exchange; RP, NP, or HIAX) in combination with MALDI-TOF MS/MS and chemical/enzymatic treatments. The remaining O-glycopeptides can be subject to β-elimination and LC-MS. The example structures, shown according to the nomenclature of the Consortium for Functional Glycomics, are from a marine snail [12]

phase contrast microscopy). For fungi/marine organisms: biological material (2–5 g) is heat-denatured for 10 min, then lyophilized overnight. Lyophilized samples are ground in liquid nitrogen in a mortar and pestle to produce a powder which is suspended in a minimal volume of water (wait briefly to avoid freezing of the added water) and transferred into a 150 ml round-bottomed flask.

3. Add formic acid [5% (v/v) final concentration] and 1 mg of pepsin (per 3 g wet weight). Incubate for 1 day at 37 °C (final volume of 5–10 ml) and centrifuge to remove insoluble material. Alternatively, add 100 mM ammonium carbonate–ammonium hydrogen carbonate buffer to a final concentration of 20–50 mM (pH 8), followed by $CaCl_2$ to a final concentration of 0.5 mM and finally thermolysin (1 mg protease per 1 g wet weight) and incubate for 2 h at 70 °C. Another alternative is trypsin, but this results in rather large glycopeptides which may not be efficiently enzymatically deglycosylated by PNGase A.

4. As required (e.g., when using thermolysin), acidify the sample with an aliquot of 10% acetic acid. Then incubate the proteolytic supernatant in a beaker with 10 packed ml of prewashed Dowex-50 W×8 for 1 h at 23 °C. Pour into a column (e.g., a Bio-Rad Econo-Pac polypropylene column). Wash the column with 2% (v/v) acetic acid to remove unbound material, and elute glycopeptides with 0.5 M ammonium acetate (pH 6). Collect 1.5 ml fractions and assess for carbohydrate by orcinol reactivity (e.g., by spotting onto TLC plates pretreated with orcinol and then developed at 90 °C for 5–10 min (*see* **Note 3**); lyophilize selected fractions and resuspend in not more than 3 ml of water.

5. Subject the sample to gel filtration on an 80-ml Sephadex G25 column (1.5×45 cm) in 0.5% (v/v) acetic acid and collect 4 ml fractions (*see* **Note 4**). Pool the orcinol-positive fractions (these should contain glycopeptides) and lyophilize.

3.2 Release of N-Glycans (See Note 5)

3.2.1 Sequential Release by PNGase F and PNGase A [6, 12]

1. Resuspend the glycopeptides in water, heat for 5 min at 95 °C (to denature remaining proteases), and subject to digestion overnight at 37 °C with 3 μL PNGase F in a final concentration of 20–50 mM NH_4HCO_3 (pH 8.0) and a volume of 200–300 μl.

2. Acidify the sample (*see* **Note 6**) and repeat the Dowex-50 W×8 chromatography step (*see* **step 4** of Subheading 3.1), collect 1.5 ml fractions and lyophilize the unbound (free N-glycans, lacking core α1,3-fucose, ready for further purification) and the bound fractions eluted with 0.5 M ammonium acetate pH 6 (remaining glycopeptides).

3. Desalt the remaining glycopeptides on Sephadex G25 (as in **step 5** of Subheading 3.1). Lyophilize orcinol-positive fractions and dissolve together in 50 mM ammonium acetate, pH 5.0. Incubate with 3 μL of PNGase A overnight at 37 °C in a total volume of 200–300 μl.
4. Repeat the Dowex-50 chromatography step (**step 4** of Subheading 3.1). The unbound fraction contains core α1,3-fucosylated glycans which are also ready for further purification below, whereas the remaining glycopeptides can be desalted (as in **step 5** of Subheading 3.1) and subject to β-elimination for O-glycan analysis [12].

3.3 Solid-Phase Extraction for Glycan Purification and Subfractionation

3.3.1 Solid-Phase Extraction and Enrichment of Hydrophobic N-Glycans [12]

1. Dilute glycan samples in 100 μl water and apply to a C_{18}-SepPak (*see* Subheading 2.1.3, **item 3**).
2. Elute most N-glycans with water (1 ml).
3. Elute further hydrophobic N-glycans with 15% MeOH (1 ml).
4. Wash column with 40% MeOH and 100% MeOH.
5. Dry by vacuum centrifugation or lyophilize.
6. Assess quality of pools by mass spectrometry.

3.3.2 Enrichment of Anionic N-Glycans by Non-porous Graphitized Carbon [8, 12]

1. Dissolve glycans in water (100–200 μl) and apply them to graphitized carbon pre-equilibrated as above (Subheading 2.1.3, **item 4**).
2. Wash the column or cartridge with water (1–2 ml).
3. Elute primarily neutral N-glycans with 40% MeCN (1 ml).
4. Elute further N-glycans with 0.1% TFA in 40% MeCN (1 ml).
5. Dry neutral and acidic-enriched pools by vacuum centrifugation or lyophilization before analysis or derivatization.
6. Assess quality of pools by mass spectrometry.

3.4 Derivatization of N-Glycans (See Note 7)

3.4.1 Reductive Amination of Glycans

Fluorophores are conjugated at the reducing terminus of the glycan, which is uniquely constituted by a reactive carbonyl moiety. Many fluorophores are available including 2-aminobenzamide (2AB), AEAB (2-amino(N-aminoethyl)benzamide) and 2-aminopyridine (2AP or PA), the latter which is described here.

1. Transfer 80 μL of fresh 2-aminopyridine (PA) solution (*see* Subheading 2.1.4, **item 1**) to the glycan sample dried in a 1.5-ml polypropylene microcentrifuge tube, apply safety cap to avoid popping, and incubate in boiling water for 15 min.
2. Continue the reaction by transferring 4 μL of a fresh cyanoborohydride/PA solution (*see* Subheading 2.1.4, **item 2**) to the sample and incubating overnight at 90 °C in an oven.

3. Dilute the sample in 1.5 ml of 0.5 % acetic acid, apply to a 30 ml Sephadex G15 column (1 × 40 cm) equilibrated in 0.5 % acetic acid, and collect 1.5 ml fractions. Transfer aliquots of fractions (80 μl) to a 96F black plate and detect fluorescence in a microtiter plate reader (excitation/emission: 320/400 nm). Pool fluorescent glycans eluting before the excess labeling reagent and lyophilize.

3.5 HPLC Separation Methods

Pyridylaminated N-glycans can be analyzed by either NP- or RP-HPLC or HIAX using an HPLC system equipped with a fluorescence detector. NP- and RP-columns are calibrated daily in terms of glucose units, using PA-labeled forms of partial dextran hydrolysates (*see* **Note 8** and Fig. 2).

3.5.1 Option 1: NP-HPLC

1. Dissolve dried sample in 50 μL of a 1:3 mixture of solvent A (10 mM ammonium formate, pH 7.0) and solvent B (95 % acetonitrile in water).
2. Inject a portion of the sample into a Tosoh-80 column equilibrated in the same 1:3 mixture.
3. Elute column at 1 ml/min as follows: 0–5 min, 75 % solvent B; 5–15 min, 75–65 % B; 15–40 min, 65 % B; 40–55 min, 65–57 %

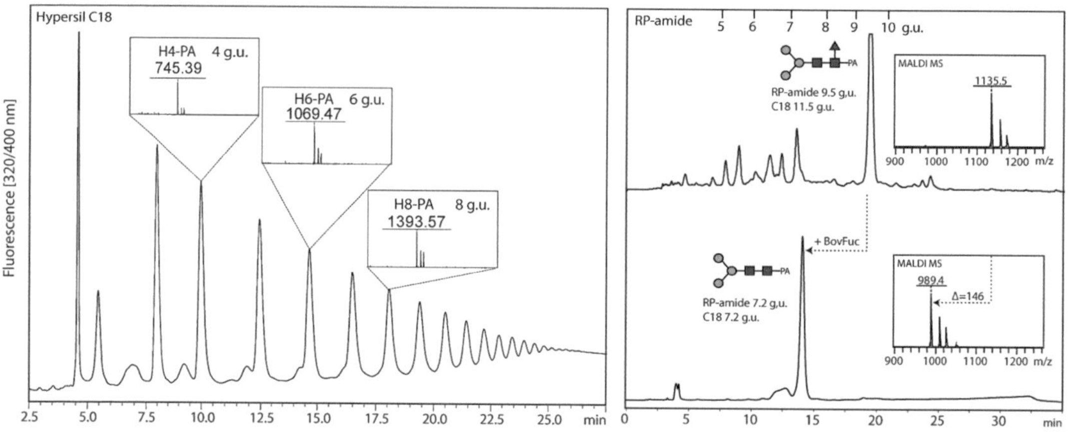

Fig. 2 Examples of RP-HPLC separations. RP-HPLC columns can be calibrated in terms of glucose units for day-to-day comparisons; the order of elution is the same as for NP-HPLC with the lowest molecular weight gluco-oligomers coming first as verified by MALDI-TOF MS of three of the standard peaks. Some modifications have also a major effect on elution times on RP-HPLC, as exemplified by the elution of Manα1,6(Manα1,3)Manβ1,4GlcNAcβ1,4(Fucα1,6)GlcNAc-PA before and after bovine α-fucosidase digestion, which results in a reduction in elution time from 9.5 to 7.2 glucose units on an RP-amide column (11.5 to 7.2 g.u. in the case of a standard C18 column); the example glycan shown is from *Trichuris suis* [17]

B; followed by a return to the starting conditions. Other variant: 1–5 min: 75%B; 5–10 min, 75–70% B; 10–15 min, 70–65% B; 15–55 min; 65–55% B; 55–56 min; 55–75%B; 56–60 min, 75% B. The remaining % (up to 100% (v/v)) derive from delivery of solvent A by the HPLC system.

4. Detect glycans by fluorescence using excitation at 310 nm and emission at 380 nm. Collect fractions based on fluorescence intensity and lyophilize prior to later analyses (*see* Subheading 3.6).

3.5.2 Option 2: RP-HPLC (See also Fig. 2)

1. Dissolve dried sample in 50 μL of water.
2. Inject portion of sample onto an RP-HPLC column (e.g., Agilent Hypersil ODS or Phenomenex Hyperclone) pre-equilibrated with 100 mM ammonium acetate (pH 4; solvent C).
3. Elute at 1.5 ml/min using a linear gradient from the starting buffer to a solution composed of 30% (v/v) MeOH (solvent D), at 1% per min (in the case of fungal samples [10], a shallower gradient was applied: 0–35 min, 0–15% D; 35–36 min, 15–35% D; 36–46 min, 35–0% D).
4. Detect glycans by fluorescence using excitation at 320 nm and emission at 400 nm. Collect fractions based on fluorescence intensity and lyophilize prior to later analyses (*see* Subheading 3.6). For fused core RP columns, the gradients may have to be adapted and the flow rate reduced to 0.8 ml/min if manually collecting fractions.

3.5.3 Option 3: 2D HPLC (See **Note 9**)

1. First fractionate by NP-HPLC (*see* Subheading 3.5.1).
2. Collect fractions, lyophilize, and identify fractions of interest by MALDI-TOF MS (*see* Subheading 3.6).
3. Subject desired fractions to RP-HPLC (*see* Subheading 3.5.2) and analyze by MALDI-TOF MS (*see* Subheading 3.6).

3.5.4 Option 4: HIAX [8, 21]

1. Dissolve sample in 10 μl water, before adding 40 μl acetonitrile; pre-equilibrate AS11 column (Dionex) with 4:1 80% acetonitrile–800 mM ammonium acetate, pH 3.85 (solvent E is 800 mM ammonia, pH 3.85, and solvent F is 80% acetonitrile).
2. Check column by injecting a mixture of PA-labeled oligomannosidic glycans (e.g., $Man_{3-9}GlcNAc_2$).
3. Inject portion of sample onto pre-equilibrated column.
4. Elute with the following gradient: 0–5 min, 99% F; 5–50 min, 99–90%F, 50–65 min; 90–80% F, 65–85 min, 80–7% F; 85–110 min, 75–65% F; 110–112 min, 65%F; 112–120 min, 65–99% F; 120–130 min, 99%F.
5. Collect (either based on fluorescence intensity or time) and dry fractions for further analyses.

3.6 MALDI-TOF and TOF-TOF Analysis (See Note 10)

1. Dried native or chemically treated (*see* Subheading 3.7.2) samples are resuspended in 5–20 μl water; exoglycosidase digested samples can normally be used directly.
2. Prepare a method blank (without analyte) to differentiate signals from background contamination (especially for exoglycosidase digests).
3. Spot 0.5 μL of the sample onto a steel MALDI target plate and vacuum dry; thereafter 0.5 μL of the matrix ATT or DHB is added and drying repeated; sometimes the signals can be improved or shifted to $[M+H]^+$ if 0.25 μl 20 mM ammonium sulfate pH 5 is co-spotted with the sample. Spot a peptide mixture (using ACH as matrix) as an external mass calibrant.
4. Analyze glycans using an MALDI-TOF-TOF in reflectron positive or negative ion modes. Acquire MALDI spectra at laser frequency of 1000 Hz and sum 1000–2000 individual spectra of each sample. Typically, glycans require higher laser power and/or higher detector gain settings than peptides. Decreasing the lens voltage, as compared to factory settings, may also improve detection of some glycans.
5. For glycan fragmentation, perform TOF-TOF MS/MS experiments by selecting precursor ions for laser-induced dissociation using the LIFT apparatus in Bruker instruments (*see* **Note 11**).

3.7 Enzymatic and Chemical Treatments

3.7.1 Exoglycosidase Digestions (See **Note 12**)

1. Resuspend pmol quantities of PA-labeled glycans in an appropriate amount of water.
2. Typically 1 μl pyridylaminated glycans and 0.5 μl 50 mM ammonium acetate (pH 5.0) are mixed with 0.2 μl enzyme (regardless of U/ml) and 0.8 μl water in a PCR tube. Incubate tubes at 37 °C overnight in an oven. If performing a number of reactions with the same enzyme, the glycosidase can be premixed with buffer and water for immediate use to reduce pipetting errors. Some glycosidases have other pH or cation requirements: e.g., β-glucuronidase requires pH 7 and α1,2/3-mannosidase requires 0.5 mM $CaCl_2$ (refer to suppliers' information as required); other glycosidases may need to be diluted before use due to contaminants.

3.7.2 Chemical Cleavage to Remove Phosphoesters, Galactofuranose, or Fucose Residues (See **Note 13**)

1. Incubate dried glycan fractions with 3 μl 48% hydrofluoric acid (HF; caution) on ice for 24–48 h in plastic microcentrifuge tubes, and then dry under a stream of dry N_2 gas or in a SpeedVac.
2. Re-evaporate from water to remove traces of HF.

3.8 Western Blotting (See Note 14)

1. Analyze crude whole cell extracts (25 μg total protein material) or purified glycoproteins of interest by Western blotting after separation by SDS-PAGE and transfer to nitrocellulose membrane using a semidry blotting apparatus.

2. Use standard Western blot procedures to probe the membrane with anti-carbohydrate antibody or lectin, such as rabbit anti-HRP (IgG) for α3-linked core fucose or β1,2-xylose, mouse anti-phosphorylcholine (TEPC-15; IgA) for phosphorylcholine, or biotin-conjugated lectins.
3. Use a relevant secondary antibody: e.g., alkaline phosphatase-conjugated forms of goat anti-rabbit antibody (for anti-HRP), goat anti-mouse IgA (for TEPC-15), or anti-biotin (or streptavidin; for lectins).
4. Color detection is with 5-bromo-4-chloro-3-indolyl phosphate/nitro blue tetrazolium or 3,3′-diaminobenzidine for alkaline phosphatase or peroxidase conjugates respectively; SigmaFAST tablets are available for both.

3.9 Glycan Data Analysis

1. Analyze MS spectra using the relevant software (e.g., flexAnalysis for Bruker instruments). Ions are typically singly charged under the MALDI conditions described, and only exact monoisotopic m/z values (i.e., the ^{12}C-only ion) should be examined. If possible, select $[M+H]^+$ ions for peak picking in positive ion mode, or $[M-H]^-$ in negative ion mode. Spectra can be exported as mzXML files if supporting information is required for publications.
2. Predict glycan compositions based on mass/charge (m/z) matching with potential glycan compositions using manual calculations or GlycoWorkbench software. Typically MALDI is accurate to ±0.3 Da; N-glycans will contain varying numbers of hexose ($\Delta m/z$ 162), *N*-acetylhexosamine ($\Delta m/z$ 203), deoxyhexose (e.g., fucose; $\Delta m/z$ 146), pentose (e.g., xylose; $\Delta m/z$ 132), sulfate/phosphate ($\Delta m/z$ 80), methylphosphate ($\Delta m/z$ 94) or zwitterionic residues (phosphorylcholine, phosphoethanolamine, *N*-methyl-2-aminoethylphosphonate or aminoethylphosphonate with respectively $\Delta m/z$ 165, 123, 121 or 107). Apply corrections for reducing terminal derivatization or permethylation. The nature of the hexose and *N*-acetylhexosamine residues depends on the species; indeed, even in fungi, hexose residues are not automatically mannose [10]. Sialic acids are rare in nonvertebrate species; their presence in invertebrate or protist samples may suggest contamination with food or media components. Western blotting data can additionally aid interpretation.
3. Predict structural models based on known biosynthetic rules, and confirm using exoglycosidase digestions (Subheading 3.7.1), chemical cleavage (Subheading 3.7.2), and MS/MS studies (Subheading 3.6). These models can be converged with findings from elution times in 1D- and 2D-HPLC studies (Subheading 3.5).

4 Notes

1. A high quality source of deionized water, to dissolve reagents and samples, is required; in some cases, the water should be bought to avoid ionic or microbial contaminants.
2. Care must be taken to consider potential contamination from food sources or media components; for instance, fetal calf serum often used in parasite cultivation contains fetuin (a sialylated glycoprotein), fungal growth media contain milk glycoproteins (casein) or mosquito larvae are reared on ground cat or fish food. This can mean that cells or unicellular organisms are kept briefly in a serum-free medium or at least adequately washed or that buffer be exchanged in order to reduce media contaminants. Also, contamination by a polyhexose series is common; to reduce the relevant signals, samples (e.g., glycopeptides) can be pre-reduced prior to N-glycan release with 100 μL of 0.5 % (w/v) sodium borohydride/2.5 % (v/v) ammonia at room temperature for 2 h followed by addition of 2.5 μL of glacial acetic acid and lyophilization [22]. The reduced material will then be inert to derivatization by reductive amination and the normal glycan release procedure then followed after adjustment to the pH required by the applied enzyme. Endogenous endoglycosidases and exoglycosidases are another potential problem, but heat inactivation before disruption of cell integrity should be sufficient to prevent glycan degradation.
3. Care should be taken when diluting or handling sulphuric acid required for the orcinol reagent (do not pour water directly into concentrated acid); for pretreating TLC plates, apply a thin layer of orcinol reagent and then dry with a hair drier.
4. Whereas sample volume is not a problem for Dowex ion exchange chromatography, in the case of gel filtration, the sample volume should be no more than 5 % of that of the gel filtration column.
5. PNGase F is the standard N-glycosidase for releasing N-glycans, but does not release N-glycans modified by Fuc α3-linked to the core GlcNAc [23]. PNGase A has broader specificity including core α3-fucosylated N-glycans, but requires small peptides for optimal activity, is more expensive and differs in its pH optimum as compared to PNGase F. Newer broad specificity enzymes are now described, but are not yet commercially available. Endoglycosidases are more specific and so are less suited for whole organism glycomics unless pretreatment to remove, e.g., oligomannosidic forms is desired. Enzymatic deglycosylation is most effective after proteolysis, especially for PNGase A. However, PNGase F is also capable of deglycosylating intact proteins, which have been denatured by SDS or urea, if SDS is first diluted in the pres-

ence of NP40, or urea is sufficiently diluted; deglycosylation can be followed by glycan recovery or by Western blotting [24]. As appropriate, glycopeptides samples may be treated with PNGase A or F alone (e.g., with PNGase A alone if core α1,3-fucosylated glycans are a major component or with PNGase F alone if, as in the case of fungi, no core α1,3-fucosylated glycans are expected); the described approach of sequential release is, however, enabling a partial sub-fractionation of complex N-glycomes. Chemical release with hydrazine is also an option, but requires thorough purification to remove contaminants interfering with mass spectrometry.

6. To prevent failure of pH-sensitive steps, such as Dowex chromatography or digestion with glycosidases, confirm proper pH by dispensing 1 μL on a strip of appropriate pH indicator paper.
7. The simplest method is to analyze released N-glycans directly by MS. However, derivatization of the reducing terminus with a fluorophore improves MS sensitivity, and allows detection of glycans by RP or NP chromatography for identification based on co-chromatography with known standards. PA is commonly used in our own laboratory and in Japan [25]; otherwise, 2AB or AA (anthranilic acid) are other commonly used fluorophores with different properties in terms of hydrophobicity and HPLC/MS sensitivity. AEAB is of interest for preparation of natural glycan arrays [26].
8. Calibrated HPLC separations can allow structure prediction based on relative elution times, because compositional isoforms may be differentially retained on selected stationary phases. For both normal and reversed phase columns, the lowest glucose units elute first (Fig. 2). In case of doubt, collected fractions containing the standards can be analyzed by MALDI-TOF MS. Anionic glycans tend to elute earlier than the corresponding neutral structures on NP- and RP-HPLC columns [8, 11]; the elution of zwitterionic phosphorylcholine-containing glycans is either relatively late on standard RP-HPLC or relatively earlier on fused core RP-HPLC columns [27]; core α1,3-fucose results in early elution and core α1,6-fucose in late elution on RP-HPLC regardless of the type of column. HPLC buffers are made from the individual components (acetic acid and ammonia) and not by merely adjusting the pH of an ammonium acetate solution prepared from the commercially available salt; the stock solutions should also be filtered before use.
9. Typically, 2D-HPLC is generally performed with NP-HPLC in the first dimension and RP-HPLC in the second [6]; however, we have occasionally done RP-HPLC followed by HIAX [10].
10. Here we focus on MALDI-TOF MS, but LC-ESI-MS is also a commonly used method in glycomic analyses and its MS^n capabilities are of interest in the case of unusual structures as

well as for cross-ring cleavages; also permethylation or perdeuteromethylation can be performed, but care must be taken with clean-up in order to avoid loss of hydrophilic sulfated or phosphoester-modified glycans. We have found ATT to be a rather robust matrix for positive and negative MALDI-TOF MS and is also suitable for glycopeptides [24]; DHB tends to favor formation of sodiated adducts of glycans. ACH is used as a matrix for the peptide standard; its chlorinated form can be also used for glycans and glycopeptides [28].

11. In our hands, best results for MS/MS are obtained with the $[M+H]^+$ form in positive mode as sodiated adducts tend to show losses from the parent ion (B-fragments), but are relatively poor in terms of Y-fragments useful for determining core modifications. High-energy collision-induced decay using argon is also possible, but reduces the parent ion signal. If a neutral glycan sample is rather concentrated, then cross-ring cleavage in negative-ion mode is possible [17].
12. Although jack bean α-mannosidase is considered nonspecific with respect to the linkage position on the underlying sugar, reaction rates vary considerably and thus times need to be extended to achieve removal of sterically constrained linkages. Some exoglycosidases are extremely sensitive to steric hindrance; furthermore, not all residues occurring in nonvertebrate glycans are susceptible to exoglycosidase digestion due either to the types of linkages or the presence of substitutions such as methylation or sulphation. If there is sufficient material, shifts in elution time by RP-HPLC offer extra information in addition to MALDI-TOF MS alone; e.g., removal of core α1,6-fucose results in a reduction in retention time (Fig. 2). On-plate digestions in a humid chamber can also be performed. Volatile buffers (e.g., ammonium acetate) are to be preferred over manufacturer-supplied buffers with citrate, phosphate, etc.
13. Hydrofluoric acid removes phosphoesters (phosphate, methylphosphate, phosphorylcholine, phosphoethanolamine, 2-aminoethylphosphonate and its N-methylated form), galactofuranose and α1,3-fucose; α1,2- and α1,4-fucose is also partially susceptible, but α1,6-fucose and α-mannose is resistant [8, 10, 12, 14]. Hydrofluoric acid is to be treated with caution; a tube of calcium gluconate gel should be to hand in case of skin contact. As for exoglycosidase digestions, if there is sufficient material, shifts in elution time by RP-HPLC offer extra information in addition to MALDI-TOF MS alone.
14. Antibodies and lectins can help screen types of glycan modifications, such as core α3- vs. α6-linked fucose when using anti-HRP or phosphorylcholine when using TEPC-15, in the samples at the glycoprotein level [7, 22]. Western blotting is readily adapted to the use of any lectin or antibody; enzyme-linked, fluorescence, or ECL detection systems can be used.

Concentrations of the antibody/lectin should be optimized by comparison with appropriate negative and positive controls; especially when using biotin-based systems, controls without the lectin (i.e., applying only the anti-biotin or streptavidin conjugate) are required. Nevertheless, interaction with a lectin is not a structural proof and can be misleading as many nonmammalian glyco-modifications have not been assessed for lectin reactivity.

Acknowledgment

This work was supported by the Austrian Fonds zur Förderung der wissenschaftlichen Forschung (FWF; grants P26662, P25058, and P23922 to A.H., K.P., and I.B.H.W.).

References

1. Schachter H (2009) Paucimannose N-glycans in *Caenorhabditis elegans* and *Drosophila melanogaster*. Carbohydr Res 344:1391–1396
2. Sabotič J, Ohm RA, Künzler M (2015) Entomotoxic and nematotoxic lectins and protease inhibitors from fungal fruiting bodies. Appl Microbiol Biotechnol 100:91–111
3. Rodrigues JA, Acosta-Serrano A, Aebi M, Ferguson MA, Routier FH, Schiller I, Soares S, Spencer D, Titz A, Wilson IB, Izquierdo L (2015) Parasite glycobiology: a bittersweet symphony. PLoS Pathog 11, e1005169
4. Samuelson J, Banerjee S, Magnelli P, Cui J, Kelleher DJ, Gilmore R, Robbins PW (2005) The diversity of dolichol-linked precursors to Asn-linked glycans likely results from secondary loss of sets of glycosyltransferases. Proc Natl Acad Sci U S A 102:1548–1553
5. Schiller B, Hykollari A, Yan S, Paschinger K, Wilson IBH (2012) Complicated N-linked glycans in simple organisms. Biol Chem Hoppe Seyler 393:661–673
6. Paschinger K, Hykollari A, Razzazi-Fazeli E, Greenwell P, Leitsch D, Walochnik J, Wilson IBH (2012) The N-glycans of *Trichomonas vaginalis* contain variable core and antennal modifications. Glycobiology 22:300–313
7. Schiller B, Makrypidi G, Razzazi-Fazeli E, Paschinger K, Walochnik J, Wilson IBH (2012) Exploring the unique N-glycome of the opportunistic human pathogen *Acanthamoeba*. J Biol Chem 287:43191–43204
8. Hykollari A, Balog CI, Rendić D, Braulke T, Wilson IBH, Paschinger K (2013) Mass spectrometric analysis of neutral and anionic N-glycans from a *Dictyostelium discoideum* model for human congenital disorder of glycosylation CDG IL. J Proteome Res 12:1173–1187
9. Hykollari A, Dragosits M, Rendić D, Wilson IBH, Paschinger K (2014) N-glycomic profiling of a glucosidase II mutant of Dictyostelium discoideum by "off-line" liquid chromatography and mass spectrometry. Electrophoresis 35:2116–2129
10. Hykollari A, Eckmair B, Voglmeir J, Jin C, Yan S, Vanbeselaere J, Razzazi-Fazeli E, Wilson IBH, Paschinger K (2016) More than just oligomannose: an N-glycomic comparison of *Penicillium* species. Mol Cell Proteomics 15:73–92
11. Kurz S, Jin C, Hykollari A, Gregorich D, Giomarelli B, Vasta GR, Wilson IBH, Paschinger K (2013) Haemocytes and plasma of the eastern oyster (*Crassostrea virginica*) display a diverse repertoire of sulphated and blood group A-modified N-glycans. J Biol Chem 288:24410–24428
12. Eckmair B, Jin C, Abed-Navandi D, Paschinger K (2016) Multi-step fractionation and mass spectrometry reveals zwitterionic and anionic modifications of the N- and O-glycans of a marine snail. Mol Cell Proteomics 15(2):573–97. doi:10.1074/mcp.M115.051573
13. Kurz S, Aoki K, Jin C, Karlsson NG, Tiemeyer M, Wilson IB, Paschinger K (2015) Targetted release and fractionation reveal glucuronylated and sulphated N- and O-glycans in larvae of dipteran insects. J Proteomics 126: 172–188

14. Yan S, Brecker L, Jin C, Titz A, Dragosits M, Karlsson N, Jantsch V, Wilson IBH, Paschinger K (2015) Bisecting galactose as a feature of N-glycans of wild-type and mutant *Caenorhabditis elegans*. Mol Cell Proteomics 14:2111–2125
15. Yan S, Jin C, Wilson IBH, Paschinger K (2015) Comparisons of *Caenorhabditis* fucosyltransferase mutants reveal a multiplicity of isomeric N-glycan structures. J Proteome Res 14: 5291–5305
16. Paschinger K, Wilson IBH (2015) Two types of galactosylated fucose motifs are present on N-glycans of *Haemonchus contortus*. Glycobiology 25:585–590
17. Wilson IBH, Paschinger K (2016) Sweet secrets of a therapeutic worm: mass spectrometric N-glycomic analysis of *Trichuris suis*. Anal Bioanal Chem 408:461–471
18. Dragosits M, Pflugl S, Kurz S, Razzazi-Fazeli E, Wilson IBH, Rendić D (2014) Recombinant Aspergillus β-galactosidases as a robust glycomic and biotechnological tool. Appl Microbiol Biotechnol 98:3553–3567
19. Dragosits M, Yan S, Razzazi-Fazeli E, Wilson IBH, Rendić D (2015) Enzymatic properties and subtle differences in the substrate specificity of phylogenetically distinct invertebrate N-glycan processing hexosaminidases. Glycobiology 25:448–464
20. Iskratsch T, Braun A, Paschinger K, Wilson IBH (2009) Specificity analysis of lectins and antibodies using remodeled glycoproteins. Anal Biochem 386:133–146
21. Neville DC, Dwek RA, Butters TD (2009) Development of a single column method for the separation of lipid- and protein-derived oligosaccharides. J Proteome Res 8:681–687
22. Pöltl G, Kerner D, Paschinger K, Wilson IBH (2007) N-Glycans of the porcine nematode parasite *Ascaris suum* are modified with phosphorylcholine and core fucose residues. FEBS J 274:714–726
23. Tretter V, Altmann F, März L (1991) Peptide-N^4-(*N*-acetyl-β-glucosaminyl)asparagine amidase F cannot release glycans with fucose attached α1→3 to the asparagine-linked *N*-acetylglucosamine residue. Eur J Biochem 199:647–652
24. Paschinger K, Gonzalez-Sapienza GG, Wilson IBH (2012) Mass spectrometric analysis of the immunodominant glycan epitope of *Echinococcus granulosus* antigen Ag5. Int J Parasitol 42:279–285
25. Hase S, Ibuki T, Ikenaka T (1984) Reexamination of the pyridylamination used for fluorescence labelling of oligosaccharides and its application to glycoproteins. J Biochem 95: 197–203
26. Song X, Xia B, Stowell SR, Lasanajak Y, Smith DF, Cummings RD (2009) Novel fluorescent glycan microarray strategy reveals ligands for galectins. Chem Biol 16:36–47
27. Yan S, Wilson IBH, Paschinger K (2015) Comparison of RP-HPLC modes to analyse the N-glycome of the free-living nematode *Pristionchus pacificus*. Electrophoresis 36: 1314–1329
28. Selman MH, Hoffmann M, Zauner G, McDonnell LA, Balog CI, Rapp E, Deelder AM, Wuhrer M (2012) MALDI-TOF-MS analysis of sialylated glycans and glycopeptides using 4-chloro-alpha-cyanocinnamic acid matrix. Proteomics 12:1337–1348

Chapter 14

High-Throughput and High-Sensitivity Mass Spectrometry-Based *N*-Glycomics of Mammalian Cells

Stephanie Holst, Gabi W. van Pelt, Wilma E. Mesker, Rob A. Tollenaar, Ana I. Belo, Irma van Die, Yoann Rombouts, and Manfred Wuhrer

Abstract

The current protocols for glycomic analysis of cells often require a large quantity of material (5–20 million cells). In order to analyze the *N*-glycosylation from small amounts of cells (≤1 million) as obtained from, for example, primary cell lines or cell sorting, and in a higher throughput approach, we set up a robust 96-well format PVDF-membrane based *N*-glycan release protocol followed by linkage-specific sialic acid stabilization, cleanup, and MALDI-TOF-MS analysis. We further evaluated the influence of PNGase F incubation time on the *N*-glycan profile.

Key words Cells, *N*-glycosylation, PVDF filter plates, PNGase F digestion, Ethyl esterification, MALDI-TOF-MS

1 Introduction

Changes in glycosylation are often found to be associated with the development of malignancies as well as infectious and autoimmune diseases [1, 2]. Current protocols for the analysis of cellular glycosylation often require several millions of cells and therefore are of limited use to decipher the glycosylation of cells that are available in minute amounts as for instance primary or stem cells, cells obtained from cell sorting, and patient-derived materials. In order to overcome this issue, we have developed and optimized a robust 96-well format sample preparation protocol followed by MALDI-TOF-MS-based profiling of cellular *N*-glycans. The method consists of the denaturation and reduction of proteins from a relatively small number of cells ($0.25–1 \times 10^6$ cells), protein immobilization on PVDF filter plates [3], enzymatic *N*-glycan release with PNGase F followed by derivatization of the glycans using a recently developed technique which enables the stabilization and linkage-differentiation of α2,6- and α2,3-linked sialic acids [4]. The resulting derivatized *N*-glycans are then easily purified using

Gordan Lauc and Manfred Wuhrer (eds.), *High-Throughput Glycomics and Glycoproteomics: Methods and Protocols*, Methods in Molecular Biology, vol. 1503, DOI 10.1007/978-1-4939-6493-2_14,

micro-hydrophilic liquid interaction chromatography (HILIC) solid-phase extraction (SPE) [5] followed by their qualitative and quantitative characterization by MALDI-TOF mass spectrometry in positive ion mode. Here, we describe this method in detail and present its application on four cancer cell lines (colorectal cancer cell lines HT29 and SW480, pancreatic cancer cell lines Pa-Tu-8988S and Pa-Tu-8988T). We also present results on the effect of variations in PNGase F incubation times on the *N*-glycomic cell line profiles. Altogether, this new method offers a sensitive, robust, and higher throughput workflow for cellular *N*-glycomics.

2 Materials

All buffers are prepared using Milli-Q (mQ) water generated from a Q-Gard 2 system (Millipore, Amsterdam, The Netherlands).

2.1 Cell Culture

1. Colorectal cancer cell lines HT29 and SW480 were obtained from the Department of Surgery, Leiden University Medical Center, Leiden, The Netherlands.
2. Pancreatic cancer cell lines Pa-Tu-8988S (PaTu-S) and Pa-Tu-8988T (PaTu-T) were obtained from the Department of Molecular Cell Biology and Immunology, VU University Medical Center, Amsterdam, The Netherlands.
3. 500 mL Hepes-buffered RPMI-1640 culture media containing l-glutamine (300 mg/L) (Invitrogen, Paisley, UK), supplemented with 5 mL penicillin (5000 IU/mL)/streptomycin (5 mg/mL) (MP Biomedicals; Santa Ana, CA), and 50 mL fetal calf serum (10% *v*/*v*) (FCS; PAA Laboratories GmbH, Pasching, Austria).
4. T75 cell culture flasks (Greiner-Bio, Alphen a/d Rijn, The Netherlands).
5. 1× PBS (B. Braun Melsungen AG, Melsungen, Germany).
6. 1× trypsin–EDTA solution: Dilute 10× trypsin–EDTA solution (5.0 g/L porcine trypsin and 2.0 g/L EDTA×4 Na in 0.9% sodium chloride; PAA Laboratories GmbH) 1:5 (*v*/*v*) in 1× PBS. For 50 mL add 5 mL 10× trypsin–EDTA solution to 45 mL 1× PBS.
7. Tryptan stain 0.4% (Life Technologies, Eugene, OR).
8. CountessTM Automated Cell-Counter (Invitrogen).
9. 15 mL conical Falcon tubes.
10. 1.5 mL Eppendorf tubes.
11. Centrifuge.
12. Incubator.

2.2 Control Sample

1. Control Visucon-F plasma pool (citrated and 0.02 M HEPES buffered plasma pool from 20 healthy human donors; (Affinity Biologicals, Ancaster, Canada).

2.3 Enzymatic Glycan Release, Derivatization, and Purification

1. MultiScreen® HTS 96 multiwell plates (pore size 0.45 μm) with hydrophobic Immobilon-P high protein-binding PVDF membrane (Millipore).
2. Conical polypropylene 96-well Nunc plates (Thermo Scientific, Roskilde, Denmark).
3. 96-well plate clear (Greiner Bio).
4. 96-well polypropylene 0.8 mL 96-deepwell plate (Greiner Bio).
5. 96-well PCR plate polypropylene (Greiner Bio).
6. Solvent resistant capmat (Thermo Scientific) or plastic seals (Thermo Scientific).
7. 15 mL Falcon tubes.
8. 70% ethanol (AZL Ziekenhuisapotheek, Leiden University Medical Center).
9. 100 mM $NaHCO_3$ buffer: Dissolve 1.68 g $NaHCO_3$ (Merck, Darmstadt, Germany) in 200 mL mQ water. Store at 4 °C for maximum 2 weeks, check pH (~pH 8).
10. 200 mM DTT: Weigh 30.85 mg dithiothreitol (DTT; Merck) and dissolve in 1 mL mQ water. Store at −20 °C.
11. Denaturation buffer: Prepare denaturation buffer master-mix of 72.5 μL 8 M guanidine hydrochloride (GuHCl; Sigma-Aldrich, Steinheim, Germany) and 2.5 μL 200 mM DTT per well (=5.8 M GuHCl/5 mM DTT). For 96 samples prepare 6960 μL 8 M GuHCl and 240 μL 200 mM DTT in a 15 mL Falcon tube. Prepare fresh each time.
12. Parafilm M (Sigma).
13. Kimberly-Clark tissue (Kimberly-Clark Professional B.V., Ede, The Netherlands).
14. *N*-Glycosidase F (PNGase F; Roche Diagnostics, Mannheim, Germany).
15. Ethylation reagent: For 10 mL ethylation reagent, sufficient for 96 samples, weigh 479.25 mg 1-ethyl-3-(3-dimethylaminopropyl) carbodiimide (EDC, Fluorochem; Hadfield, UK) into a 15 mL Falcon tube. Weigh 337.80 mg 1-hydroxybenzotriazole (HOBt) hydrate (Sigma-Aldrich) into a second 15 mL Falcon tube. Dissolve each reagent separately in 5 mL 100% ethanol (Merck) (=0.5 M each). Mix both reagents (0.25 M EDC/0.25 M HOBt). When prepared

in excess, store at −20 °C in a glass bottle or as aliquots in 2 mL Eppendorf tubes (*see* **Note 1**).

16. Trifluoroacetic acid (TFA) (Sigma-Aldrich).
17. HPLC SupraGradient acetonitrile (ACN) (Biosolve, Valkenswaard, The Netherlands).
18. Pipette tips 20 μL (Rainin/ Mettler-Toledo, Tiel, The Netherlands).
19. Cotton-thread.
20. 85% ACN/15% mQ water solution (*v*/*v*): For 1 L measure 850 mL ACN in a cylinder, pour into a clean 1 L glass bottle. Measure 150 mL mQ water in a cylinder. Add the mQ water to the ACN and mix well. Label and store for maximum 1 month at 4 °C. Bring to room temperature for 30 min prior to use.
21. 85% ACN/14% mQ water/1% TFA (*v*/*v*) solution: For 1 L measure 850 mL ACN in a cylinder, pour into a clean 1 L glass bottle. Measure 140 mL mQ water in a cylinder and add to the ACN. Measure 10 mL TFA and add to the ACN/water mix. Mix well, label and store for maximum 1 month at 4 °C. Bring to room temperature for 30 min prior to use.
22. Centrifuge.
23. Horizontal shaker.
24. 60 °C oven.
25. 37 °C oven.
26. Buckets/containers suited for the use of 12-channel pipettes.

2.4 MALDI-TOF-MS

1. Micro-focus MALDI-TOF-MS target plate; here an AnchorChip plate (Bruker Daltonics, Bremen, Germany).
2. Prepare 1 M NaOH solution by adding 53 μL 50% NaOH solution (Sigma-Aldrich) to 947 μL mQ water in a glass vial. Store at room temperature.
3. Super DHB matrix (2-hydroxy-5-methoxy-benzoic acid and 2,5-Dihydroxybenzoic acid, 1:9; Sigma-Aldrich): Prepare matrix for MALDI-TOF-MS by weighing 5 mg superDHB into a 1.5 mL Eppendorf tube. Resuspend the powder in 999 μL 50% ACN (500 μL ACN + 499 μL mQ water). Add 1 μL 1 M NaOH in mQ water to achieve a final concentration of 1 mM NaOH. Store at 4 °C.
4. Mass calibration standard; here a peptide calibration standard (Bruker Daltonics).
5. MALDI-TOF mass spectrometer; here a UltrafleXtreme (Bruker Daltonics).

3 Methods

3.1 Cell Culture

Pancreatic cancer cell lines PaTu-S and PaTu-T are cultured as described in Belo *et al.* [6] in high-glucose DMEM culture medium with 10% FCS and antibiotic supplement. For colorectal cancer cell lines, a detailed protocol is given:

1. Culture cells in T75 flask with 14 mL Hepes-buffered RPMI-1640 culture medium in an incubator at 37 °C with 5% CO_2 in humidified air. Culture cells to a confluence of 80% under sterile conditions.
2. Harvest the cells by removing the medium and washing adherent cells twice with 2 mL 1× PBS. Remove medium and PBS by glass pipette using an aspiration system.
3. Detach the cells by trypsinizing with 2 mL 1× trypsin–EDTA solution in 1× PBS for approx. 5 min at 37 °C.
4. Stop trypsin activity by adding 5 mL medium and transfer cells to conical 15 mL Falcon tube.
5. Pellet cells at 300×*g* for 5 min.
6. Remove the supernatant and resuspend cells in 5 mL 1× PBS by carefully pipetting up and down (*see* **Note 2**)
7. Take 10 μL of cell suspension into a separate Eppendorf tube and add 10 μL tryptan blue staining solution. Count cells using a CountessTM Automated Cell-Counter.
8. Aliquot cells from the Falcon tube to 2.0×10^6 cells per mL of 1× PBS in 1.5 mL Eppendorf tubes and label tubes.
9. Pellet cells for 3 min at 1000×*g*, remove supernatant and wash cells twice with 500 μL 1× PBS for 3 min at 1000×*g*. Remove supernatant and store pellets at −20 °C until further use (*see* **Note 3**).

3.2 Enzymatic N-Glycan Release

PVDF-membrane based release protocol using 96-well filter plates is modified from Burnina *et al.* [3]:

1. Resuspend cell pellets containing two million cells in 100 μL mQ water each and homogenize 30 min in a sonication bath (*see* **Note 4**).
2. As a control, bring 5–10 μL pooled human Visucon plasma to 25 μL with mQ water (*see* **Note 5**).
3. Precondition PVDF membrane with 200 μL 70% ethanol, followed by 2× 200 μL 100 mM $NaHCO_3$ on a vacuum manifold (*see* **Note 6**).
4. Dip the plate gently on a Kimberly-Clark tissue to remove residual solvent and place on top of a 96-well NUNC plate as washing plate (*see* **Note 7**).

5. Load cell line samples (in technical replicates) with 25 μL/well (=500,000 cells) onto preconditioned PVDF membrane. Add 25 μL/well diluted human plasma control as well as 25 μL/well mQ water as blank in empty wells (*see* **Note 8**).
6. Add 75 μL of premixed denaturation buffer per well using a multistep pipette. Incubate protected from evaporation using a Tupper box containing a moistured tissue as incubation chamber for 30 min in a 60 °C stove, seal plate additionally by surrounding with Parafilm (*see* **Note 9**).
7. Remove the plates (PVDF filter plate on washing plate) from the incubation box and shake on a horizontal shaker for 5 min.
8. Next, remove non-bound material by centrifugation for 1 min at 500 × *g* (*see* **Note 10**).
9. Wash the membrane with 200 μL mQ by shaking 3 min on a horizontal shaker prior to centrifugation for 1 min at 500 × *g*. Repeat this washing step once more, followed by washing with 200 μL 100 mM $NaHCO_3$ through centrifugation for 1 min at 500 × *g* (*see* **Note 10**).
10. Remove residual solvent from the bottom of the filter by gently dipping on a Kimberly-Clark tissue (*see* **Note 11**)
11. Place the PVDF filter plate on a clear 96-well collection plate.
12. For the *N*-glycans release, add 50 μL 100 mM $NaCHO_3$ and 1 μL (=1 mU) PNGase F to each well. Incubate samples for either 3 h or 6 h or overnight (O/N) at 37 °C in a moistured incubator box to prevent evaporation (*see* **Note 12**).
13. After the incubation with PNGase F, recover glycans by centrifuging 2 min at 1000 × *g* (*see* **Note 13**).
14. Store samples at −20 °C or use directly for further processing.

3.3 Sialic Acid Derivatization

Sialic acid derivatization using ethyl esterification was adapted from Reiding *et al.* [4] and is explained in detail for human plasma *N*-glycans in [7].

1. If stored at −20 °C, thaw samples on a horizontal shaker.
2. Prepare polypropylene 96-deepwell plate with 100 μL ethylation reagent using a multistep pipette.
3. Mix glycan-containing sample. Pipet 20 μL of sample carefully into ethylation reagent using a multichannel pipette (*see* **Note 14**). Do not mix!
4. Seal with solvent resistant capmat or sticker to prevent evaporation. Beware of potential contamination by plastic sticker.
5. Incubate for 1 h in a 37 °C oven.

6. Add 100 μL 100% ACN to each well using a multistep pipette.
7. Incubate for 15 min at −20 °C to precipitate possibly remaining proteins (*see* **Note 15**).

3.4 Micro-HILIC SPE Purification

Micro-HILIC SPE purification of glycans was modified from Selman *et al.* [5]:

1. Prepare 4 mm cotton-thread tips by cutting cotton threads in 4 mm pieces using a clean scalpel and a clean glass petri dish. Insert single threads into 20 μL pipette tips. Apply 50 kPa air pressure to push cotton thread in the tips (*see* **Note 16**).
2. Prepare a 96-well PCR plate with 10 μL mQ water per well for the elution using a multistep or 12-channel pipette.
3. Pour mQ water, 85% ACN, and 85% ACN/1% TFA separately into clean buckets/containers suited for the use of 12-channel pipettes. Label containers with the containing solution and cover ACN-containing buckets.
4. Remove samples from the freezer and let them reach room temperature.
5. Precondition 12 cotton-tips by pipetting 3×20 μL mQ water using a 2–20 μL multichannel pipette, followed by 3×20 μL 85% ACN.
6. Load sample by pipetting carefully up and down 50-times. Do not pipette precipitate.
7. Wash unbound material by pipetting 3×20 μL 85% ACN/1% TFA, followed by 3×20 μL 85% ACN.
8. Elute glycans in 10 μL mQ water by pipetting up and down 15-times. Make sure all liquid is retrieved from the pipette tip.
9. Store purified samples at −20 °C or use directly for MALDI-TOF-MS analysis.

3.5 MALDI-TOF-MS(/MS)

1. Spot 5 μL of sample from the HILIC-SPE elution onto an AnchorChip target plate and let dry at room temperature (ca. 2–3 h). Vortex matrix and co-crystallize dried samples with 0.5 μL matrix (5 mg/mL superDHB/50%ACN/1 mM NaOH).
2. Spot 0.5 μL peptide calibration standard with 0.5 μL matrix onto in-between spots of the AnchorChip plate.
3. Record MALDI-TOF-MS spectra in positive reflectron mode on a Bruker UltrafleXtreme mass spectrometer over a range of *m*/*z* 900–5000 using random walk across 200 μm diameter, 200 shots per raster, and accumulation of a total of 10,000 shot at 1000 Hz frequency. An exemplary mass spectrum is shown in Fig. 1 (*see* **Note 17**).

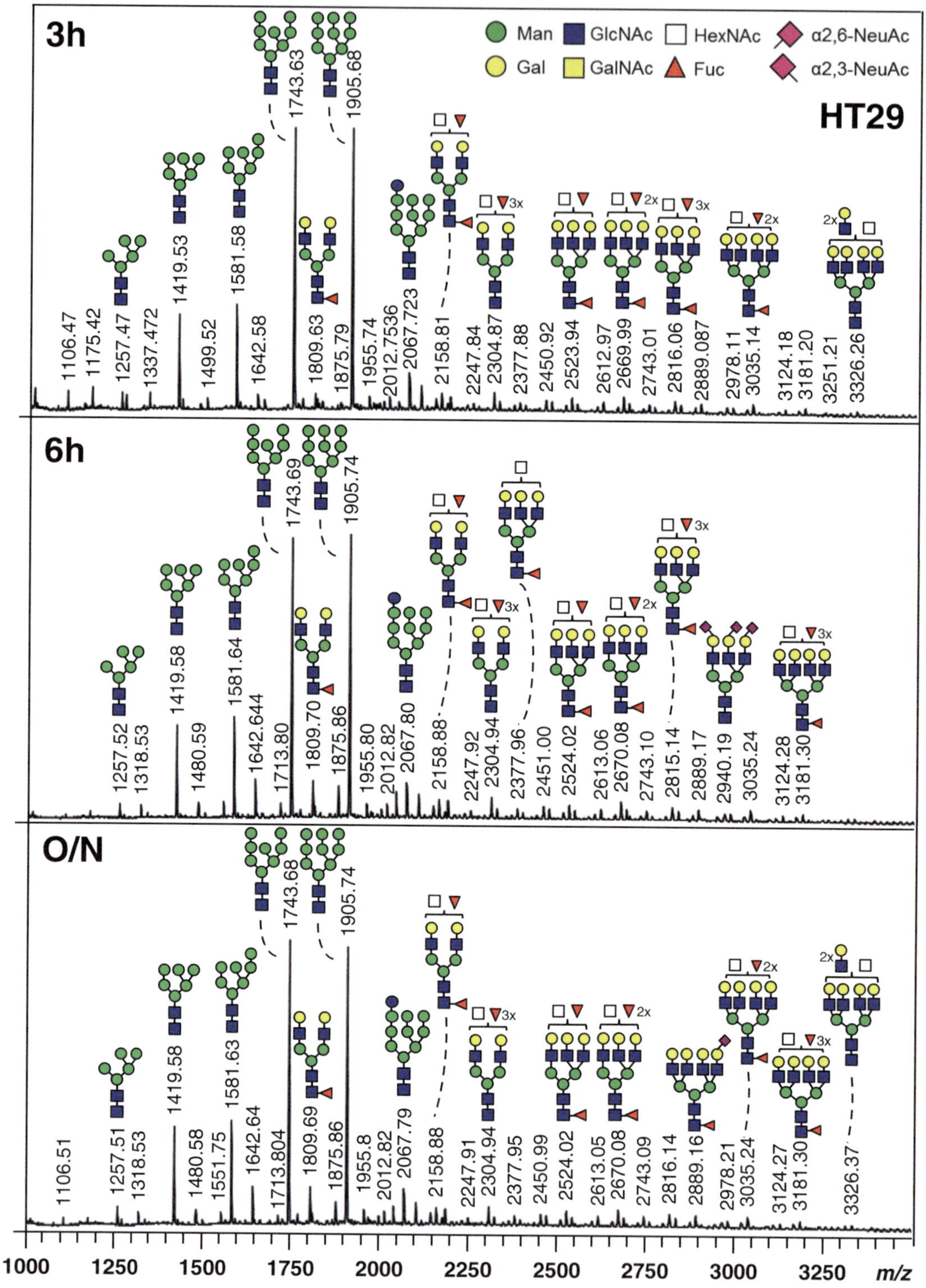

Fig. 1 Exemplary mass spectra of *N*-glycans released from colorectal cancer cell line HT29 by 3 h, 6 h, or O/N digestion with PNGase F. To ensure a complete glycan release and good quality vs. high-throughput, we tested different PNGase F incubation times (3 h vs. 6 h vs. O/N) using the protocol described above. Released, derivatized, and purified *N*-glycans were measured by MALDI-TOF-MS on a Bruker UltrafleXtreme mass spectrometer in positive reflectron mode. Spectra are shown in relative intensity and main peaks are annotated with glycan compositions. The presence of isomers cannot be excluded. Notably, the spectra do not differ largely between the three incubation times. The average absolute intensities of mass spectra, however, increased from 3 h to O/N incubation (not shown) which is advantageously for the analysis of low sample amounts

4 Notes

1. It is important to dissolve the chemicals completely before mixing them together.
2. The amount of PBS to resuspend the cells can vary depending on the amount of cells or on the size of the cell pellet.
3. The supernatant needs to be removed carefully to prevent sample loss. For some cells $1000 \times g$ may not be enough to result in a pellet, increase g-power accordingly.
4. It is essential to dissolve and disrupt the cell pellet properly. This leads to the presence of also intracellular precursors which increases the abundance of high-mannose type *N*-glycans. Other homogenizing methods may be used, e.g., the MPBio FastPrep with Matrix Z beads, which is suited for higher volumes. Different cell amounts and dilution volumes can be used, but including the denaturation mix should not exceed 200 μL/well.
5. Instead of human plasma any other standard protein(s) can be used.
6. It is crucial that the membrane from this moment on does not completely dry out.
7. To prevent leaking of the sample, no droplet should be left on the bottom side of the filter. The washing plates can be reused.
8. The cell amount loaded per well can be varied between 250,000 and 1 million cells, depending on the cell size—overloading will cause clogging of the filter. Chose for a randomized plate design, especially for cohorts. Sample and denaturation solution can as well be mixed prior to loading onto the filter plate.
9. In case different sample volumes are used, adapt denaturation buffer volume to keep correct concentrations. During the incubation it is important to prevent evaporation. For that any other suited device can be used.
10. In case residual solution remains on the filter, carefully pipette the solution out of the wells and discard, do not damage the membrane. Empty washing plate each time or use a 96-deep well plate.
11. To prevent leaking of the sample, no droplet should be left on the bottom side of the filter. Take care that the PVDF-filter plate and the collection plate are aligned to prevent mixing up of samples.
12. The buffer and PNGase F can be pre-premixed, but thorough mixing is essential to obtain an uniform solution. The incubation time with PNGase F can be varied, but needs to be evaluated for different samples. *See* the influence of incubation time discussed below (Figs. 1 and 2).

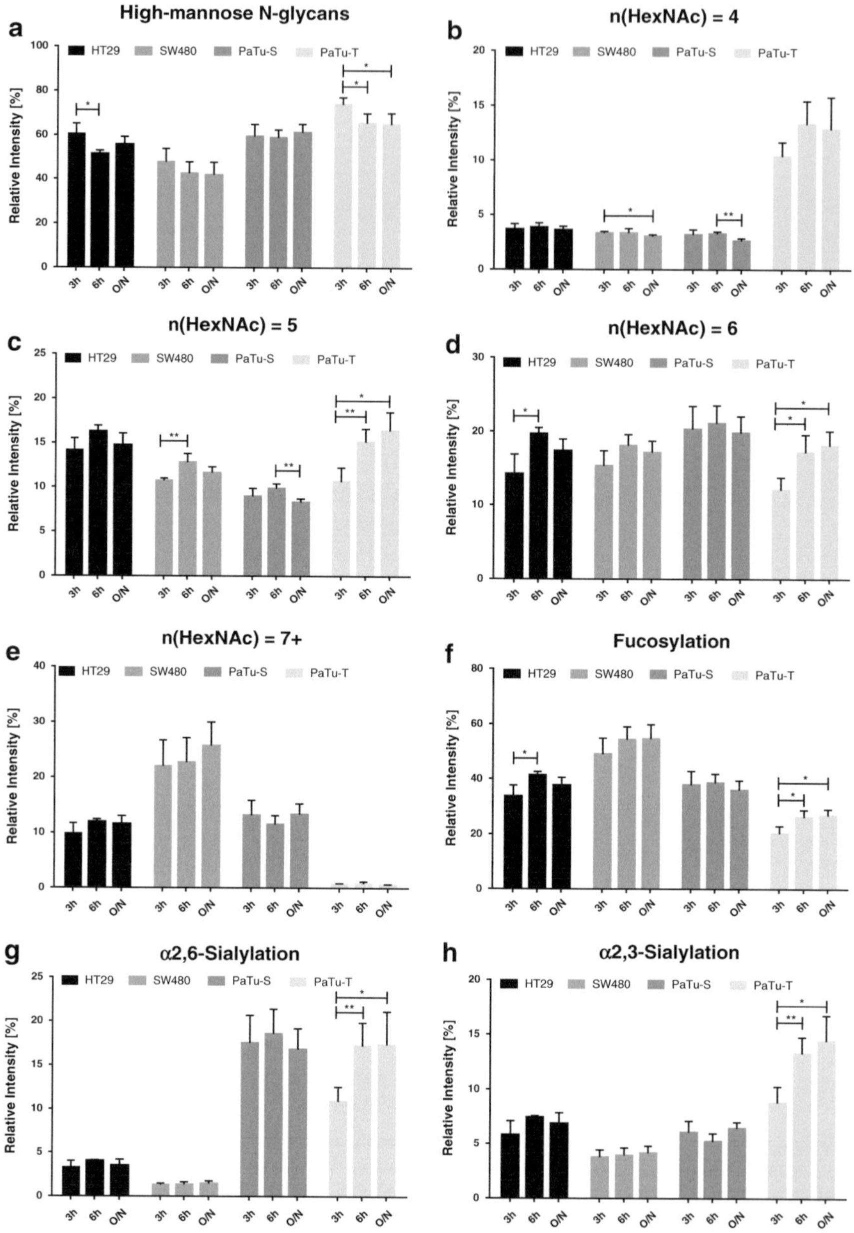

Fig. 2 Relative abundances of derived *N*-glycan traits after different PNGase F incubation times. Data from the MALDI-TOF-MS measurements was processed and analyzed as described elsewhere [8]. Shortly, an average sum spectrum over all obtained MS-spectra was generated in order to define a composition list for targeted data extraction of each spectrum using the in-house developed software Massy Tools V. 0.1.8.0 [9]. Spectra were smoothed, baseline subtracted, and internally recalibrated. Background was dynamically detected and subtracted. Data was revised based on several quality parameters for analytes (ppm error, signal/noise, quality score) as well as spectra (total intensity, fraction of analytes above S/N 6). Revised data was rescaled to 100% per spectrum and derived glycosylation traits were calculated by summing relative intensities of glycans corresponding to a trait. A *t*-test was performed to evaluate significant differences between each incubation time

13. In case of filter clogging, collect also possible residual solution from the filter.
14. Be careful when handling ethanol—always cover the ethylation reagent to prevent evaporation.
15. The incubation with ACN at −20 °C can be prolonged if needed.
16. Alternatively, cotton from cotton pads can be used to pack 200 μL pipette tips to increase the working/washing volume to 150 μL, elution can be performed in 10 μL to 30 μL mQ water.
17. Other MALDI target plates as well as mass spectrometers can be used.

Acknowledgements

We thank B. C. Jansen and K. R. Reiding for in-house developed Python scripts and data analysis support. This work was supported by the European Union (Seventh Framework Programme HighGlycan project, grant number: 278535).

References

1. Drake RR, Ball L (2015) Glycosylation and cancer, vol 126. Advances in cancer research. Elsevier, London, UK. doi:10.1016/S0065-230X(15)00022-6
2. Maverakis E, Kim K, Shimoda M, Gershwin ME, Patel F, Wilken R, Raychaudhuri S, Ruhaak LR, Lebrilla CB (2015) Glycans in the immune system and The Altered Glycan Theory of Autoimmunity: a critical review. J Autoimmun 57:1–13. doi:10.1016/j.jaut.2014.12.002
3. Burnina I, Hoyt E, Lynaugh H, Li H, Gong B (2013) A cost-effective plate-based sample preparation for antibody N-glycan analysis. J Chromatogr A 1307:201–206. doi:10.1016/j.chroma.2013.07.104
4. Reiding KR, Blank D, Kuijper DM, Deelder AM, Wuhrer M (2014) High-throughput profiling of protein N-glycosylation by MALDI-TOF-MS employing linkage-specific sialic acid esterification. Anal Chem 86(12):5784–5793. doi:10.1021/ac500335t
5. Selman MH, Hemayatkar M, Deelder AM, Wuhrer M (2011) Cotton HILIC SPE microtips for microscale purification and enrichment of glycans and glycopeptides. Anal Chem 83(7):2492–2499. doi:10.1021/ac1027116

Fig. 2 (continued) (3 h vs. 6 h, 3 h vs. O/N, 6 h vs O/N). Bar graphs were generated in GraphPad Prism Version 6.02 (GraphPad Software, Inc., La Jolla, CA) and show the means with standard deviation as error bars. Significance is displayed as **p*-value < 0.05, ***p*-value < 0.01. The relative abundances of the calculated traits show only minor differences between 3 h of incubation with PNGase F and overnight incubation for cell lines HT29, SW480, and PaTu-S, making this protocol suited to perform within 1 day. The cell line PaTu-T, however, shows more variation and an overnight *N*-glycan release is recommended. When applying this method to a novel cell type, the optimal incubation time of PNGase F to obtain complete digestion should be investigated for the specific sample used

6. Belo AI, van der Sar AM, Tefsen B, van Die I (2013) Galectin-4 reduces migration and metastasis formation of pancreatic cancer cells. PLoS One 8(6), e65957. doi:10.1371/journal.pone.0065957
7. Reiding KR, Lonardi E, Hipgrave Ederveen AL, Wuhrer M (2016) Ethyl esterification for MALDI-MS analysis of protein glycosylation. Methods Mol Biol 1394:151–162. doi:10.1007/978-1-4939-3341-9_11.
8. Holst S, Deuss AJM, van Pelt GW, van Vliet SJ, Garcia-Vallejo JJ, Koeleman CAM, Deelder AM, Mesker WE, Tollenaar RA, Rombouts Y, Wuhrer M (2016) N-glycosylation profiling of colorectal cancer cell lines reveals association of fucosylation with differentiation and CDX1/villin mRNA expression. Mol Cell Proteomics 15(1):124–140. doi:10.1074/mcp.M115.051235
9. Jansen BC, Reiding KR, Bondt A, Hipgrave Ederveen AL, Palmblad M, Falck D, Wuhrer M (2015) MassyTools: a high throughput targeted data processing tool for relative quantitation and quality control developed for glycomic and glycoproteomic MALDI-MS. J Proteome Res 14(12):5088–5098

Chapter 15

Analysis of *N*-Glycosylation of Total Membrane Proteins

Tamara Pavić and Olga Gornik

Abstract

Glycosylation of membrane proteins plays a crucial role in various physiological events, including intercellular recognition and intermolecular interactions on the cell surface (Gornik et al., Biochim Biophys Acta 1820:1318–1326, 2012). To study composition and function of *N*-glycans on membrane proteins one has to have an efficient and reproducible analytical method, which includes protein extraction and analysis of glycans. In this chapter we provide an analytical approach that includes cloud-point extraction (CPE) of total membrane proteins with the non-ionic detergent Triton X-114 and subsequent analysis of their *N*-glycans using hydrophilic interaction liquid chromatography (HILIC)–UPLC/HPLC. The protocol presented here can be used for parallel analysis of both membrane and intracellular proteins.

Key words *N*-Glycans, Membrane proteins, Cloud point extraction, Triton X-114, Protein enrichment, HILIC-UPLC

1 Introduction

Due to the crucial role of glycans attached to membrane proteins in many physiological processes [1], such as cell–cell communication [2], development and differentiation [3], immunity and infection, the analysis of glycosylation of membrane proteins has become an important topic. Many different analytical approaches, including different protein preparation methods, have been used for this purpose, but when it comes to the analysis of total membrane proteins, they all lack good reproducibility and reliability [4]. The usual approach for studying cell glycosylation includes homogenization of the total cell. This approach offers the total glycan profile of the cell, but also includes glycans from soluble glycoproteins, as well as glycans from proteins whose glycosylation process did not finish yet. Thus, it does not result with the reliable and objective situation from which further conclusions on the role and importance of certain glycan structures can be deduced.

Gordan Lauc and Manfred Wuhrer (eds.), *High-Throughput Glycomics and Glycoproteomics: Methods and Protocols*, Methods in Molecular Biology, vol. 1503, DOI 10.1007/978-1-4939-6493-2_15, © Springer Science+Business Media New York 2017

Membrane proteins show low abundance and poor solubility in aqueous solutions, so their enrichment must be carried out prior to glycan analysis [5, 6]. In this chapter we present the procedure based on cloud-point extraction (CPE) of total membrane proteins using detergent Triton X-114 and subsequent analysis of their *N*-glycans using hydrophilic interaction—ultra performance liquid chromatography (HILIC-UPLC/HPLC) of fluorescently labeled molecules.

CPE is a detergent based enrichment method for hydrophobic analytes [5, 7]. It is based on the phenomenon that, at concentrations above the critical micellar concentration (CMC), non-ionic polyoxyethylene detergents form clear micellar solutions in water. At low concentrations detergent binds to the cell membrane by partitioning into the lipid bilayer, which is then disrupted and lysed upon the increase of detergent concentration above CMC. With the increase in temperature, the micellar molecular weight also increases and the solution turns turbid, it reaches its "cloud point." At this point the solution separates into two different phases: detergent-rich phase and aqueous phase [7]. In this way, membrane proteins are relatively easily separated and enriched from hydrophilic proteins in biological matrices. Triton X-114 is a member of non-ionic polyoxyethylene detergent family, containing seven to eight oxyethylene units condensed on the hydrophobic octylphenyl residue. Solution of Triton X-114 reaches its cloud point at about 20 °C, thus fractioned proteins are not degraded or denaturated in any way [7].

HILIC-UPLC is reliable and robust, semiquantitative method, optimized for the analysis of glycans [8]. Since glycome, of either protein mixture or a single protein, consists of larger number of different and structurally complex glycans, their analysis requires the method which can offer good separation based on their physical and chemical properties.

2 Materials

Prepare all solutions using ultrapure water (prepared by purifying deionised water to attain a sensitivity of 18 MΩ cm at 25 °C) and analytical grade reagents. Prepare and store all reagents at room temperature (unless indicated otherwise). Diligently follow all waste disposal regulations when disposing waste materials.

2.1 Protein Extraction

1. Phosphate-buffered saline (PBS): 137 mM NaCl, 2.7 mM KCl, 10 mM Na_2HPO_4, 1.8 mM KH_2PO_4 pH 7.4. To prepare 1 L of 10× PBS dissolve 80 g NaCl, 2 g KCl, 14.4 g Na_2HPO_4, and 2.4 g KH_2PO_4 in 800 mL of H_2O. Adjust the pH to 7.4 with HCl and then add H_2O to 1 L. Store 10× PBS at room temperature. From this stock solution you can easily prepare

5× PBS by dissolving the 10× stock twice (e.g., to 100 mL of 10× PBS add 100 mL of water) or 1× PBS (e.g., to 100 mL of 10× PBS add 900 mL of water).

2. Tris–HCl stock solution: 1 M Tris–HCl, pH 7.4. Weigh 12.11 g Tris, dissolve in 70 mL of water. Set the pH to 7.4 using HCl and make up to 100 mL with water.
3. NaCl solution: 1.5 M NaCl. Weigh 8.77 g NaCl and dissolve in 100 mL of water.
4. EDTA solution: 0.1 M EDTA. Weigh 3.72 g EDTA (EDTA-$Na_2 \times 2H_2O$) and dissolve in 100 mL of water.
5. Triton X-114 stock solution: around 10% Triton X-114. Commercially available Triton is not pure enough for the analytical process and thus must be subjected to the following procedure. Dissolve 10 g of Triton X-114 in 490 mL of 10 mM Tris–HCl pH 7.4, 150 mM NaCl (make by mixing 5 mL 1 M Tris–HCl, 50 mL 1.5 M NaCl, and 435 mL water). Keep the mixture at 4 °C until complete dissolution of Triton X-114 and incubate clear solution at 37 °C. Condensation of the detergent will occur and the mixture will separate overnight into a large aqueous phase (poor in detergent) and a smaller phase enriched in detergent. Discard the aqueous phase and replace it with the new 490 mL of 10 mM Tris–HCl pH 7.4, 150 mM NaCl. Repeat the whole procedure twice and use the third detergent phase as Triton X-114 stock solution (concentration of Triton X-114 in this detergent-rich phase is now around 10%).
6. Triton lysis buffer: 10 mM Tris–HCl pH 7.4, 150 mM NaCl, 1 mM EDTA, 1% (v/v) Triton X-114 in PBS. To make 10 mL of Triton lysis buffer mix 0.1 mL 1 M Tris–HCl pH 7.4, 1 mL 1.5 M NaCl, 0.1 mL 0.1 M EDTA, 1 mL Triton X-114 stock solution, and PBS up to 10 mL. At the end add one tablet of protease inhibitor cocktail (*see* **Note 1**).
7. Sucrose cushion solution: 6% (w/v) sucrose, 10 mM Tris–HCl pH 7.4, 150 mM NaCl and 0.06% Triton X-114. To prepare 10 mL of sucrose cushion solution weigh 0.6 g sucrose and dissolve it in a mixture of 0.1 mL 1 M Tris–HCl, 1 mL 1.5 M NaCl, 0.06 mL Triton X-114 stock solution, and PBS up to 10 mL.

2.2 Glycan Analysis

1. SDS solution: 1.33% SDS (w/v). Weigh 1.33 g SDS and dissolve in 100 mL water (*see* **Note 2**).
2. Igepal-CA630 solution: 4% Igepal-CA630 (v/v). Take 4 mL of Igepal-CA630 and dissolve in 96 mL of water (*see* **Note 3**).
3. Ethanol: 70% ethanol. Take 300 mL of 95% ethanol and add 100 mL of water. By this you will get 400 mL of 70% ethanol.

4. Acetonitrile: 96% acetonitrile. To prepare 100 mL of 96% acetonitrile take 96 mL of 100% acetonitrile and add 4 mL of water. Store at 4 °C.
5. PNGase F: Peptide *N*-glycosidase F (Promega, 10 U/μL). Store at 4 °C.
6. Glycan labeling mixture: 2-aminobenzamide (19.2 mg/mL) and 2-picoline borane (44.8 mg/mL) in DMSO and acetic acid 30:70, v/v. This labeling mixture *must always be prepared fresh*! Calculate the number of samples and prepare adequate volume of labeling mixture. Amounts given here are for one sample (*see* **Note 4**). Per sample, dissolve 0.48 mg of 2-aminobenzamide and 1.12 mg of picoline-borane in 25 μL 30% glacial acetic acid and DMSO (85:15, v/v; this you prepare by mixing 7.5 μL of acetic acid with 17.5 μL of DMSO).
7. Solvent A stock: 2 M ammonium formate, pH 4.4. To prepare 1 L add 77 mL of formic acid to 600 mL of water. Adjust pH *on the ice block*, using magnetic stirrer by adding ammonium and waiting for the temperature to decrease to 22 °C. At the end, adjust the final volume with water to 1 L.

   To be used in chromatographic analysis, the working solution must be prepared (from the Stock A)!

   To prepare working solution for UPLC, you must mix 50 mL of Stock A with 950 mL of water.

   To prepare working solution for HPLC, you must mix 25 mL of Stock A with 975 mL of water.
8. Solvent B: LC-MS grade 100% acetonitrile.

3 Methods

Carry out all procedures at room temperature unless specified otherwise. Schematic work flow is given in Fig. 1.

3.1 Isolation of Membrane Proteins Using Triton X-114

3.1.1 Triton X-114 Phase Partitioning

1. Homogenize cca 4×10^6 cells in the presence of 1 mL of Triton lysis buffer and protease inhibitor cocktail (*see* **Note 5**), to prevent the protein degradation, using ultrasonic processor (4 cycles, 15–20 s) (*see* **Note 6**). Incubate samples at 4 °C for 1 h and clarify cell lysate by centrifugation (30 min, 10,000 × *g*, 4 °C) (*see* **Note 7**).
2. Overlay the clear supernatant on 200 μL of sucrose cushion and incubate at 37 °C during 20 min. Clouding of the solution will occur (*see* **Note 8**). Centrifuge samples for 3 min at 400 × *g*, 37 °C. Two phases (detergent-rich and aqueous, detergent-poor) will be formed.

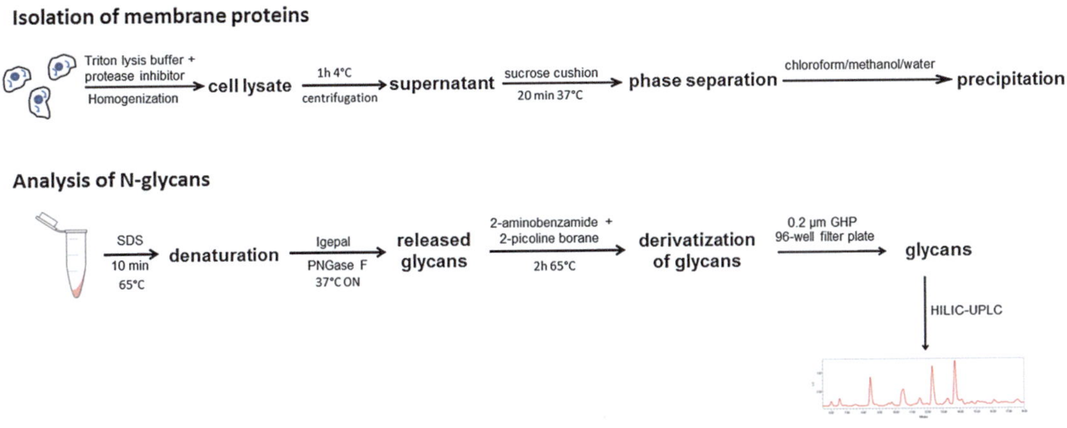

Fig. 1 Work flowchart representing the steps of the enrichment of total membrane proteins and HILIC (hydrophilic interaction liquid chromatography)-UPLC analysis of their *N*-glycans

3. Transfer the aqueous phase to a fresh tube and keep it on ice (*see* **Note 9**).
4. Resuspend the detergent phase in 500 μL of cold PBS (*see* **Note 10**) and repeat the phase separation once again. Pool this aqueous phase with initial one and re-extract by adding 50 μL of Triton stock solution to perform the phase separation as described previously (*see* **Note 11**).

3.1.2 Protein Precipitation

Since presence of Triton X-114 interferes with enzymatic deglycosylation with *N*-glycosidase F, an amidase, also known as PNGase F, which cleaves between innermost GlcNAc of *N*-glycan and asparagine residue on protein, it must be removed prior to glycan analysis. The best method is precipitation with organic solvents. The method of choice, which is described below, is the chloroform–methanol–water precipitation (*see* **Note 12**).

1. Add four times the sample volume of methanol to each sample (prepared as described in Subheading 3.1.1) and vortex well.
2. Sequentially, add twice the initial sample volume of chloroform to each sample and vortex well.
3. At the end, add three times of the initial sample volume of water, vortex vigorously and centrifuge at 9000 × *g* for 1 min at 4 °C. After centrifugation the proteins are located in the liquid interphase.
4. Remove the aqueous top layer, add additional three volumes of methanol, vortex samples and centrifuge again for 2 min at 9000 × *g*, 4 °C to pellet the proteins.
5. Remove the supernatant as much as possible, without disturbing the precipitate, and leave samples to air-dry (*see* **Note 13**).

3.2 Analysis of N-linked Glycans Using HILIC-UPLC or HPLC

3.2.1 Glycan Release and Labeling

1. Resuspend pellets by adding 30 μL of 1.33% SDS (w/v) and incubate at 65 °C for 10 min.
2. Subsequently, add 10 μL of 4% Igepal-CA630 and 1.25 mU of PNGase F in 10 μL of 5× PBS to each sample. Incubate the samples overnight at 37 °C for *N*-glycan release.
3. Now prepare labeling mixture as indicated in Subheading 2. Always prepare this mixture FRESH (*see* **Note 14**).
4. Add 25 μL of labeling mixture to each sample, followed by 2 h incubation at 65 °C.
5. Add 700 μL of ACN to each sample in order to bring it to 96% ACN (v/v) (*see* **Note 15**).
6. Prepare the 0.2 μm GHP 96-well filter plate! For this you need to prewash the plate with 200 μL of 70% ethanol (v/v) and then 200 μL of water, followed by equilibration with 200 μL of 96% ACN (v/v) (*see* **Note 16**).
7. Transfer the samples to the prepared plate by using multichannel pipette. Remove the solvent using vacuum (*see* **Note 17**).
8. Wash loaded samples 5× with 200 μL of 96% ACN (v/v) (*see* **Note 18**) following the same procedure for loading and removing the solution as described in Subheading 2 (preparation of the filter plate on vacuum manifold). Elute glycans with 2×90 μL water and store at −20 °C until chromatographic analysis.

3.2.2 Hydrophilic Interaction Liquid Chromatography (HILIC)-UPLC/HPLC

Fluorescently labeled *N*-glycans can then be separated by hydrophilic interaction chromatography on an UPLC (or a HPLC) instrument with fluorescence detector set with excitation and emission wavelengths of 250 and 420 nm, respectively.

If using the UPLC, labeled *N*-glycans can be separated on a BEH Glycan chromatography column, 150×2.1 mm i.d., 1.7 μm BEH particles, with 100 mM ammonium formate, pH 4.4, as a solvent A and acetonitrile as a solvent B. Separation method uses linear gradient of 75–53% acetonitrile (v/v) at flow rate of 0.561 mL/min in a 25 min analytical run. Before injection maintain samples at 5 °C, while the separation temperature is 25 °C. Integrate the chromatographic areas (peaks). The amount of glycans in each peak can be expressed as percentage of total integrated area (*see* **Note 19**). *See* Fig. 2 for an example of the resulting chromatogram.

If using HPLC, labeled *N*-glycans can be separated on a 250 mm i.d. 5 mm particle packed TSKgel Amide 80 column at 30 °C with 50 mM ammonium formate, pH 4.4, as a solvent A and acetonitrile as a solvent B. Separation method uses linear gradient of 65–53% acetonitrile (v/v) at flow rate of 0.8 mL/min in a 60 min analytical run (*see* **Note 19**).

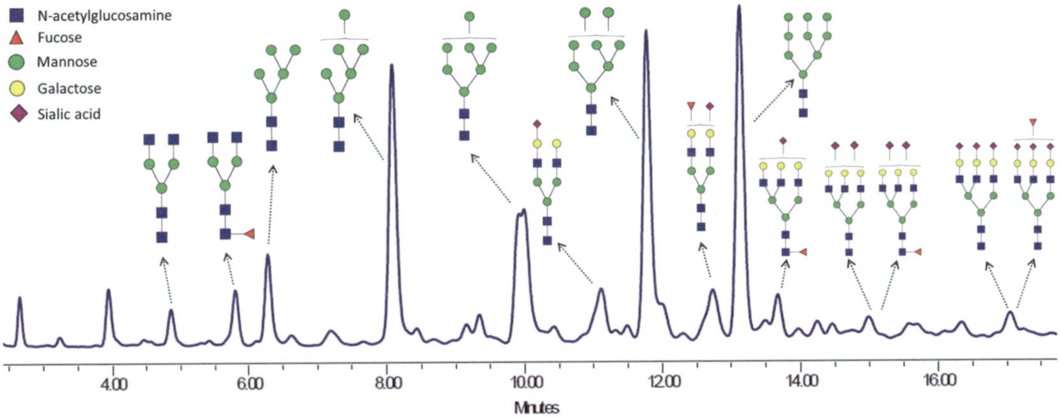

Fig. 2 Example of the HILIC (hydrophilic interaction liquid chromatography)-UPLC chromatogram of *N*-glycans released from total membrane proteins of THP-1 cells. Proteins were extracted using Triton X-114 based protocol described in this chapter

4 Notes

1. In described preparation protocol used protease inhibitor cocktail was in tablet form and sufficient for 10 mL of solution (purchased from Roche Diagnostics, Mannheim, Germany). Please note that other manufacturers may require different preparation protocols for their inhibitors. In this particular case, depending on the number of samples, i.e., on the volume of prepared Triton lysis buffer, you can add a whole tablet of protease inhibitor cocktail (sufficient for 10 mL of solution) or make a stock solution by dissolving a whole tablet in 1.5 mL of PBS (7× concentrated) and use appropriate volume. Stock solution should be kept at −20 °C, and it is stable for at least 12 weeks.
2. When making SDS solution you can raise the temperature on a magnetic stirrer to for example 60 °C to help dissolving. SDS solution should be kept at room temperature or at 37 °C, since the detergent precipitates out of the solution at 4 °C.
3. We recommed cutting off the pipette tip end to make the orifice wider when pipetting Igepal-CA630. To help dissolving you can put the mixture at 65 °C. Keep the prepared solution at 4 °C.
4. When calculating the amounts needed for your sample number, calculate for ten samples more than the actual number to account for the pipetting errors.
5. Triton lysis buffer should be cooled down to 4 °C to ensure the detergent is in its soluble state.
6. Keep your samples on ice during ultrasonic processing to prevent overheating.

7. Cool down the centrifuge to 4 °C prior sample centrifugation.
8. If clouding of the solution doesn't occur after 20 min, leave the samples for additional 10–15 min at 37 °C.
9. While transferring the aqueous phase to the fresh tube be careful not to disturb the detergent phase. It is better to leave some of the aqueous phase on top of the detergent phase, since the latter will be resuspended in cold PBS and phase separation will be performed once more.
10. Be sure to keep your PBS solution cold (store it at 4 °C), since the detergent-rich phase could only be resuspended in cold PBS.
11. When pipetting Triton stock solution cut off the end of the pipette tip to make the orifice wider, since the stock solution is very viscous.
12. All organic solvents should be cooled down to at least 4 °C before use. It is highly recommendable to carry out the procedure of protein precipitation on ice, in order to decrease the temperature and therefore achieving better protein precipitation.
13. Alternatively, if a vacuum concentrator is available, to speed up the process, you can dry your samples in it instead of air-drying.
14. It is extremely important to always prepare fresh labeling mixture. Once prepared, the mixture should be stable for an hour. Be sure to prepare it in the fume hood.
15. Allow your samples to cool down before adding the acetonitrile (approximately 30 min). Also, keep the acetonitrile at 4 °C before you add it to the samples. By doing so you decrease the evaporation of volatile acetonitrile and ensure that your samples are brought to 96% ACN (v/v).
16. This is done in a way that the filter plate is put on the vacuum manifold (according device instructions), solvents are then added sequentially to each well using multichannel pipette and then removed using vacuum. The second solvent is added when the first one is removed etc.
17. At this point be very diligent with the pressure, i.e., vacuum should not exceed 2 inHg and you should stop it as soon as filter plate membrane becomes dry.
18. 96% ACN (v/v) should also be cooled down to 4 °C prior to use.
19. The chromatography system can be calibrated using an external standard of hydrolyzed and 2-AB-labeled glucose oligomers from which the retention times for the individual glycans can be converted to glucose units (GU). Glycans can be then analyzed on the basis of their elution positions and measured in glucose units, which can be compared to reference values in NIBRT's "GlycoBase v3.2" database available at http://glycobase.nibrt.ie for structure assignment.

References

1. Gornik O, Pavić T, Lauc G (2012) Alternative glycosylation modulates function of IgG and other proteins - Implications on evolution and disease. Biochim Biophys Acta 1820(9):1318–1326. doi:10.1016/j.bbagen.2011.12.004
2. Taniguchi N, Miyoshi E, Gu J, Honke K, Matsumoto A (2006) Decoding sugar functions by identifying target glycoproteins. Curr Opin Struct Biol 16(5):561–566. doi:10.1016/j.sbi.2006.08.011
3. Haltiwanger RS, Lowe JB (2004) Role of glycosylation in development. Annu Rev Biochem 73:491–537. doi:10.1146/annurev.biochem.73.011303.074043
4. Ito H et al (2016) Comparison of analytical methods for profiling N- and O-linked glycans from cultured cell lines: HUPO Human Disease Glycomics/Proteome Initiative multi-institutional study. GlycoconjJ33:405–415.doi:10.1007/s10719-015-9625-3
5. Shevchenko G, Sjo MOD, Malmstro D, Wetterhall M, Bergquist J (2010) Cloud-point extraction and delipidation of porcine brain proteins in combination with bottom-up mass spectrometry approaches for proteome analysis. J Proteome Res 9(8):3903–3911. doi:10.1021/pr100116k
6. Donoghue PM, Hughes C, Vissers JPC, Langridge JI, Dunn MJ (2008) Nonionic detergent phase extraction for the proteomic analysis of heart membrane proteins using label-free LC-MS. Proteomics 8(18):3895–3905. doi:10.1002/pmic.200800116
7. Bordier C (1981) Phase separation of integral membrane proteins in Triton X-114 solution. J Biol Chem 256(4):1604–1607
8. Huffman JE et al (2014) Comparative performance of four methods for high-throughput glycosylation analysis of immunoglobulin G in genetic and epidemiological research. Mol Cell Proteomics 13(6):1598–1610. doi:10.1074/mcp.M113.037465

Chapter 16

HILIC-UPLC Analysis of Brain Tissue *N*-Glycans

Thomas Klarić and Ivan Gudelj

Abstract

Extraction of *N*-glycans from intact tissue presents a unique set of challenges which makes it a relatively laborious and time-consuming process in comparison to other sample types, such as plasma. Here we present an approach designed for the extraction, purification, and labeling of free *N*-glycans from brain tissue. Using this method, up to 16 samples can be processed at once which translates to an output of 48 samples per week when rounds of extraction are staggered. Moreover, although intended for brain tissue, the method could easily be adapted to other tissue types as well. The protocol involves several stages. First, the tissue is homogenized and total proteins are isolated using chloroform–methanol extraction. The proteins are then deglycosylated using the Peptide *N*-Glycosidase F (PNGase F) enzyme in a reaction lasting two days. The released *N*-glycans are subsequently cleaned up from the reaction mixture using a centrifugal filter device and dried overnight. Next, the *N*-glycans are resuspended, labeled with 2-aminobenzamide (2-AB) and once again cleaned up using a filter plate. The purified *N*-glycans are released from the filter using ultrapure water and are then ready for analysis by for hydrophilic interaction ultra performance liquid chromatography (HILIC-UPLC).

Key words Brain tissue, *N*-glycans, Glycomics, Glycosylation, HILIC-UPLC, 2-aminobenzamide

1 Introduction

The field of glycomics is rapidly expanding [1] and high-throughput analysis of large numbers of samples is becoming increasingly common [2]. Isolation of *N*-glycans from plasma is relatively straightforward and is routinely performed in a high-throughput format [3]. Indeed, methods for the preparation, labeling and analysis of *N*-glycans from plasma have been streamlined to allow the simultaneous processing of hundreds of samples. However, this is still not the case for *N*-glycan analysis of tissue samples which is why this approach is often limited to smaller-scale studies. Improving the efficiency and throughput of this method would enable these types of studies to be upscaled thereby increasing their output and statistical power. We have developed a method specifically designed for the extraction, purification and labeling of free *N*-glycans from brain tissue for HILIC-UPLC analysis. The first step in this method is

Gordan Lauc and Manfred Wuhrer (eds.), *High-Throughput Glycomics and Glycoproteomics: Methods and Protocols*, Methods in Molecular Biology, vol. 1503, DOI 10.1007/978-1-4939-6493-2_16, © Springer Science+Business Media New York 2017

tissue homogenization and total protein isolation using chloroform–methanol extraction [4], followed with 2-day PNGase F deglycosylation. The released *N*-glycans are subsequently purified using a centrifugal device, dried overnight and labeled with 2-AB. After labeling, clean up using a filter plate is performed and the purified *N*-glycans are then ready for HILIC-UPLC analysis. Sixteen samples can be processed with this method at once which, although still not at the level of high throughput analysis, represents a significant improvement upon existing methods [5–10] in terms of scaling and allows for the rapid processing of batches of samples which number in the hundreds. Moreover, when compared to the methods which use direct application of PNGase F on tissue [11] this method is more reproducible (average coefficient of variation: 3.93%), and therefore more convenient for comparison studies.

2 Materials

Prepare all solutions using ultrapure water (prepared by purifying deionized water to attain a sensitivity of 18 MΩ·cm at 25 °C) and analytical grade reagents. Prepare and store all reagents at room temperature (unless indicated otherwise). Diligently follow all waste disposal regulations when disposing of waste materials.

2.1 Tissue Homogenization and Protein Extraction

1. Cold methanol.
2. Cold chloroform.
3. Cold 50% methanol solution: 50% methanol (v/v) in ultrapure water.
4. Lysing matrix D (MP Biomedicals): 1.4 mm ceramic spheres (Vickers hardness 800).
5. Parafilm.
6. FastPrep®-24 Instrument (MP Biomedicals).
7. Ultrasonic water bath.
8. Savant SC210A SpeedVac vacuum concentrator (Thermo Scientific).

2.2 Deglycosylation and Cleanup

1. Phosphate buffered saline (PBS): Prepare 10× PBS stock by weighing NaCl (80.028 g), Na_2HPO_4 (13.832 g), KH_2PO_4 (2.964 g), and KCl (1.976 g). Take a 1 L glass beaker, fill the beaker with 800 mL of distilled water (dH_2O), add the chemicals, a magnetic stirring bar, cover the beaker with Parafilm, and put the beaker on a magnetic stirrer—stir until completely dissolved (*see* **Note 1**). When completely dissolved, transfer the unfinished buffer into a 1 L measuring flask, add dH_2O up to 1 L, cover the flask with Parafilm and shake the buffer by

holding the flask with both hands. Filter the buffer through a bottle top 0.2 μm Supor filter into a clean 1 L glass bottle. Check the pH of the buffer (pH 6.6–6.8). To make 1× PBS, add 100 mL of 10× PBS into a 100 mL measuring flask and transfer into a 1 L glass beaker. Add dH_2O up to 0.9 L. Adjust the pH to 7.4 with 1 M NaOH. Transfer the unfinished buffer into a 1 L measuring flask and add dH_2O up to 1 L. Filter the buffer through a bottle top 0.2 μm Supor filter into a clean 1 L glass bottle. Store at 4 °C.

2. 10% (w/v) sodium dodecyl sulfate (SDS) in ultrapure water: Store at 37 °C.
3. 2-mercaptoethanol (βME).
4. 10% (w/v) IGEPAL CA-630 (Sigma-Aldrich) in ultrapure water: Store at 4 °C.
5. 10% (w/v) sodium azide (NaN_3) in ultrapure water.
6. PNGase F (Promega, 10 U/μL): Store at 4 °C.
7. Parafilm.
8. Thermomixer comfort (Eppendorf).
9. Amicon® Ultra 2 mL Centrifugal Filter Devices with Ultracel-30 membrane (Merck Millipore): Nominal Molecular Weight Limit (NMWL) 30 kDa.
10. Savant SC210A SpeedVac vacuum concentrator (Thermo Scientific).

2.3 Labeling and Cleanup

1. 2-AB labeling solution: The amount required for one sample is 50 μL 30% acetic acid in DMSO, 0.96 mg 2-AB, and 2.24 mg 2-picoline borane (PB). Upscale as required (prepare a master mix with 10% extra so as not to run out). First prepare the 30% acetic acid in DMSO solution. Then weigh out the required mass of 2-AB and PB in separate tubes (*see* **Note 2**). Next, add the acetic acid–DMSO mixture to the 2-AB and mix well by vortexing until completely dissolved. Add the entire volume to the PB and mix well by vortexing until completely dissolved.
2. Cold 100% acetonitrile (ACN): Store at 4 °C.
3. Cold 96% (v/v) ACN in ultrapure water: Prepare fresh and store at 4 °C.
4. 70% (v/v) ethanol in ultrapure water: Prepare fresh.
5. 1 mL AcroPrep™ GHP membrane 96-well filter plates (Pall): 0.2 μm membrane.
6. 2 mL plate stand.
7. 96-well collection plate (200 μL well volume) and caps.
8. Vacuum manifold.

2.4 HILIC-UPLC Analysis

1. Solvent A: 100 mM ammonium formate, pH 4.4: To make a 2 M stock solution, add 300 mL of ice cold ultrapure water to a 500 mL glass beaker. Add 37.75 mL of formic acid to the ice cold water. Add a magnetic stirring bar to the glass beaker and stir on a magnetic stirrer. Insert an electrode of the pH meter into the glass beaker and check the pH (which should be <2). Slowly add cca. 75 mL of ammonia solution to the diluted formic acid until the pH reaches 4.2. Leave at room temperature overnight. Adjust the pH to 4.4 with ammonia solution. Transfer ammonium formate, pH 4.4, into a 500 mL volumetric flask and add ultrapure water to the final volume of 500 mL. Filter the prepared 2 M ammonium formate, pH 4.4, through a 0.2 μm PES membrane into a dark bottle and store at room temperature. To make a 100 mM working solution, transfer 50 mL of 2 M ammonium formate into a 1 L volumetric flask and add ultrapure water to the final volume of 1 L.
2. Solvent B: LC-MS grade 100% ACN.
3. Acquity UPLC Glycan BEH Amide column, 130 Å, 1.7 μm, 2.1 mm × 150 mm.
4. Waters Acquity UPLC H-class system consisting of a quaternary solvent manager, sample manager and a fluorescence (FLR) detector set with excitation and emission wavelengths of 250 and 428 nm, respectively.

3 Methods

Carry out all procedures at room temperature unless otherwise specified.

3.1 Preparation

1. Place ultrapure water, methanol, chloroform, and PBS in the refrigerator and allow to cool down to 4 °C (*see* **Note 3**).
2. Make a 50% (v/v) methanol in water solution by mixing equal volumes of ultrapure water and methanol. Make enough to process all of the samples in the batch. Typically, it takes approximately 7 mL of 50% methanol to process one sample. Store the solution at 4 °C.
3. Label your tubes.

3.2 Homogenization of Brain Tissue

1. This protocol works with fresh or fresh frozen (unfixed) brain tissue. Place the brain tissue into a 2 mL microfuge tube and add Lysing Matrix D (*see* **Note 4**). The quantity of lysing matrix used depends on the starting amount of tissue. Using approximately 5–6 beads per 25 mg of tissue is sufficient to achieve complete homogenization (*see* **Note 5**).

2. Add 1 mL of cold ultrapure water to the tube, close the lid and seal tightly with Parafilm.
3. Place the tubes in the FastPrep®-24 Instrument (making sure that the instrument is balanced) and homogenize the tissue using the following settings: Speed 6, Time 40 s. The solution should turn a pale milky color.
4. Transfer the supernatant (leaving the beads behind) into a clean 15 mL tube.
5. Rinse the microfuge tube with 1 mL cold methanol to collect remaining homogenate, vortex briefly and transfer the solution to the same 15 mL tube.
6. Rinse the microfuge tube once more with 0.75 mL cold methanol, vortex briefly and transfer the solution to the same 15 mL tube. Discard the microfuge tube with the beads.
7. Vortex the 15 mL tube, seal it with Parafilm and place it in an ultrasonic water bath (*see* **Note 6**).
8. Sonicate for 15 min.

3.3 Protein Extraction

1. Remove the tube from the water bath, remove the Parafilm, add 3.25 mL of cold chloroform to the tube and vortex well.
2. Tightly seal the lid with Parafilm and sonicate for 15 min.
3. Remove the Parafilm and centrifuge the tube at 4500 × *g* for 30 min. After centrifugation, separation of the phases should be apparent. The proteins are located at the interface of the aqueous phase (top) and the organic phase (bottom).
4. Carefully remove and discard the aqueous phase using a 1 mL pipette, being careful not to disturb the protein layer (*see* **Note 7**).
5. To the remaining organic phase, add an equal volume of cold 50% methanol solution and vortex well.
6. Repeat **steps 2–5**. Additional cycles of methanol–chloroform extraction help to clean up the proteins.
7. Tightly seal the lid with Parafilm and sonicate for 15 min.
8. Remove the Parafilm and centrifuge the tube at 4500 × *g* for 30 min.
9. Carefully remove and discard the aqueous phase using a 1 mL pipette, being careful not to disturb the protein layer.
10. To the remaining organic phase, add an equal volume of cold 100% methanol solution and vortex well. Addition of pure methanol causes precipitation of the proteins.
11. Centrifuge the tube at 4500 × *g* for 30 min. After centrifugation, a protein pellet should be visible at the bottom of the tube.
12. Carefully remove and discard the supernatant using a 1 mL pipette, being careful not to disturb the protein pellet.

13. Add 0.5 mL cold methanol to the 15 mL tube.
14. Carefully transfer the pellet and methanol solution from the 15 mL tube into a 2 mL microfuge tube using a 1 mL pipette (*see* **Note 8**).
15. Add a further 0.5 mL cold methanol to the 15 mL tube, rinse the sides of the tube to collect any remaining pieces of the protein pellet and transfer to the same microfuge tube.
16. Centrifuge the tube at 13,000 × *g* for 15 min.
17. Carefully remove and discard the supernatant using a 1 mL pipette, being careful not to disturb the protein pellet.
18. Add 1 mL cold methanol to the tube and vortex to resuspend the pellet.
19. Centrifuge the tube at approximately 13,000 × *g* for 15 min.
20. Repeat **steps 17–19** another two times.
21. Carefully remove and discard the supernatant using a 1 mL pipette, being careful not to disturb the protein pellet.
22. Place the open tubes into a vacuum concentrator and dry for 10 min (*see* **Note 9**).

3.4 Deglycosylation and Cleanup

1. Resuspend the dried protein pellet in 1 mL cold PBS.
2. Add 150 μL 10% SDS (final concentration 1.3% w/v) and 6 μL βME (final concentration 0.5% v/v). Denaturation of proteins and elimination of disulphide bonds is required for effective deglycosylation by PNGase F.
3. Vortex the solution and incubate in the Thermomixer comfort at 95 °C for 10 min with gentle shaking.
4. Incubate on ice for 5 min (*see* **Note 10**).
5. Add 15 μL 10% NaN3 (final concentration 0.1% w/v) and 200 μL 10% IGEPAL CA-630 (final concentration 1.5% w/v).
6. Add 0.5 μL PNGase F enzyme (5 U), tightly seal the tube with Parafilm and vortex well.
7. Incubate overnight at 37 °C.
8. The following day repeat **steps 6** and **7** (*see* **Note 11**).
9. The following day prepare the Amicon® Ultra 2 mL Centrifugal Filter Devices by adding 1 mL of ultrapure water to the filter device and centrifuging for 5 min at 2450 × *g*.
10. Discard the flow-through, add a further 1 mL of ultrapure water and centrifuge once more for 5 min at 2450 × *g*.
11. Discard the flow-through and transfer the entire volume of the deglycosylation reaction from the microfuge tube to the filter device (*see* **Note 12**). Do not discard the microfuge tube.
12. Centrifuge for 20 min at 4000 × *g*. Do not discard the flow-through; this contains the free *N*-glycans.

13. Rinse the microfuge tube in which the deglycosylation reaction took place with 500 μL ultrapure water to collect remaining traces of the reaction mixture and transfer the solution to the filter device.
14. Centrifuge for 10 min at 4000 × *g*. Do not discard the flow-through; this contains the free *N*-glycans.
15. Repeat **step 13**.
16. Centrifuge for 15 min at 4000 × *g*. Do not discard the flow-through; this contains the free *N*-glycans.
17. Remove the filtrate collection tube, which contains the flow-through, from the filtration device and place it into a vacuum concentrator to dry for 2-4 h (*see* **Note 13**).
18. Transfer the solution from the filtrate collection tube into a 2 mL microfuge tube. Do not discard the filtrate collection tube.
19. Rinse the filtrate collection tube with 200 μL ultrapure water to collect remaining traces of *N*-glycans and transfer the solution to the microfuge tube. Do this twice.
20. Place the microfuge tube in into a vacuum concentrator and incubate overnight or until completely dry (*see* **Note 14**).

3.5 Labeling and Cleanup

1. Resuspend the *N*-glycans in 100 μL ultrapure water. If desired, the procedure can be upscaled to a high-throughput format at this point by transferring the resuspended *N*-glycans to a 96-well plate. The subsequent labeling and cleanup steps can be carried out in 96-well plates.
2. Prepare the 2-AB labeling solution and add 50 μL to each sample. Mix well by pipetting.
3. Seal the plate with an adhesive plate seal and incubate for 10 min at room temperature with gentle shaking.
4. Incubate the plate for 2 h at 65 °C.
5. Remove the plate from the incubator and cool at room temperature for 30 min.
6. Prepare the 1 mL AcroPrep™ GHP membrane 96-well filter plate by pipetting 200 μL of 70 % ethanol to each well. Vacuum to waste (*see* **Note 15**).
7. Pipette 200 μL of ultrapure water to each well and vacuum to waste.
8. Pipette 200 μL of cold 96 % ACN to each well and vacuum to waste.
9. Prepare the *N*-glycan samples for cleanup by adding 800 μL of cold 100 % ACN to each sample. Mix well by pipetting and transfer the entire solution to the AcroPrep™ GHP membrane filter plate.
10. Incubate for 2 min.

11. Vacuum to waste.
12. Pipette 200 μL of cold 96% ACN to each well vacuum to waste.
13. Repeat the previous step three times.
14. Place the AcroPrep™ GHP membrane filter plate on top of a 2 mL stand and pipette 200 μL of cold 96% ACN to each well of the GHP plate.
15. Centrifuge the GHP plate for 5 min at 165 × *g*. Make sure the centrifuge is balanced.
16. Place the AcroPrep™ GHP membrane filter plate on top of a 96-well collection plate.
17. Pipette 90 μL of ultrapure water to each well of the GHP plate and incubate the plate for 15 min at room temperature with gentle shaking.
18. Centrifuge the plate for 5 min at 165 × *g* to collect the first fraction of the eluate into the collection plate.
19. Repeat **steps 17** and **18** to collect the second fraction of the eluate.
20. Seal the plate with plastic caps and store at −20 °C.

3.6 HILIC-UPLC Analysis

1. Prepare sample solution by resuspending 15 μL of fluorescently labeled *N*-glycans in 35 μL of ACN.
2. Maintain the sample solution at 10 °C before injection and the column at 25 °C during separation.
3. Turn on the lamp of the FLR detector (*see* **Note 16**).
4. Equilibrate the system to starting conditions (0.561 mL/min, 27% solvent A, 73% solvent B) by gradually increasing the flow rate by 0.1 mL/min from 0.1 mL/min to 0.561 mL/min. Make sure the pressure ripple delta value is ≤ 10 psi before each increase.
5. Inject 40 μL of the prepared solution and run the samples at flow rate of 0.561 mL/min using a linear gradient of 27–29.5% solvent A in the first 15 min and 29.5–38.7% for the next 80 min.

4 Notes

1. Raise the temperature on the magnetic stirrer (to around 60 °C) to help dissolving.
2. PB should be stored at 4 °C. Before weighing, allow the PB to reach room temperature.
3. To save time, place the solutions in the refrigerator the day before and allow them to cool overnight.

4. As a general rule, the more starting tissue, the better the results. We routinely get excellent results with as little as 25 mg of brain tissue, though even less could probably be used.
5. **Steps 1–4** should be performed as quickly as possible and with cold solutions so as to minimize the activity of proteases and glycosidases which may be released by homogenization. After the addition of methanol, enzymes will be denatured and lose their activity so haste is no longer as critical.
6. It is important to tightly wrap the Parafilm around the lid of the tube as sometimes sonication can cause the lids to loosen which results in leakage of the solution into the water bath.
7. It is not essential to completely remove all of the aqueous solution; a small amount can remain behind. It is better to leave a small amount of aqueous phase than to disturb the protein layer.
8. We recommend cutting off the end of the pipette tip to make the orifice wider.
9. If a vacuum concentrator is not available, samples can also be dried by leaving them on the bench at room temperature with the lids open for 30–60 min as methanol is quite volatile and evaporates quickly. A vacuum concentrator simply speeds up the process.
10. Alternatively, place the tubes at −20 °C for 5 min.
11. In our experience, we find deglycosylation to be more effective with an extended incubation and the addition of fresh enzyme.
12. We recommend using wide orifice pipette tips for this step.
13. The purpose of this step is to reduce the volume of the flow-through solution to below 1.5 mL so that it can be transferred to a 2 mL microfuge tube which is more convenient for downstream steps. Be careful not to let the solution evaporate completely or else it will be more difficult to resuspend the *N*-glycans and transfer them.
14. At this point, the procedure can be paused if desired. The dried *N*-glycans can be stored at −20 °C until ready for labeling.
15. Make sure the vacuum does not exceed 2 inHg and do not let the membrane become overly dry. It is a good idea to turn off the suction between vacuuming steps (i.e., while pipetting the next solution).
16. For optimal results, wait for 30–60 min for the FLR detector to stabilize before acquiring data.

Acknowledgements

The research leading to these results has received funding from the European Union Seventh Framework Programme (FP7 2007-2013) under grant agreement n° 291823 Marie Curie FP7-PEOPLE-2011-COFUND (The new International Fellowship Mobility Programme for Experienced Researchers in Croatia-NEWFELPRO). This chapter has been written as a part of a project "A spatio-temporal analysis of glycosylation in the human brain (Human Neuroglycome)" which has received funding through NEWFELPRO project under grant agreement n° 34.

References

1. Committee on Assessing the Importance and Impact of Glycomics and Glycosciences (2012) Transforming glycoscience: a roadmap for the future. The National Acadamies Press, Washington, DC
2. Zoldoš V, Horvat T, Lauc G (2013) Glycomics meets genomics, epigenomics and other high throughput omics for system biology studies. Curr Opin Chem Biol 17:34–40
3. Trbojević Akmačić I, Ugrina I, Štambuk J, Gudelj I, Vučković F, Lauc G et al (2015) High-throughput glycomics: optimization of sample preparation. Biochemistry (Mosc) 80:934–942
4. Wessel D, Flügge UI (1984) A method for the quantitative recovery of protein in dilute solution in the presence of detergents and lipids. Anal Biochem 138:141–143
5. Manzi AE, Norgard-Sumnicht K, Argade S, Marth JD, van Halbeek H, Varki A (2000) Exploring the glycan repertoire of genetically modified mice by isolation and profiling of the major glycan classes and nano-NMR analysis of glycan mixtures. Glycobiology 10:669–689
6. Chen Y-J, Wing DR, Guile GR, Dwek RA, Harvey DJ, Zamze S (1998) Neutral N-glycans in adult rat brain tissue. Complete characterisation reveals fucosylated hybrid and complex structures. Eur J Biochem 251:691–703
7. Zamze S, Harvey DJ, Chen Y-J, Guile GR, Dwek RA, Wing DR (1998) Sialylated N-glycans in adult rat brain tissue. A widespread distribution of disialylated antennae in complex and hybrid structures. Eur J Biochem 258:243–270
8. Ji IJ, Hua S, Shin D, Seo N, Hwang JY, Jang I-S et al (2015) Spatially-resolved exploration of the mouse brain glycome by tissue glyco-capture (TGC) and nano-LC/MS. Anal Chem 87:2869–2877
9. Parker BL, Thaysen-andersen M, Solis N, Scott NE, Larsen MR, Graham ME et al (2013) Site-specific glycan-peptide analysis for determination of N-glycoproteome heterogeneity. J Proteome Res 12:5791–5800
10. Torii T, Yoshimura T, Narumi M, Hitoshi S, Takaki Y, Tsuji S et al (2014) Determination of major sialylated N-glycans and identification of branched sialylated N-glycans that dynamically change their content during development in the mouse cerebral cortex. Glycoconj J 31:671–683
11. Hu Y, Zhou S, Khalil SI, Renteria CL, Mechref Y (2013) Glycomic profiling of tissue sections by LC-MS. Anal Chem 85:4074–4079

Chapter 17

Automated Integration of a UPLC Glycomic Profile

Anna Agakova, Frano Vučković, Lucija Klarić, Gordan Lauc, and Felix Agakov

Abstract

Ultra-performance liquid chromatography (UPLC) is the established technology for accurate analysis of IgG Fc N-glycosylation due to its superior sensitivity, resolution, speed, and its capability to provide branch-specific information of glycan species. Correct and cost-efficient preprocessing of chromatographic data is the major prerequisite for subsequent analyses ranging from inference of structural isomers to biomarker discovery and prediction of humoral immune response from characterized changes in glycosylation. The complexity of glycomic chromatograms poses a number of challenges for developing automated data annotation and quantitation algorithms, which frequently necessitated manual or semi-manual approaches to preprocessing, most notably to peak detection and integration. Such procedures are meticulous and time-consuming, and may be a source of confounding due to their dependence on human labelers. Although liquid chromatography is a mature field and a number of methods have been developed for automatic peak detection outside the area of glycomics analysis, we found that hardly any of them are suitable for automatic integration of UPLC glycomic profiles without substantial modifications. In this chapter, we illustrate practical challenges of automatic peak detection of UPLC glycomics chromatograms. We outline a robust, semi-supervised method ACE (Automatic Chromatogram Extraction) for automated alignment and detection of glycan peaks in chromatograms, developed by Pharmatics Limited (UK) in collaboration with Genos Limited (Croatia). Application of the tool requires minimal human interference, which results in a significant reduction in the time and cost of IgG glycomics signal integration using Waters Acquity UPLC instrument (Milford, MA, USA) in several human cohorts with blind technical replicas.

Key words Liquid chromatography, Peak detection, Automatic alignment, Glycomics, Data analysis

1 Introduction

Recent years have witnessed impressive developments in 'omics technologies, leading to an exploding amount of collected 'omics data. However, increased efforts of the academic community and life science industries to use the technological advancements for improving effectiveness of drug development and precision medicine have so far led only to moderate successes. For instance, despite the 50-fold increase in investments in R&D by big pharmas from 1980 to 2010, drug approval rates remained largely unchanged [1],

Gordan Lauc and Manfred Wuhrer (eds.), *High-Throughput Glycomics and Glycoproteomics: Methods and Protocols*, Methods in Molecular Biology, vol. 1503, DOI 10.1007/978-1-4939-6493-2_17,

and a recent survey showed that productivity may be even lower than previous estimates [2]. A study conducted by Bayer revealed that only 21% of published results on potential drug targets were in line with their in-house findings, citing inappropriate analysis and inadequate experimental control among key causes for the observed lack of reproducibility [3]. Statistical analysis and experimental control may be particularly challenging when data acquisition is based on liquid chromatography (used as a stand-alone method or applied in a tandem with other separation technologies), where a broad range of environmental variations may lead to shifts and amplifications of observed and extracted expressions.

1.1 Glycomic Chromatograms

In case of glycomic techniques based on high-(or ultra-)performance liquid chromatography, the asynchronous nature of glycomics chromatograms commonly necessitates manual or semi-manual signal preprocessing (such as sequence alignment and peak detection) prior to subsequent statistical analyses. Such (semi-)manual procedures can be justifiable in situations when chromatographic profiles are highly heterogeneous, prone to complex nonlinear trends, heteroscedastic noise, or complex edge effects, or when shapes of the characterized peaks are complex and not readily conforming to simple parametric approximations. However, when applied to large quantities of data, such (semi-) manual procedures are slow, expensive, and prone to subjective judgement. Although there have been numerous developments in chromatography for automating preprocessing of chromatograms, we found that many of them are not readily applicable to sequence alignment and peak detection from UPLC glycomic chromatograms using the Waters Acquity instrument (*see* Subheading 3.2) due to the high complexity of glycan profiles. We found that glycomic chromatograms vary in length, scale, and pattern distributions across samples, which significantly complicated applications of automated glycan annotation methods. In particular, by using plasma samples collected from several human cohorts of several thousand people, we noticed that:

1. Glycosylation reaction time varies significantly across samples, ranging from 20 to 40 min depending on a sample;
2. Shapes and areas of each peak vary substantially across samples even for technical replicas;
3. Some glycan peaks are weak or nonexistent for some but not other runs, which made it difficult to separate them from noise;
4. There are complex spikes of nonbiological origin at the beginning and end of the process;
5. Different trends of nonbiological origin are often present in the observed patterns and needed to be removed to obtain accurate estimates of glycan peak intensities.

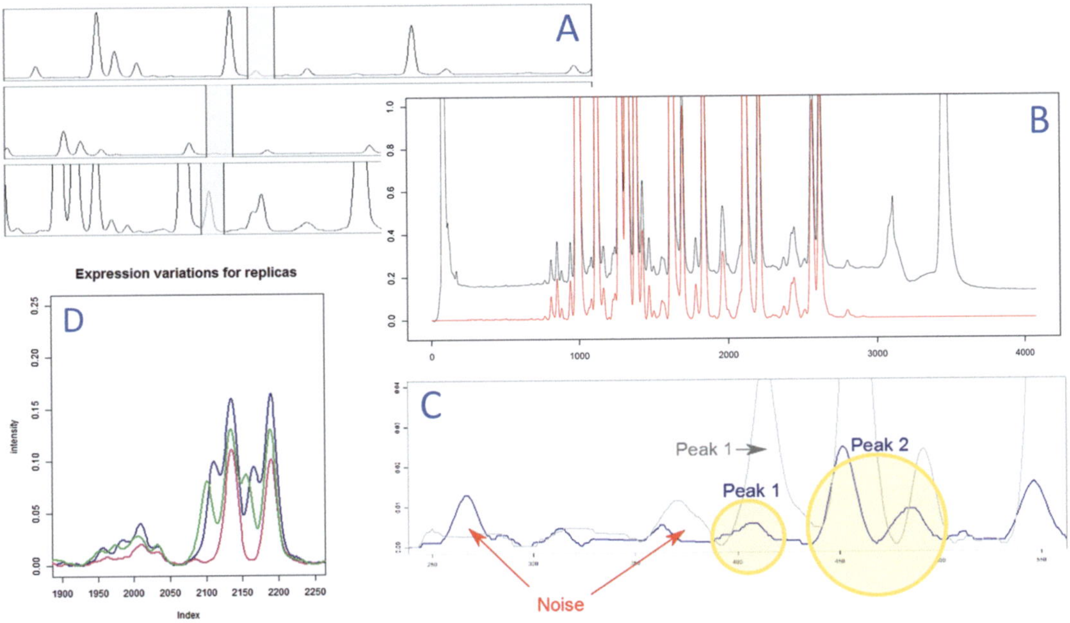

Fig. 1 Typical IgG glycan profiles collected by Genos Ltd using the Waters Acquity UPLC instrument for human plasma samples. (**a**) Different runs of the instrument result in shifts and amplifications of chromatographic patterns (the target peak is highlighted in light gray, the *y*-axes are fixed to have the same scale after removing trends and edge effects. (**b**) Nonlinear trends and edge effects of technical origin (upper and lower for the raw and preprocessed series respectively). (**c**) Chromatograms for two manually co-aligned technical replicas. Note that amplitudes and areas corresponding to some peaks are lower than amplitudes and areas of some of the noisy spikes for some, but not other chromatograms. (**d**) Co-aligned chromatogram fragments for three technical replicas; note differences in the scaling and shapes (two well-separated peaks in one chromatogram vs more complex multi-modal shapes in other chromatographic patterns). The presence of the irregular shifts, amplifications, edge effects, noisy nonlinear trends, and complex peak shapes complicate automatic detection of glycan peaks

Figure 1 demonstrates some of these challenges, which complicate automatic peak detection and integration of glycomic profiles from chromatograms.

The standard procedure for integration of glycomic profiles deployed by Genos Ltd until recently was to use the Empower Data Chromatography software developed by Waters Corporation, which required semi-manual post-processing of the results, costing several person-weeks of experienced researchers' time per cohort of a few thousand samples. Motivated by the low throughput and high costs of the procedure, we set out to develop a generic semi-supervised approach to automated peak detection and integration. The approach uses a small number of fully annotated sequences (for examples, those produced semi-manually) to be used as training data, and learns to mimic peak detection by iterative applications of sequence alignment and outlier detection algorithms using features extracted from time series chromatograms.

Although the approach should be generally applicable to peak detection using disparate separation technologies, so far it was tested only for IgG glycans extracted by Waters UPLC platform from several human cohorts, where it compared favorably with open-source peak detection techniques and with the semi-manual procedure industrially deployed by Genos, at a fraction of the time expenditure.

1.2 Automatic Peak Detection

The majority of the existing peak detection algorithms developed for liquid chromatography and other separation technologies operate directly on unlabeled time series chromatograms. The detection procedure usually comprises smoothing (for example, by using moving averages, Gaussian filters, or wavelet transforms), trend removal, and peak extraction. The last step is often performed by applying sets of thresholds on peak intensities or widths, or by fitting mathematical models to peak shapes using a variety of locally smooth functions. For example, Vivó-Truyols et al. [4] suggest to detect peaks by using combinations of the estimated 1st- and 2nd-order derivatives of locally smoothed signals with width-based heuristics; Yu and Peng [5] use bi-Gaussian mixture models and statistical model selection to integrate partially overlapping peaks for LC-MS. Numerous related algorithms were developed outside the field of the liquid chromatography; *see* for example Yang et al. [6] for a thorough review. These algorithms are intrinsically *unsupervised*, that is, they do not attempt to reproduce *labeled* peak annotations. Instead, they make specific assumptions about the geometry of peak profiles, which often requires comprehensive and time-consuming parametrization and semi-manual fine-tuning, and may still not adequately model the complexity of signals and noise due to restrictions of the parametric representations. Because the problem of peak detection is decoupled from the problem of sequence alignment, these methods are not well suited for producing the consistent number and ordering of the extracted peaks, which significantly complicates interpretation and analysis of the annotated data. Additionally, the methods do not easily accommodate complex effects of signal or noise intensities on peak shapes, cannot easily exploit prior knowledge about peak expressions or collocations, and cannot easily reconstruct annotations provided by human technicians or by applications of carefully fine-tuned (semi-manual) techniques.

One way to facilitate consistency of peak extraction procedures for multiple time series is by solving the sequence alignment problem, which minimizes pairwise distances between the aligned sequences. The problem can be addressed by warping methods, which typically use dynamic programming to map a target sequence to a reference by the local stretching or squeezing of the time axes for the two time series. Two common sequence alignment algorithms are dynamic time warping (DTW) and correlation optimized warping (COW) [7]. The methods are widely used for comparing speech sequences, but have also found applications in chromatography [8, 9]. One example is the improved COW-based chromatogram aligner—Chromaligner [10], which uses preliminary segmentation prior to alignment, but requires careful tuning of the segment lengths. There are multiple modifications of these methods varying in the choice of the minimized distance functions or preprocessing methods. The method suggested in Bork et al. [11] uses DTW of the signal derivatives to prevent mapping of a single point of one chromatogram onto a large subsection of the other. Other extensions include the OBI-warp algorithm [12], Center-Star Multiple alignment by Pairwise Partitioned Dynamic Time Warping (CEMAPP-DTW), and multiple alignment by Bidirectional Best Hits Peak Assignment and Clique Extension (BIPACE) recently suggested in Hoffman et al. [13] for automated matching and grouping of multiple peaks. Empirically, we found that warping algorithms such as DTW and COW do not provide accurate annotations of glycomic peaks in our data, and often led to unintuitive mappings (which is consistent with some previous observations in the literature [14, 15]). This is particularly the case for chromatograms with large peak shifts. On the other hand, these methods are often accurate for shorter, approximately well co-aligning time series segments, so they are included as auxiliary components of our approach.

2 Materials

To test the performance of the algorithm we used 1500–3000 samples from each of the three human cohorts: the Orkney Complex Disease Study ORCADES [16] supported by the Chief Scientist Office of the Scottish Government, the Royal Society, the MRC Human Genetics Unit, Arthritis Research UK, and the European Union framework program 6 EUROSPAN project (Contract no. LSHG-CT-2006-018947); TwinsUK funded by the Wellcome Trust, European Community's Seventh Framework Programme (FP7/2007-2013), with the support from the National Institute for Health Research (NIHR)-funded BioResource, Clinical Research Facility and Biomedical Research Centre based at Guy's and St Thomas' NHS Foundation Trust in partnership with King's College London; and a proprietary cohort of Chinese patients.

3 Methods

3.1 IgG Purification and Labeling

1. The IgG is isolated from plasma samples using 96-well protein G monolithic plates (BIA Separations, Ajdovščina, Slovenia). 50–90 μl of plasma is diluted 10× with 1× PBS pH 7.4 and applied to the protein G plate. IgG is eluted with 0.1 M formic acid (v = 1 mL; Merck, Darmstadt, Germany) and neutralized with 1 M ammonium bicarbonate (Merck).

2. Purified IgG is first denatured with the addition of 30 μL 1.33% SDS (w/v) (Invitrogen, Carlsbad, CA, USA) and 10-min incubation at 65 °C. Subsequently, 10 μL of 4% Igepal-CA630 (Sigma-Aldrich, St. Louis, MO, USA) and 1.25 mU of PNGase F (ProZyme, Hayward, CA, USA) in 10 μL 5× PBS are added to the samples and incubated overnight at 37 °C to release N-glycans.
3. The released N-glycans are labeled with 2-aminobenzamide (2-AB). The labeling mixture is freshly prepared by dissolving 2-AB (Sigma-Aldrich) in DMSO (Sigma-Aldrich) and glacial acetic acid (Merck) mixture (85:15, v/v) to a final concentration of 48 mg/mL
4. A volume of 25 μL of labeling mixture is added to each N-glycan sample in the 96-well plate. 25 μL of freshly prepared reducing agent solution (106.96 mg/ml 2-picoline borane (Sigma-Aldrich) is added, and the plate is sealed using adhesive tape.
5. Mixing is achieved by shaking for 10 min, followed by 2 h incubation at 65 °C. Samples (in a volume of 100 μL) are brought to 80% ACN (v/v) by adding 400 μL of ACN (J.T. Baker, Phillipsburg, NJ, USA). Free label and reducing agent are removed from the samples using HILIC-SPE.
6. An amount of 200 μL of 0.1 g/mL suspension of microcrystalline cellulose (Merck) in water is applied to each well of a 0.45 μm GHP filter plate (Pall Corporation, Ann Arbor, MI, USA). Solvent is removed by application of vacuum using a vacuum manifold (Millipore Corporation, Billerica, MA, USA). All wells are prewashed using 5× 200 μL water, followed by equilibration using 3× 200 μL acetonitrile–water (80:20, v/v). The samples are loaded to the wells.
7. The wells are subsequently washed 7× using 200 μL acetonitrile–water (80:20, v/v). Glycans are eluted with 2× 100 μL of water and combined eluates are stored at −20 °C until usage.

3.2 Hydrophilic Interaction Chromatography (HILIC)-UPLC

1. Fluorescently labeled N-glycans are separated by HILIC on a Waters Acquity UPLC instrument (Milford, MA, USA) with fluorescence detector set with excitation and emission wavelengths of 330 and 420 nm, respectively. The instrument is controlled by Empower 2 software, build 2145 (Waters, Milford, USA).
2. Labeled N-glycans are separated on a Waters BEH Glycan chromatography column, 100×2.1 mm i.d., 1.7 μm BEH particles, with 100 mM ammonium formate, pH 4.4, as solvent A and acetonitrile as solvent B.
3. Separation method used linear gradient of 75–62% acetonitrile (v/v) at flow rate of 0.4 ml/min in a 25 min analytical

run. Samples are maintained at 5 °C before injection, and the separation temperature is 60 °C. The system is calibrated using an external standard of hydrolyzed and 2-AB labeled glucose oligomers from which the retention times for the individual glycans are converted to glucose units.

3.3 ACE: Automatic Integration of Glycan Profiles for UPLC

The key motivation for our method, automatic chromatogram extraction (ACE), is to combine the high flexibility of manual annotations with cost-effectiveness and speed of automatic procedures. The algorithm takes into account *relative locations of multiple adjacent peaks for multiple time series* in a batch of samples. This is achieved by a *semi-supervised* approach that tries to predict locations of intensity peaks by using both labeled and unlabeled time series data. Initially, the algorithm applies an active learning technique to a batch of unlabeled chromatograms in order to select a small set of sequences for manual or semi-manual annotations (data dictionary construction). Then ACE iteratively applies sequence alignment together with a local peak detection procedure. During each iteration of the detection step, the algorithm attempts to identify incorrectly labeled peak patterns, and then refines the current annotations. The probability that a peak pattern is detected correctly is set to be proportional to the likelihood of that pattern under the empirical distribution defined over the dictionary samples, which takes into account probability estimates of the neighboring peaks or peak patterns. The alignment and extraction procedures are repeatedly applied to each contiguous segment by using different signal representations at each iteration.

The key feature of ACE is that in contrast to common sequence detection and peak alignment approaches that aim to use a single peak model or a single optimization technique to annotate time series, it uses a *cascade of multiple techniques* via an iterative procedure. Note that in order to resolve the problem of chromatographic time shifts we want an annotation algorithm to omit extraneous time segments by mapping them to a single point or a short interval on a reference panel. However, the algorithm should avoid mapping useful segments to a single point, as this would lead to a loss of useful information. In practice, it is challenging to determine optimal parameters of warping procedures and balance their ability to eliminate shifts and to control the risk of losing useful signals by applying a single universally optimal approach. ACE draws motivations from *boosting* methods [17] of supervised classification, where the challenging problem of constructing a single accurate classifier is replaced by the sequential fitting of multiple "weak classifiers." Each such classifier focuses on a simple prediction task, typically uses a subset of features, and aims to correct errors of other weak classifiers fitted at previous iterations of the learning procedure, which often leads to superb performance in a number of applications (for example, in object detection and

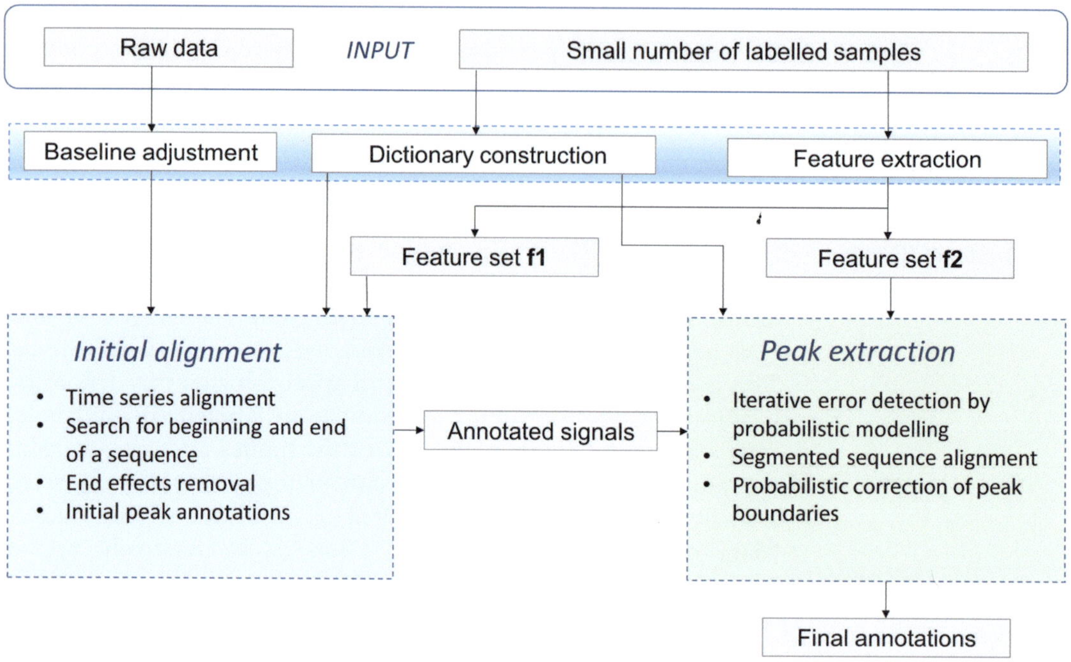

Fig. 2 Block diagram of the ACE method. Note the three distinct logical blocks: Preprocessing (comprising baseline adjustment, dictionary construction, and feature extraction), Initial alignment aimed at detecting boundaries of the informative segments, and Peak extraction aimed at extracting individual glycan peaks

machine vision [18]). By using an analogous strategy, *ACE can detect and correct many of its errors*, which empirically resulted in accurate annotations of IgG glycomic profiles. A useful feature of the approach is that it is guaranteed to generate a fixed ordering of the extracted peaks consistent with annotations from the data dictionary, which often facilitates interpretation of the signals. If the dictionary is constructed by manual or semi-manual approaches, ACE will learn to mimic the manual labeling prior to applying it to new sequences.

For the sake of space and clarity to glycobiologists, we omit mathematical and implementation details, and only provide a sketch of the automated integration algorithm and some vignette illustrations for IgG glycomic profile annotations. The algorithm contains three key steps: preprocessing (including baseline adjustment, dictionary construction, and feature extraction), initial alignment, and iterative peak extraction as outlined below; its block-diagram is shown on Fig. 2.

3.3.1 Baseline Adjustment

The purpose of this step is the initial smoothing and trend removal:

1. Smoothing by approximate Gaussian filtering.
2. Trend fitting by consecutive applications of the flat discriminative mixture of the local B-spline and linear regression. Learning the degree of smoothness by tenfold cross-validation.

3. Constrained baseline correction to ensure non-negativity of the adjusted baseline (*see* Fig. 1b).
4. Normalization (for example, by the total area).
5. Repeating **steps 1–4** for all raw sequences in the batch.
6. Note: other approaches can be used (*see* for example Yang et al. [6]).

3.3.2 Dictionary Construction and Labeled Data Generation

As ACE is a semi-supervised method, it needs a small set of manually labeled reference samples representative of the cohort. To identify a small set of signals representative of the cohort, we use:

1. Probabilistic consensus clustering of chromatograms; learning the number of clusters by cross-validation.
2. Construction of the *data dictionary Dict* by selecting sequences closest to the cluster centroids from **step 2** according to the L_2 distance. We refer to such sequences as *dictionary* samples. In the considered cohorts, the dictionary typically consisted of only 4–7 entries.
3. Construction of the *reference panel Dtr* by selecting additional sequences from the learned clusters to augment the chosen dictionary construction *Dict*. Within each cluster, the reference samples maximize an approximation of the amount of information between the selected and the discarded sequences by using a modification of the principal variables method [19].
4. Passing IDs of the selected reference sequences to technologists for (semi-)manual annotations.
5. Note: $Dict \subseteq D_{tr}$. Empirically, we found that 30–40 labeled reference samples from a batch are usually sufficient for accurate annotations of the remaining samples for batches of 1500–3000 individuals. Additionally, our empirical results show that ACE performs well even when the labeled set comes from a different cohort, as long as the platform and the measurement procedure stays unaltered (*see* Subheading 3.2).

3.3.3 Feature Extraction and Density Estimation

1. Extraction of two sets of feature vectors for edge effect removal $\{f_1\}$ and peak detection and refinement $\{f_2\}$ using the annotated sequences *Dtr*. The sets of features include peak widths and intensities, parameters of curve models, and offset times (full list omitted), necessary for performing the initial alignment and peak refinement.
2. Choosing a representation for the joint density $p(\{f_1\}, \{f_2\})$. For example, for the generalized Markov chains, we get

$$p(\{\boldsymbol{f}_1\},\{\boldsymbol{f}_2\} \mid D_{tr}) = \prod_{k=1}^{M} p\left(\boldsymbol{f}_1^{\{(k)\}}, \boldsymbol{f}_2^{\{(k)\}} \mid \boldsymbol{f}_1^{\{(k-1)\}}, \boldsymbol{f}_2^{\{(k-1)\}}\right)$$
$$\approx \prod_{j=1}^{F}\prod_{k=1}^{M} p\left(f_{1j}^{(k)} \mid f_{1j}^{(k-1)}\right) p\left(f_{2j}^{(k)} \mid f_{2j}^{(k-1)}\right)$$

where $k = 1..M$ corresponds to each of the M annotated peak patterns, and $j = 1..F$ corresponds to each of the extracted boundary or peak features. We suggest heavy-tail parameterizations of the conditional densities, e.g., ref. [20].

3. Fitting $p(\{f_1\}, \{f_2\})$ to the labeled samples D_{tr} by maximizing the likelihood.
4. Note: we considered several parameterizations of $p(\{f_1\}, \{f_2\})$ including the fully factorized approximations, Markov chains, and hidden Markov models, with several choices of the emission and transition probabilities. The hidden Markov models resulted in the most robust performance across the cohorts; the factorized parameterization was least robust, but most computationally efficient.

3.3.4 End Effects Removal and the Initial Alignment

The goal of this step is to remove edge effects via the initial alignment of the unlabeled sequences with the dictionary. This is achieved by co-alignment of new unlabeled time series with typical time series—dictionary entries.

1. Sequence leveling by a proprietary intensity harmonization algorithm.
2. For each sequence $x_i \in D$ in the unlabeled batch D, identification of the best match between x_i and dictionary entry $y_j \in Dict$, and extraction of the resulting boundary features $\{f_1^{i,j}\}$. The time series are can be matched even if they do not line up in the time domain.
3. Error checking by computing the probability of observing the resulting boundary features $p\left(\left\{f_1^{i,j}\right\} | D_{tr}\right)$ under the model trained in Subheading 3.3.3. Repeating **steps 2** and **3** with different distance definitions until this probability exceeds a specified threshold, or until the maximal number of iterations has been exceeded. The procedure corrects sequence boundaries and provides initial peak annotations (*see* Fig. 3).
4. Note: The sequence leveling method (**step 1** of this procedure, details to be published elsewhere) was inspired by analog processing in acoustics. Empirically, it more than halved the number of errors at the first application of the initial alignment, thus significantly speeding up removal of the edge effects.

3.3.5 Detection and Refinement of Individual Peaks

This step identifies and corrects any mislabeled peak boundaries within the informative part of the signal. The probabilistic error detection and sequence alignment procedures are analogous to the initial alignment described in Subheading 3.3.4, but use peak features f_2 defined over *adjacent peak patterns* rather than boundary features f_1. For an efficient detection and correction of peak boundaries, it is crucial to take into account peaks or peak patterns

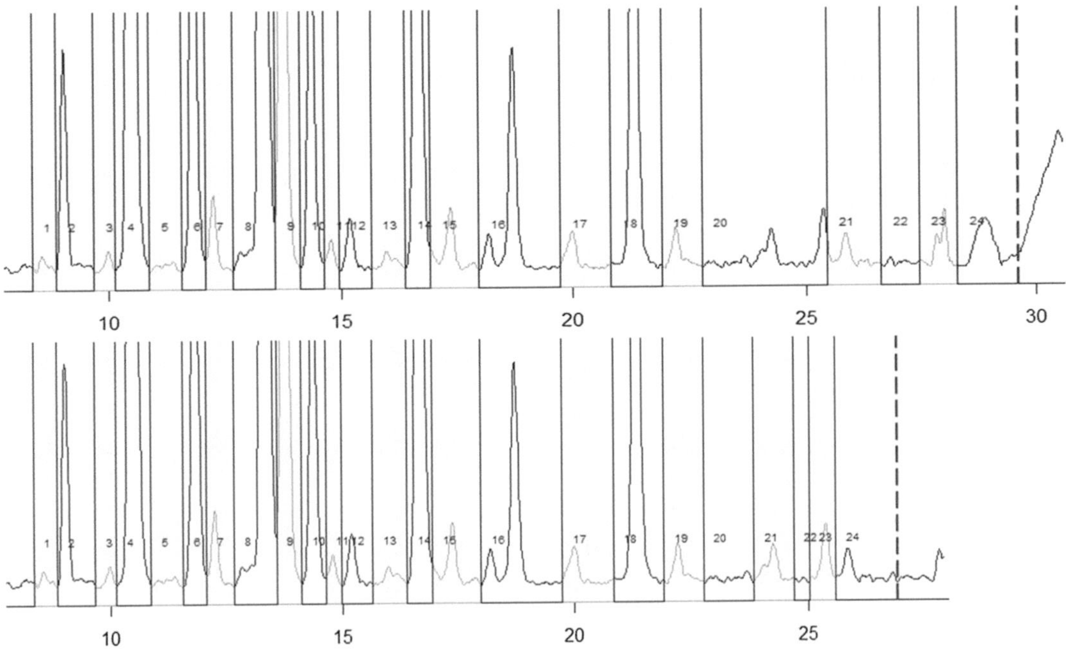

Fig. 3 Illustration of the corrected edge effects. *Top*: an example of a wrong identification of the end of a sequence, where the end spikes are erroneously labeled as glycan peaks. *Bottom*: consecutive correction of the edge effect in one of subsequent iterations after the error detection check

adjacent to the currently processed peak and distinguish the following scenarios:

- The current peak is likely to be annotated incorrectly (i.e., it is in the low-likelihood region under the trained model $p(\{f_1\},\{f_2\} \mid D_{tr})$, but is surrounded by peaks that are likely to be annotated correctly (i.e., those lying in the high-probability regions under the trained model);
- There are multiple incorrectly labeled peaks near the currently labeled peak;
- Incorrect annotations are in the middle or end of each sequence, etc.

The refinement is by a probabilistic search for local optima. Additional explanations are provided in **Note 1**. Comparisons with unsupervised and semi-manual procedures are discussed in **Notes 2–3**. Limitations and advantages of ACE are highlighted in **Note 4**.

3.3.6 Final Boundary Adjustment

At this point, the annotation is nearly complete. A potential problem lies in the labeling of the first and last boundary. The accuracy of their annotations can be strongly affected by the edge noise, which sometimes has more variation (versatile structure and patterns) than the noise inside the informative part, and therefore has a large effect on sequence mapping. This noise resembling

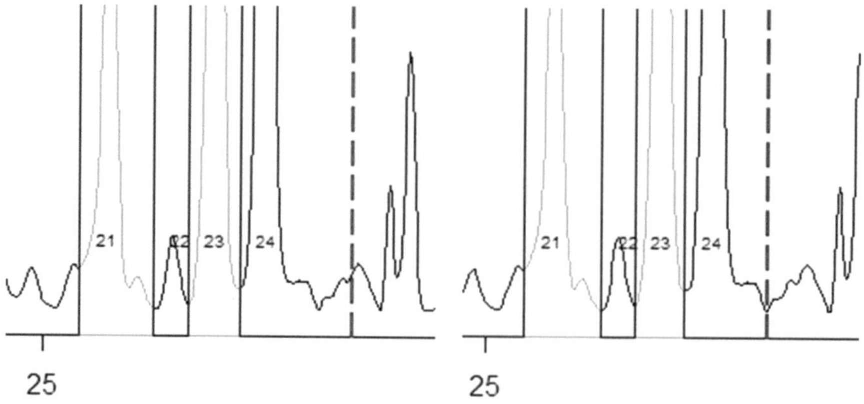

Fig. 4 Final adjustment of the ending peaks immediately followed by spurious spikes. *Left* plot—before, *right*—after the adjustment

additional peaks (Fig. 4) can be more efficiently removed by setting the widths of the first and the last peaks to be close to their statistical medians under D_{tr}, and by making adjustments to the boundaries by local methods applied to smoothed signals.

4 Notes

1. *Error detection.* Under decoupled probability models of peak features, error detection can be performed by checking whether features of the detected peaks (such as peak widths) have high probabilities under the corresponding marginal distributions computed from the labeled data *Dtr*. This approach suffers from *outliers* and *shifts*. The problem with the outliers occurs when correctly labeled peaks that are abnormally wide or narrow under *Dtr* are mislabeled by the decoupled error detection procedure (Fig. 5, Left). The problem with the shifts occurs when both boundaries of a peak are mislabeled, but the width is typical under *Dtr* (Fig. 5, Right). Both problems can be addressed by considering richer probabilistic models that take into account dependencies between the neighboring peaks or peak patterns.

   The consecutive steps of the error detection procedure analyze each mislabeled peak jointly with its neighbors, ensuring that false negatives (that is, mislabeled peaks erroneously marked as correct) are marked for correction, and false positives (that is, correctly labeled peaks erroneously marked as incorrect) are ignored. For example, if a peak is marked as erroneous, but a few peaks adjacent to it are correct, then the target peak is more likely to be correct. The alternating "correct" and "wrong" indicators may denote shifts, so the whole segment is considered for relabeling. Once mislabeled bound-

Fig. 5 Limitations of decoupled error detection models. *Left*: An outlier peak, which is wider than average, is incorrectly detected as being mislabeled by the error detection procedure (*dashed line*: correct annotation, *solid line*: automatic annotation). *Right*: Erroneously shifted boundaries are not detected for the peak in the middle, despite the abnormally wide or narrow widths of the adjacent peaks. This figure demonstrates that error correction in peak x_i may be facilitated by inclusion of information about peaks x_{i-1} and x_{i+1}

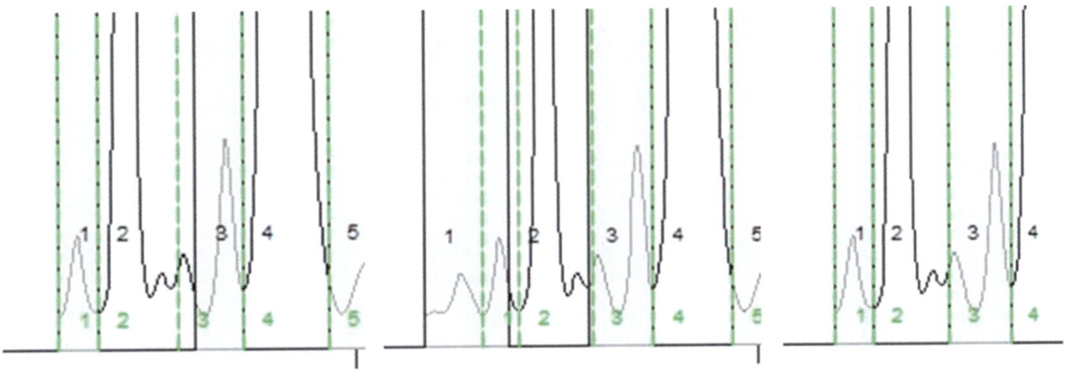

Fig. 6 An illustration of three consecutive steps of error detection and correction. *Left to right*: (1) mislabeled boundary between peaks 2 and 3 is detected; (2) iterative error correction is applied and improves boundary 2–3, but wrongly shifts boundary 1–2; (3) tuning restores boundary 1–2 (*dashed line*: correct annotation, *solid line*: annotation by ACE). The figure motivates the need for combining sequence alignment with a local search at the tuning step

aries are detected, the marked sections of signals containing errors are passed through another alignment step with a different parametrization. After several iterations, the segments are merged back to the rest of the sequence, which is vaguely analogous to boosting in supervised classification. Final tuning of the boundaries is based on the marginal distribution of f_2 under *Dtr*, and on the probability of the peaks being in a local minimum. Figure 6 illustrates consecutive error correction by ACE. Figure 7 gives a step-by-step illustration of the ACE annotation procedure for a particularly challenging trailing segment of one of IgG chromatograms.

2. *Comparison with unsupervised peak detection*. We compared ACE with the recent unsupervised FastChrom algorithm [21]. As FastChrom required careful and time-consuming fine-tuning, we only tested it for 10 randomly chosen samples from UK's ORCADES cohort [16]. Our goal was to identify 24

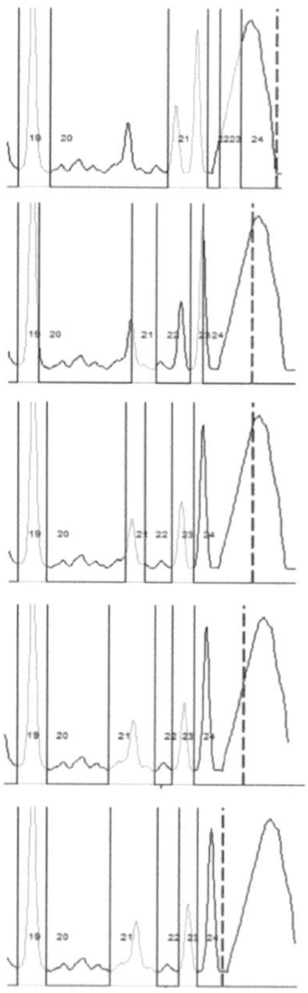

Initial alignment approximates the end of the informative part of a sequence (dashed line), but includes a spurious wide peak of non-biologic origin. This sequence does not pass the statistical check and needs to be aligned again using a different distance metric.

After finding a more appropriate distance metric, the alignment is slightly improved. This plot demonstrates the sequence after the first error detection check. Some of the internal boundaries are not yet correctly identified. The error detection procedure correctly labels peaks 20-24 for re-labeling.

Boosting corrects the boundaries of peaks 19, 23 and improves the annotation for 20, 21 and 23. The end of the sequence (dashed line), although mislabeled, is kept unchanged. At this intermediate step the ending is considered to be "correct" and is left for re-labelling at a later step.

Statistical tuning further improves peaks 20, 21, 22 and 24.

Cutting of the noisy peak by uptrend removal step of the post-processing procedure. The final annotation of all the glycan peaks is correct.

Fig. 7 A step-by-step example of chromatogram annotation by ACE, with the focus on the last four peaks of the UPLC IgG glycomic profile

intensity peaks routinely produced by the conventional semi-manual procedure deployed by Genos. By using the default parameter settings, FastChrom produced 9–18 peaks. After some careful fine-tuning, FastChrom produced 24 peaks for 3, and 18–23 peaks for 7 out of 10 samples. The inconsistency in the number and order of the detected peaks significantly complicates their characterization, as each peak could encode different biological signals for any new sample. A number of other publicly available approaches based on unsupervised peak detection [6] behave qualitatively similarly.

3. *Comparison with a semi-manual peak detection procedure.* We compared ACE with a traditional integration algorithm deployed by Genos and based on Waters Empower software, where each chromatogram was manually corrected to maintain the same intervals of integration for all the samples. The chromatograms were separated into 24 peaks. In the considered IgG cohorts, ACE accurately reconstructed the semi-manual labeling. Figures 8 and 9 show the very similar scatter plots and correlation profiles of the areas under the

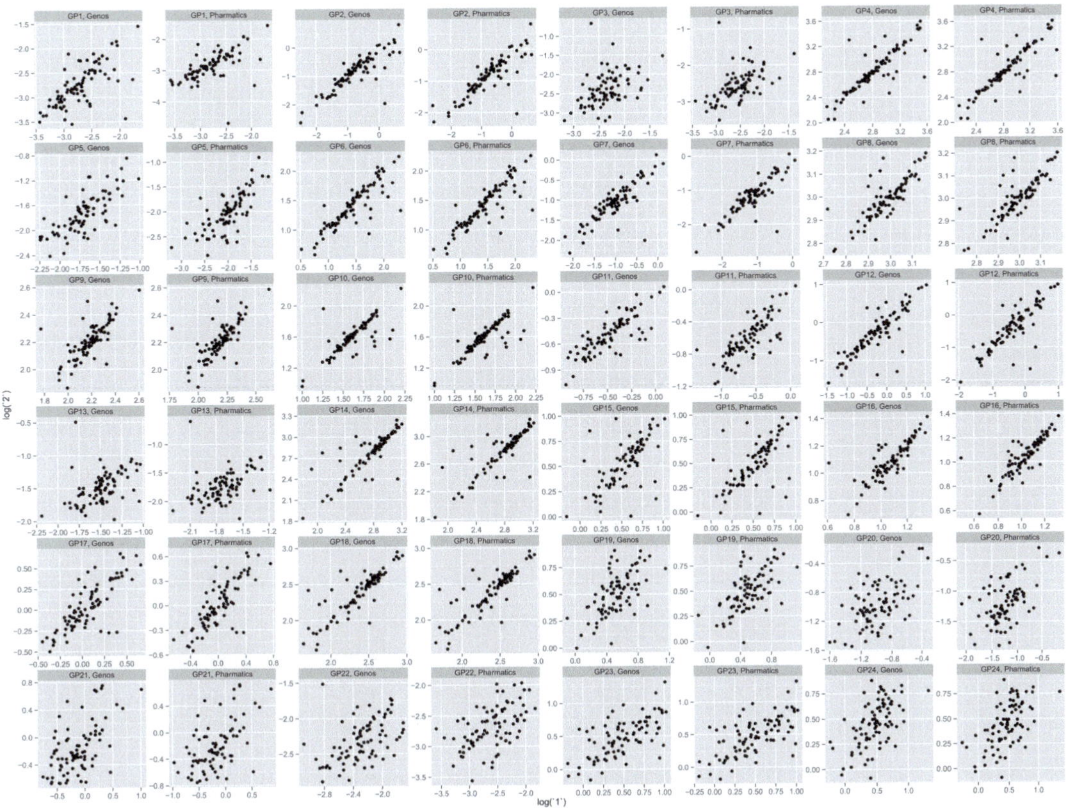

Fig. 8 Scatter plots of the areas under the 24 IgG glycan peaks extracted by the semi-manual annotation procedure deployed by Genos (via Waters Empower Chromatography Software), next to automatic annotations produced by Pharmatics ACE tool, for 84 replicate pairs. The plot shows very close correspondence between the semi-manual and the automatic procedures. The semi-manual annotation of 3000 samples takes between 5 and 6 person-weeks. The very similar labeling resulting from the automatic procedure takes ~3 h on a basic laptop

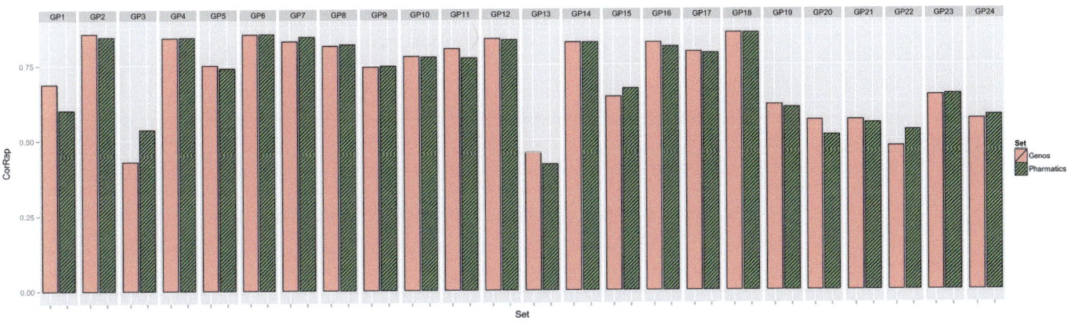

Fig. 9 Correlations between the areas under 24 IgG glycan peaks for the semi-manual (*solid color*, Genos) and automatic (*striped*, Pharmatics) annotations for 84 replicate pairs. The nearly indistinguishable correlations for both methods indicate that the automatic data-driven algorithm ACE has learned to mimic the manual annotations at a significant reduction in time and costs

glycan peaks for pairs of replicated samples produced by the semi-manual procedure (6 person-weeks for 3000 samples) and ACE (3 machine hours for 3000 samples). Correlations between the physical locations of the peak boundaries detected by the semi-manual procedure and ACE were even higher and ranged between 0.96 and 0.99 for ORCADES.

4. *ACE Advantages and Limitations.* ACE is an efficient machine learning-based tool for automatic chromatogram extraction, which:
   (a) Produces the consistent number and ordering of the peaks
   (b) Needs minimal fine-tuning and adjustments
   (c) Handles complex peak shapes and automatically removes noise
   (d) Produces an optimal match to (semi-)manually labeled signals
   (e) Needs little labeled data, and can even be used without it[1]
   (f) Learns to detect and correct common errors
   (g) Saves time and costs, producing 1000+ annotations in 1 h on a basic laptop (Intel i5 4th Gen, 8GB DDR3 RAM, SSD).

ACE's main limitation is its reliance on labeled data, and the assumption that each segment of a new time series is well represented in the automatically constructed reference panel. If labeled data contains errors, ACE will learn to mimic them. Additionally, if new time series are not adequately represented in the reference panel, or if such representations are ambiguous (which may be the case for lower-intensity signals), the algorithm may need to be extended to allow for the automatic expansion of the dictionary. Although so far ACE has only been tested for automatic integration of IgG glycomic profiles for UPLC platforms, we expect the method to be generally applicable.

Acknowledgements

Pharmatics and Genos acknowledge partial support of this work by EU FP7 MIMOmics. F.A. thanks Yurii Aulchenko and Lennart Karssen for useful discussions.

[1] Data patterns from one cohort will typically be representative of other cohorts. In this case, the old labeled data can still be used without any noticeable drop in performance for labeling of a new batch of chromatographic data. However, it is recommended to provide manually labeled set for each new cohort by following the procedure of **step 1** of the algorithm.

References

1. Paul SM, Mytelka DS, Dunwiddie CT et al (2010) How to improve R&D productivity: the pharmaceutical industry's grand challenge. Nat Rev Drug Discov 9:203–214. doi:10.1038/nrd3078
2. Hay M, Thomas DW, Craighead JL et al (2014) Clinical development success rates for investigational drugs. Nat Biotechnol 32:40–51. doi:10.1038/nbt.2786
3. Prinz F, Schlange T, Asadullah K (2011) Believe it or not: how much can we rely on published data on potential drug targets? Nat Rev Drug Discov 10:712. doi:10.1038/nrd3439-c1
4. Vivó-Truyols G, Torres-Lapasió JR, Van Nederkassel AM et al (2005) Automatic program for peak detection and deconvolution of multi-overlapped chromatographic signals: Part II: peak model and deconvolution algorithms. J Chromatogr A 1096:146–155
5. Yu T, Peng H (2010) Quantification and deconvolution of asymmetric LC-MS peaks using the bi-Gaussian mixture model and statistical model selection. BMC Bioinformatics 11:559. doi:10.1186/1471-2105-11-559
6. Yang C, He Z, Yu W (2009) Comparison of public peak detection algorithms for MALDI mass spectrometry data analysis. BMC Bioinformatics 10:4. doi:10.1186/1471-2105-10-4
7. Tomasi G, van den Berg F, Andersson C (2004) Correlation optimized warping and dynamic time warping as preprocessing methods for chromatographic data. J Chemom 18:231–241
8. Wang CP, Isenhour TL (1987) Time-warping algorithm applied to chromatographic peak matching gas chromatography/Fourier transform infrared/mass spectrometry. Anal Chem 59:649–654
9. Clifford D, Stone G (2012) Variable penalty dynamic time warping code for aligning mass spectrometry chromatograms in R. J Stat Softw 47:1–17
10. Wang S-Y, Ho T-J, Kuo C-H, Tseng YJ (2010) Chromaligner: a web server for chromatogram alignment. Bioinformatics 26:2338–2339
11. Bork C, Ng K, Liu Y et al (2013) Chromatographic peak alignment using derivative dynamic time warping. Biotechnol Prog 29:394–402
12. Prince JT, Marcotte EM (2006) Chromatographic alignment of ESI-LC-MS proteomics data sets by ordered bijective interpolated warping. Anal Chem 78:6140–6152
13. Hoffmann N, Keck M, Neuweger H et al (2012) Combining peak- and chromatogram-based retention time alignment algorithms for multiple chromatography-mass spectrometry datasets. BMC Bioinformatics 13:214. doi:10.1186/1471-2105-13-214
14. Keogh EJ, Pazzani MJ (2001) Derivative dynamic time warping. In: SDM. SIAM, pp 5–7
15. Ratanamahatana CA, Keogh E (2005) Three myths about dynamic time warping data mining. In: Proc. SIAM Int. Conf. Data Min. SDM'05. SIAM, pp 506–510
16. McQuillan R, Leutenegger A-L, Abdel-Rahman R et al (2008) Runs of homozygosity in European Populations. Am J Hum Genet 3:359–372
17. Freund Y, Schapire R, Abe N (1999) A short introduction to boosting. J Jpn Soc Artif Intell 14:1612
18. Viola P, Jones M (2001) Rapid object detection using a boosted cascade of simple features. In: Comput. Vis. Pattern Recognit. 2001 CVPR 2001 Proc. 2001 IEEE Comput. Soc. Conf. On. IEEE, pp I-511
19. McCabe GP (1984) Principal variables. Technometrics 26:137–144
20. Balkema AA, De Haan L (1974) Residual life time at great age. Ann Probab 2:792–804
21. Johnsen LG, Skov T, Houlberg U, Bro R (2013) An automated method for baseline correction, peak finding and peak grouping in chromatographic data. Analyst 138:3502–3511

Chapter 18

Databases and Associated Tools for Glycomics and Glycoproteomics

Frederique Lisacek, Julien Mariethoz, Davide Alocci, Pauline M. Rudd, Jodie L. Abrahams, Matthew P. Campbell, Nicolle H. Packer, Jonas Ståhle, Göran Widmalm, Elaine Mullen, Barbara Adamczyk, Miguel A. Rojas-Macias, Chunsheng Jin, and Niclas G. Karlsson

Abstract

The access to biodatabases for glycomics and glycoproteomics has proven to be essential for current glycobiological research. This chapter presents available databases that are devoted to different aspects of glycobioinformatics. This includes oligosaccharide sequence databases, experimental databases, 3D structure databases (of both glycans and glycorelated proteins) and association of glycans with tissue, disease, and proteins. Specific search protocols are also provided using tools associated with experimental databases for converting primary glycoanalytical data to glycan structural information. In particular, researchers using glycoanalysis methods by U/HPLC (GlycoBase), MS (GlycoWorkbench, UniCarb-DB, GlycoDigest), and NMR (CASPER) will benefit from this chapter. In addition we also include information on how to utilize glycan structural information to query databases that associate glycans with proteins (UniCarbKB) and with interactions with pathogens (SugarBind)

Key words Glycobioinformatics, Oligosaccharide, Omics, HPLC, NMR, Mass spectrometry, Databases

1 Introduction

The success of bioinformatics in providing breakthroughs in life science research has been gradual, and it is hard to say when it all began. Publically accessible life science databases, with their symbiotic connection to the open Internet, turned the concept into reality at the end of the last century. At the same time, the need to query databases with intelligent software increased. Genomic databases are currently the dominating resource, and the population of these databases with genome sequences has been boosted by

Electronic supplementary material: The online version of this chapter (doi:10.1007/978-1-4939-6493-2_18) contains supplementary material, which is available to authorized users.

Gordan Lauc and Manfred Wuhrer (eds.), *High-Throughput Glycomics and Glycoproteomics: Methods and Protocols*, Methods in Molecular Biology, vol. 1503, DOI 10.1007/978-1-4939-6493-2_18, © Springer Science+Business Media New York 2017

analytical developments starting from Sanger sequencing to what is now known as NextGen sequencing that allows sequencing of the genome (DNA) and transcriptome (mRNA). These nucleotide databases (GenBank, European Nucleotide Archive, and DNA Data Bank of Japan) are now invaluable resources for the life sciences. Subsequently, with the improvement in biological mass spectrometric ionization techniques such as ESI and MALDI, large-scale protein identification (proteomics), utilizing nucleotide databases, became possible. Other -omics such as metabolomics and, as is the case for this chapter, glycomics/glycoproteomics have been more challenging because the molecules are secondary gene products that cannot be predicted from the genomic databases. Instead, the generation of glycan structural databases relies on data from the glycoanalytical field, and on the development of glycobioinformatics standards and symbolic nomenclature to allow the transition from the glycocommunity to the wider life science arena. Early attempts to provide glycostructural databases include CarbBank [1], EUROCarbDB [2], and GlycoSuiteDB [3], and their legacy whereby they are now integrated into the UniCarb family [4, 5], GlycomeDB [6], GlycoBase [7], and GRITS (www.grits-toolbox.org). With the realization that understanding post-translational modifications is key to describing the functional molecules that operate in every cell in an organism, glycodatabases and associated tools have become a prerequisite for life science research, not only for addressing glycobiology specifically, but for overall understanding of cell systems biology. The integration of various -omics data will be required for modulating an in silico cell (Fig. 1).

Glycomics databases that are available today can roughly be divided into analytical databases that can be queried and compared with data to predict structures, and databases that can be queried by the resultant glycan structural information to obtain associated metadata (including binding partners, associated glycosyltransferases and glycosidases, attached proteins, lipids, related organisms and diseases, and prediction of 3D structures). Glycosylation information can also be found in more generic life science databases such as UniProt, Protein DataBase, and Protein Atlas. Glycomics has been provided with a separate tab in ExPASy (www.expasy.org/glycomics) that includes both internal and external links to databases and tools. Without claiming to describe every resource that is available we provide a table (Table 1) that contains many glycodatabases. In addition we also describe in detail a selection of tools that can be used to interpret glycomic data generated by well-established analytical methods, including MS, NMR, and HPLC. The focus of this chapter is on the application of bioinformatics to the *N*-linked and *O*-linked glycans on mammalian glycoproteins. This chapter also includes examples of how to indicate the way glycan structures interact with pathogens, and how to find proteins on which a particular glycan structure(s) is expressed. The methods described here are devoted to interpretation of data generated

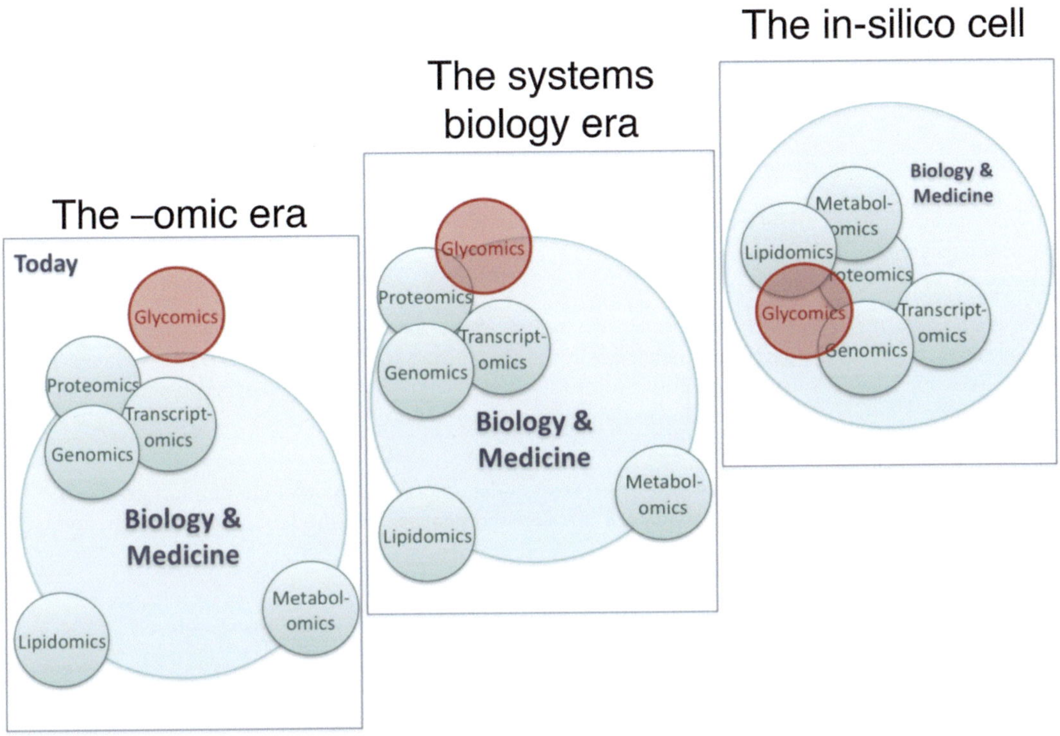

Fig. 1 The future of integration of -omics research in biology and medicine in order to understand the interplay between molecules of life

by glycomic/glycoproteomic analyses. There are other chapters and literature (e.g., Hayes et al. [8] and Campbell et al. [9]) that specifically address the workflows for generation of data that can be used for the tools presented in this chapter. We provide datasets as supplementary material to this chapter for the reader to test the different query functions of databases and tools described herein.

2 Materials

2.1 From Analytical MS Data to Monosaccharide Composition: GlycoMod

GlycoMod uses experimentally determined MS data to predict, for glycans released from a submitted peptide or protein sequence, the monosaccharide composition of free or derivatized *N*- and *O*-glycans (*see* **Note 1**), as well as the monosaccharide composition of glycans on glycopeptides from proteolytically digested proteins. The prediction is based on calculation of the combination of monosaccharide masses with some rules restricting unlikely combinations [10].

Online website: web.expasy.org/glycomod/

Type of data: MS peak list can be submitted from both negative and positive ion mode mass spectrometry.

Table 1
List of databases relevant for glycomics and glycoproteomics

Science databases that inclusively contain glycoinformation
The Universal Protein Resource (*UniProt*) is a resource for protein sequence with annotations, www.uniprot.org. *Glycodatabase related*: Predicted and confirmed sites of glycosylation within protein amino acid sequences. *References*: Farriol-Mathis et al. [40]
The Protein Data Bank (*PDB*) is a database containing three-dimensional data from large biological molecules, such as proteins and nucleic acids, www.rcsb.org. *Glycodatabase related*: Information about localization of *N*-and *O*-linked glycosylation sites as well as ligand localization of carbohydrate-binding proteins. *References*: Berman et al. [41]
The Gene Expression Omnibus (*GEO*) gene expression data repository, www.ncbi.nlm.nih.gov/geo. *Glycodatabase related*: Expression profiles of glycorelated genes in cell models, animal models, and clinical samples. *Reference*: Barett et al. [42]
The Human Proteome Atlas is a tissue, cell line, and cancer tissue map of the proteome, www.proteinatlas.org. *Glycodatabase related*: Glycoprotein and glycogene localization in tissues, as well as subcellular localization. *References*: Uhlén et al. [43]
LipidBank is a database of natural lipids, www.lipidbank.jp. *Glycodatabase related*: Literature data and experimental evidence (e.g., MS, UV, IR, NMR) of glycolipids. *References*: Watanabe et al. [44]
LIPID MAPS contains chemical representation and annotations of biologically relevant lipids, www.lipidmaps.org. *Glycodatabase related*: Chemical representations of glycoplipids from consortium and literature data. *Reference*: Sud et al. [45]
Exclusive glycodatabases
The Consortium for Functional Glycomics (*CFG*) is an international initiative providing glycomic services and databases, www.functionalglycomics.org. *Glycodatabase related*: Data from the screening of the consortium's glycan array platform, glycogene microarray of tissues and cells, MALDI-MS screening of glycans from mouse and human tissue. It also contains information about histology, immunology, hematology, and metabolism/behavior of glyco-related gene mouse knockouts. The "CFG Molecule Pages" integrate data generated from the CFG screening with public data about glycosyltransferases, glycan structures, and glycan-binding proteins. Glycan array data from the consortium can be mined by GlycoPattern (glycopattern.emory.edu) [46].
The UniCarb KnowledgeBase (*UniCarbKB*) is a curated database of information on the glycan structures of glycoproteins reported in scientific literature, www.unicarbkb.org. *Glycodatabase related*: Current version contains 3238 structures from the literature, including information about taxonomy, tissue, and associated protein. *Reference*: Campbell et al. [29]

(continued)

Table 1
(continued)

The UniCarb-DB database is an assembly of glycosequences with associated MS fragmentation data, www.unicarb-db.org. *Glycodatabase related*: Current version contains more than 600 spectra from published work, including information about structure, taxonomy, tissue, and associated protein. *Reference*: Hayes et al. [4]
GlycomeDB is a metadatabase of carbohydrate sequences, www.glycome-db.org. *Glycodatabase related*: Database includes structures from CFG, KEGG, GLYCOSCIENCES.de, BCSDB, and CarbBank [1], as well as information about species and associated aglycon. *References*: Ranzinger et al. [6]
GlyTouCan, International glycan sequence repository, www.glytoucan.org. *Glycodatabase related*: Allows registration of a glycan structure with minimal metadata to obtain a unique accession number. *References*: Aoki-Kinoshita et al. [47]
Kyoto Encyclopedia of Genes and Genomes (*KEGG*) is a database of genome sequences and other high-throughput data, www.genome.jp/kegg. *Glycodatabase related*: www.genome.jp/kegg/glycan. Pathways for glycan biosynthesis and metabolism, roles of glycans in cellular processes, glycans in cancer, glycosyltransferase reactions and relation to diseases, carbohydrate binding proteins, The KEGG GLYCAN structure database contains structures from CarbBank [1], literature, and from structures present in KEGG pathways. *Reference*: Kanehisa et al. [48], Hashimoto et al. [49]
The Japan Consortium for Glycobiology and Glycotechnology DataBase (*JCCGB*) is a metadatabase that allows one to simultaneously search in glyco related databases, www.jcggdb.jp. *Glycodatabase related*: The current version includes GGDB (GlycoGene Database, www.jcggdb.jp/rcmg/ggdb), LfDB (Lectin Frontier Database, www.jcggdb.jp/rcmg/glycodb/LectinSearch), GlycoPOD (Glyco-Science Protocol Online Database, www.jcggdb.jp/GlycoPOD), GlycoProtDB (GlycoProtein Database, www.jcggdb.jp/rcmg/gpdb), GMDB (Glycan Mass Spectral Database, www.jcggdb.jp/rcmg/glycodb/Ms_ResultSearch), LipidBank, GlycoEpitope, GALAXY (Glycoanalysis by the three axes of MS and chromatography), KO mice DB (Functional Glycomics with KO Mice Database, www.metadb.riken.jp/metadb/db/Glycomics_mouse), Glycoforum, PACDB (Pathogen Adherence to Carbohydrate Database, www.jcggdb.jp/search/PACDB.cgi), GDGDB (Glyco-Disease Genes Database), TuMaRdb (Tumor Markers Reference Database), JCGGDB Reports, GlycosidaseDB, and JIGSDB (JCGG Integrated Glycan Structures Database). *Reference*: Maeda et al. [50]
The GlycoEpitope is a database of 173 common names of glycobiological epitopes associated with known oligosaccharide sequence, www.glycoepitope.jp.
GALAXY is a database of glycan sequences. Glycoanalysis using MS and chromatography, www.glycoanalysis.info. *Glycodatabase related*: Elution profiles of PA-labeled *N*-linked oligosaccharides analyzed by different LC-methods. *Reference*: Tomiya et al. [51]
GlycoDomain Viewer is a graphic interface that displays a database of *O*-linked and *N*-linked glycosylation sites on proteins which are based on UniProt annotation and in-house discovery, glycodomain.glycomics.ku.dk. *Glycodatabase related*: Database of *O*-linked sites has been used in NetOGlyc neural network predictions of mucin type GalNAc O-glycosylation sites in mammalian proteins (www.cbs.dtu.dk/services/NetOGlyc). *Reference*: Steentoft et al. [52]

(continued)

Table 1
(continued)

GLYCOSCIENCES is a database of carbohydrate sequences from literature that provides taxonomy, linking to PDB, NMR and 3D modules, www.glycosciences.de. *Glycodatabase related*: Database built on CarbBank [1] which is updated by including glycans from PDB. It also includes the MonosaccharideDB, a database of monosaccharides (www.monosaccharidedb.org). *Reference*: Lütteke et al. [53]
Glycobase (*Lille*) is a glycan sequence database which includes taxonomy and NMR data, glycobase, www.glycobase.univ-lille1.fr/base.
Glycobase (*Dublin*) is a database of glycan sequences that includes experimental HPLC data, taxonomy, MS and literature references. glycobase, www.nibrt.ie/glycobase/show_nibrt.action. *Glycodatabase related*: Current version v3.2 include 655 sequences. *Reference*: Campbell et al. [7]
Carbohydrate Structure Database (*CSDB*) is a merged database of Bacterial (BCSDB), and Plant and Fungal (PFCSDB) databases, www.csdb.glycoscience.ru. *Glycodatabase related*: Structures of glycans and NMR data from literature.
Glyco3D is a portal for databases covering three-dimensional features of monosaccharides, disaccharides, oligosaccharides, polysaccharides, glycosyltransferases, lectins, monoclonal antibodies against carbohydrates, and glycosaminoglycan-binding proteins, www.glyco3d.cermav.cnrs.fr/home.php. *Glycodatabase related*: The database contains three-dimensional data on oligosaccharides (BiOligo), lectins (Lectin3D), glucosaminoglycan-binding proteins with ligands (GAG), monoclonal antibodies with carbohydrate ligands (mAbs), glycosyltransferases (GT), and polysaccharides (Polysac). *Reference*: Pérez et al. [54]
The Carbohydrate-Active enZYmes Database (*CAZy*) contains genomic, structural and biochemical information on glyco-enzymes, www.cazy.org. *Reference*: Lombard et al. [55]
The Unipep database contains theoretical N-linked glycosylation sites in UniProt that have been mapped to LC-MS data, www.unipep.org. *Reference*: Zhang et al. [56]
GLYCAM libraries contain predicted 3D structures of oligosaccharides. Structures are available for *N*-linked, *N*-acetyllactosamine building blocks, and from the CFG-array, glycam.org/Pre-builtLibraries.jsp. *Reference*: Kirschner et al. [57]
SugarBind Database contains literature references on lectin adhesins of viral and bacterial pathogens and biotoxins, and their known carbohydrate ligands, www.sugarbind.expasy.org. *Reference*: Mariethoz et al. [32]
ECODAB (*Escherichia coli* O-antigen database) contains O-polysaccharide structures of *E. coli* lipopolysaccharides, putative, predicted, or confirmed glycosyltransferase functions, ^{1}H and ^{13}C NMR chemical shift data, flippase and polymerase sequences, and literature references, www.nevyn.organ.su.se/ECODAB. *Reference*: Rojas-Macias et al. [22]
EK3D is a manually curated database of K antigens corresponding to various *E. coli* serotypes, and their modeled 3D structures, www.iith.ac.in/EK3D. *Reference*: Kunduru et al. [58]

Test data for:

N-linked glycopeptides: *N*-linked glycopeptide of human alpha-HS-glycoprotein (P02765), reduced and alkylated with iodoacetamide. $[M+H]^+$ ion of m/z 3394.4495 (±0.05 Da)

Reduced, released *O*-linked oligosaccharides: $[M-H]^-$ ions of m/z 384.1, 675.1, 966.2, and 1331.2 (±0.3 Da)

Released AB-labeled *N*-linked glycans: $[M-H]^-$ ions of m/z 1581.92, 1744.00, and 1906.01 (±0.5), mass of 2-aminobenzamide 136.06 Da

2.2 From Structure to Predicted MS/MS Data: GlycoWorkbench

GlycoWorkbench is designed to facilitate the annotation of mass spectrometry data while presenting corresponding glycan structures [11, 12]. A particular feature of this software is that it provides a friendly interface for drawing various glycans of different types, labeling, and charge states. The theoretical MS/MS fragment of a selected structure can be computed.

Website for downloading tool: code.google.com/p/glycoworkbench/ (static); github.com/alternativeTime/glycoworkbench (current)

Manual: http://bit.ly/1Q3pAiI

Type of data: MS/MS peak lists of glycans

Test MS/MS data: A peak list of a PA-labeled *N*-linked glycan with $[M-H]^-$ ion of m/z 1311.08 and a composition of five hexoses and two *N*-acetylhexosamines (Supplementary Table 1).

2.3 From Analytical MS and MS/MS Data to Structure: UniCarb-DB

UniCarb-DB [4] provides access to a collection of LC-MS/MS *N*- and *O*-linked glycan fragments released from glycoproteins and complemented with experimental metadata. UniCarb-DB supports the comparison of users' experimental spectra against the experimental spectra in the database. Therefore, UniCarb-DB can be used as reference to aid manual annotation and development of more precise software tools to analyze and interpret spectrometric data.

Online website: www.unicarb-db.org

Type of data: MS/MS fragment peak list from bar spectra together with pseudomolecular ion mass (*see* **Note 2**)

Test MS data for MS/MS search: MS/MS peak lists of a reduced *N*-linked oligosaccharide with [M−H] ion of m/z 1056.48 (Supplementary Table 2).

2.4 From NMR Data to Structure: CASPER

The computer program CASPER (Computer Assisted SPectrum Evaluation of Regular polysaccharides) was originally developed to facilitate structural elucidation of polysaccharides using input sugar and methylation analyses together with *unassigned* ^{13}C NMR data [13]. Subsequently, ^{1}H and 2D NMR data were made possible as input to the program [14–16] (which was also enhanced to include oligosaccharides [17]), in the analysis of unknown structures. CASPER is freely available online at the website www.casper.organ.su.se/casper/ [18].

CASPER is used in two modes: (1) prediction of ^{1}H and ^{13}C NMR chemical shifts using a given structure as input or (2) structural determination of a glycan using unassigned ^{1}H and/or ^{13}C NMR data as well as components [19, 20] and linkage positions when known, e.g., as suggested by bioinformatics information in the database ECODAB [21, 22]. Besides NMR chemical shift predictions and structural elucidation based on the constituent sugar residues, substituents such as *O*-acetyl, phosphoryl, sulfuryl, and pyruvyl groups can be added. Glycosides functionalized at the reducing end by amino acids, methyl, propyl, 2-azidoethyl, and trifluoroacetyl-*para*-aminophenylethyl groups are also available as well as sugar residues reduced to their corresponding alditols [23, 24].

2.5 From HPLC/UHPLC/CE Data to N-linked Structure: GlycoBase

The National Institute of Bioprocessing Research and Training (NIBRT) experimental database, GlycoBase is an experimental resource that allows users to conduct structural assignments based on standardized liquid chromatography retention times, expressed as glucose unit (GU) values. Further confirmation of structures can be achieved using exoglycosidase digestions followed by further chromatographic analysis to determine monosaccharide sequence and linkage.

On-line open access website: glycobase.nibrt.ie

Type of data: Chromatographic data: Released glycans labeled with 2-aminobenzamide with GU values and relative percent areas. A description of how data can be generated is found in Chapters 4 and 10 and in references [8, 25–27]. Test data can be found in the table in Fig. 7. 8-aminopyrene-1,3,6-trisulfonic acid-labeled glycans can be analyzed on CE using the same rationale.

2.6 Validation of Monosaccharide Sequence and Linkages with the Aid of Exoglycosidase Digestion: GlycoDigest

One of the most commonly used methods for determining the sequence and structure of oligosaccharides is the enzymatic analysis of oligosaccharides using highly specific exoglycosidases, either sequentially or in array form. GlycoDigest is a web-application and standalone tool that simulates the behavior of these exoglycosidase digestions on released *N*- and *O*-linked oligosaccharides. The tool is based on a set of rules acquired from expert knowledge and supporting experimental evidence provided by NIBRT's GlycoBase, and can be used to tailor a panel of exoglycosidases to unambiguously confirm oligosaccharide structure.

Online website: www.glycodigest.org

Type of data: A description of the enzymes supported and a user guide can be accessed at www.glycodigest.org and in reference [28].

2.7 Literature Report on Structure and Association with Protein: UniCarbKB

UniCarbKB is an international collaboration that develops open source glycobioinformatics databases to support glycomics and the emerging technology of glycoproteomics [29]. It is a curated collection of information on well-characterized glycan structures derived from glycoproteins, biological fluids, and tissues. Although UniCarbKB provides annotated entries for many species, its primary focus is on proteins from mammalian systems of distinct taxonomic groups to ensure the presence of high quality annotations for representative members of all protein families. The database provides contextual information for *N*- and *O*-glycan structures attached to proteins and where known, maintains the commonly lost connection between a glycan structure and its attached protein(s), as annotated in UniProtKB [30] and neXtProt [31]. In addition, an entry is supplemented with descriptions of its biological source, supporting reference and experimental methods employed to determine the glycan structure.

Online website: www.unicarbkb.org.

2.8 Association of Structure with Microbial Interactions: SugarBindDB

SugarBindDB is a curated database that covers knowledge of glycan binding of human pathogen lectin adhesins. A set of three inseparable components constitutes the core information that is stored, i.e., the pathogenic agent, a lectin adhesin, and a glycan ligand. Each of these entities is named with as much precision as possible: taxonomic designation for pathogen agent, protein name for lectin, and epitope name for ligand. Furthermore, each glycan-binding event is backed with at least one published reference. SugarBindDB is cross-linked to protein and glycan-related resources such as UniProtKB and UniCarbKB. It is tightly bound to the latter via a substructure search tool that maps each ligand to full structures when it is available. Other search, navigation, and visualization tools are implemented to consult, compare, and bring out information on glycan binding. For instance, SugarBindDB can be used to identify protein partners interacting via glycans as shown in Subheading 3.8.

Online website: sugarbind.expasy.org [32]

3 Methods

3.1 From Analytical MS Data to Monosaccharide Composition: GlycoMod

The most straightforward use of MS analysis for glycoconjugates is to determine monosaccharide compositions based on experimentally determined masses; and GlycoMod can be used for this purpose on both released glycans and glycopeptide MS data. GlycoMod is also cross-linked to UniCarbKB where predicted compositions are linked to possible glycan sequences reported in the literature and found in the database. This allows the combinatorial possibilities of monosaccharides to be narrowed down to previously seen and reported structures in biological systems.

3.1.1 Searching GlycoMod Using Oligosaccharides and Glycopeptides

1. Launch the GlycoMod website (Fig. 2, numbers below refer to this figure)
2. Export peak lists of singly charged pseudomolecular ions from mass spectrometric software (*see* **Note 3**).
3. Paste the peak list of pseudomolecular ions (1) together with information on type of adducts and ion mode (2) and MS accuracy (3) into the website (*see* test data in Subheading 2.1).
4. Enter information into the website about the type of glycoconjugate you are analyzing:
   (a) *N*- or *O*-linked
   (b) Glycopeptides, free oligosaccharides, reduced oligosaccharides (4) (for the *O*-linked test data) and derivatized oligosaccharides (for the *N*-linked test data)
   (c) Glycoprotein sequence or UniProt ID, type of protease, and cysteine modification (5) (for glycopeptide test data insert P02765, trypsin and iodoacetamide, respectively)
   (d) Reducing end derivatization mass (6) (for the *N*-linked test data 136.06)
   (e) Type of monosaccharides (7) (*see* **Note 4**)
5. Submit the search
6. For each match, compositions are displayed that match submitted data. UniCarbKB database can be used initially to narrow down the possibilities. The data will be provided as compositions along with MS error and masses of displayed glycoforms (*see* **Note 5**).
7. For glycopeptide data, the result should be: 1736.65 (glycoform mass), −0.003 (Δmass (Dalton)), $(HexNAc)_2$ (Deoxyhex ose$)_3$ + $(Man)_3(GlcNAc)_2$ (structure), 1656.796 (peptide mass [M]), and VCQDCPLLAPLNDTR (peptide sequence)
8. For the AB-labeled *N*-linked test data the compositions with UniCarbKB entries should be according to Table 2.
9. For the reduced *O*-linked test data, the compositions with UniCarbKB entries (in addition to compositions that are less likely to occur in nature) should be according to Table 3.

3.2 From Structure to Predicted MS/MS Data: GlycoWorkbench

GlycoWorkbench is designed for rapid structure drawing, theoretical fragment computation, and association of theoretical fragments with experimental MS/MS peak lists, as well as other many useful functions.

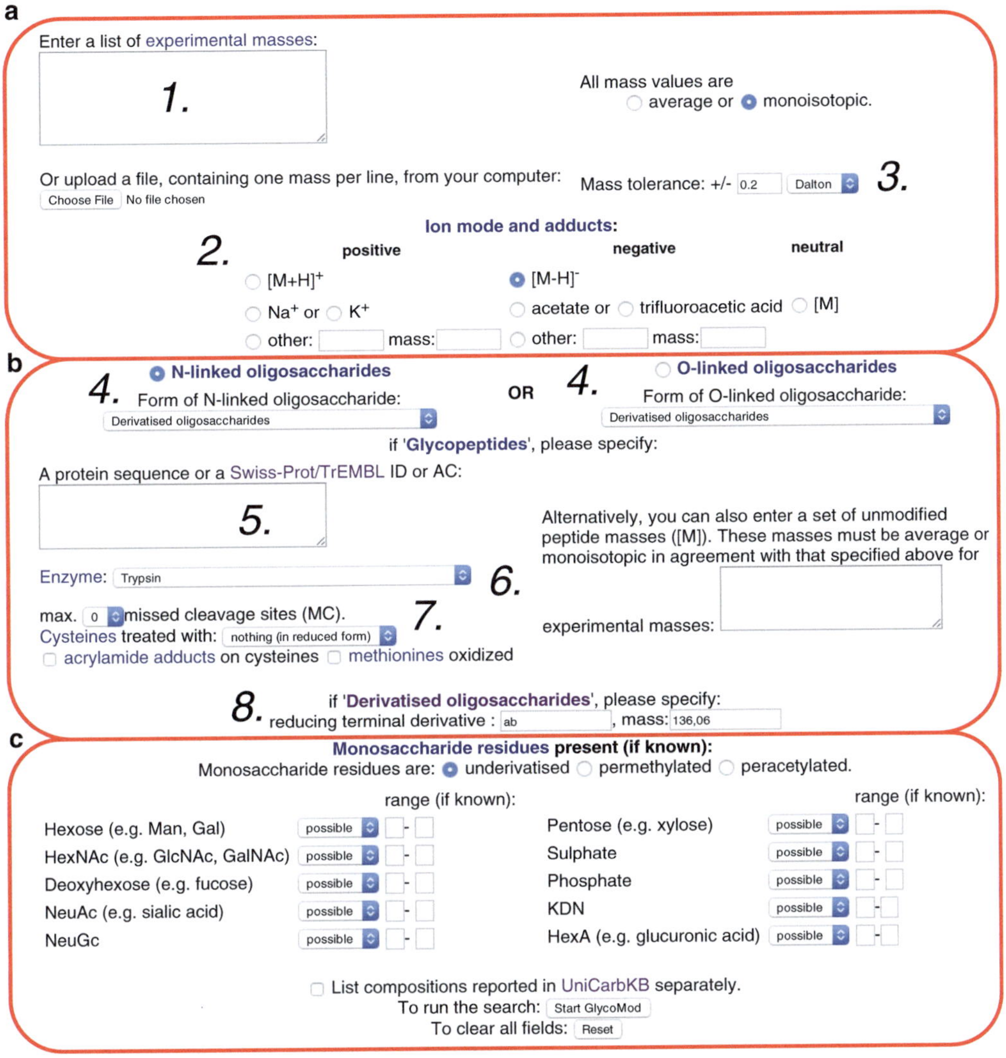

Fig. 2 Input page from Glycomod, showing: (**a**) area for input of MS parameters; (**b**) input of glycoconjugate metadata; and (**c**) monosaccharide compositional data. *Numbers* refer to the text in Subheading 3.1.1

Table 2 Results from *N*-linked test data using GlycoMod

Composition	Submitted [M – H]	Glycoform mass (Da)	Δmass (Da)
$(HexNAc)_2$ $(Deoxyhexose)_1 + (Man)_3(GlcNAc)_2$	1581.92	1444.534	0.318
$(Hex)_1$ $(HexNAc)_2$ $(Deoxyhexose)_1 + (Man)_3(GlcNAc)_2$	1744.00	1606.587	0.345
$(Hex)_2$ $(HexNAc)_2$ $(Deoxyhexose)_1 + (Man)_3(GlcNAc)_2$	1906.01	1768.64	0.302

Table 3 Results from *O*-linked test data using GlycoMod

Composition	Submitted [M – H]	Glycoform mass (Da)	Δmass (Da)
$(Hex)_1$ $(HexNAc)_1$	384.1	365.132	–0.05
$(Hex)_1$ $(HexNAc)_1$ $(NeuAc)_1$	675.1	656.228	–0.146
$(Hex)_1$ $(HexNAc)_1$ $(NeuAc)_2$	966.2	947.323	–0.141
$(Hex)_2$ $(HexNAc)_2$ $(NeuAc)_2$	1331.2	1312.455	–0.273

1. Start GlycoWorkbench and draw the putative glycan structure. Go to *Structure*>*Add structure*, and choose the type of glycan to draw on canvas. The structure can be further modified by other functional tools in *Structure* drop-down list, such as *Add residue* and *Add terminal*.
2. To set the theoretical mass, go to *Tools*>*Mass options of all structures*, where theoretical mass can be changed according to different settings.
3. To paste the MS/MS peak list to GlycoWorkbench (Fig. 3a), copy the list of fragments *m/z* and intensity from software such as Microsoft Excel, or directly copy from MS software. (*see* **Note 6**)
4. To calculate the theoretical fragmentation ions of the drawn structure, select *Tools*>*Fragments*>*Compute fragments for selected structures*. The window of *Fragment options* appears. After selecting glycosidic and cross-ring cleavage, click *OK* (Fig. 3b). The theoretical fragmentation list will appear in right panel (Fig. 3c).
5. To associate experimental MS/MS peak list with a computed peak list, go to *Tools*>*Annotation*>*Annotate peaks with fragments from selected structures*. A *Fragment option* window will appear. After setting the parameter, click *OK*. Then the associated peak list will appear in right panel (Fig. 3d).

3.3 From Analytical MS and MS/MS Data to Structure: UniCarb-DB

Different searches have been implemented for retrieving data from UniCarb-DB:

(a) Taxonomy/Tissue, where entries are selected based on their biological context,

(b) Mass and composition. As the name indicates, the search can be performed by mass and/or composition. Additional parameters can be entered to make the query more specific.

(c) MS/MS search, or peak-matching tool, retrieves and scores the peak lists in the database according to their similarity to a

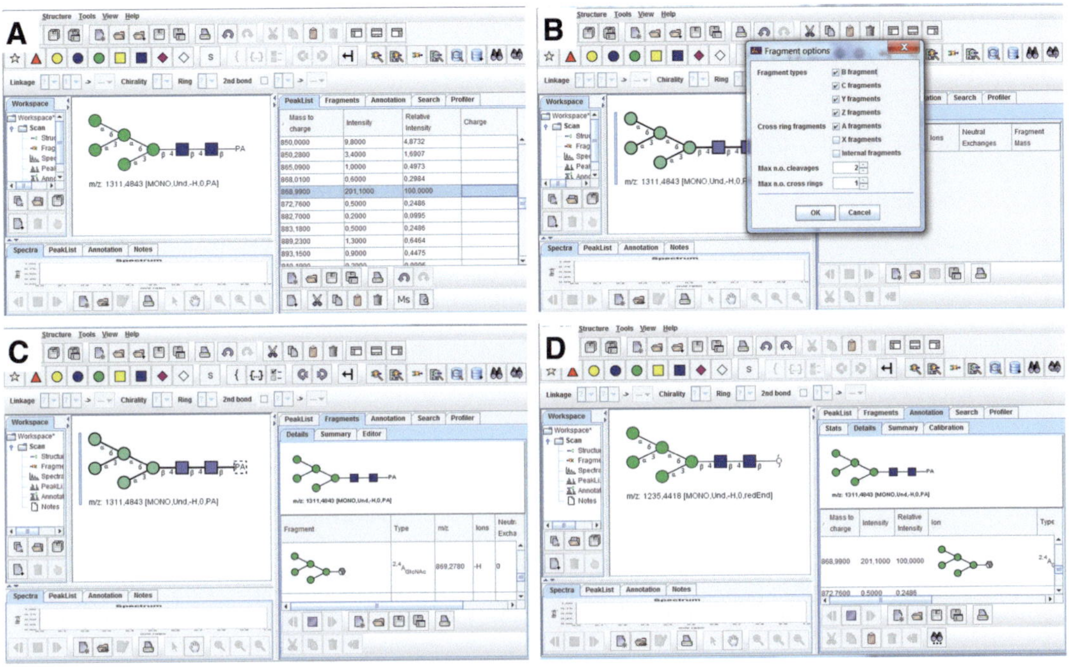

Fig. 3 Annotation of experimental MS/MS peak list using GlycoWorkbench. (**a**) Input experimental MS/MS peak list into GlycoWorkbench; (**b**, **c**) Computing theoretic glycosidic and cross-ring cleavage using GlycoWorkbench; (**d**) Associating experimental MS/MS peak list with computed fragment of selected structure

target peak list. The minimum input data is a peak list in plain text separated by a tab stop character (Fig. 4a). These data can be copied and pasted directly into the search page from software such as Microsoft Excel, or directly copied from MS software.

3.3.1 Mass and Composition

1. Upon accessing UniCarb-DB on www.unicarb-db.org click on the "Get Started" button at the bottom of the page.
2. Select the option "Search" from the menu located at the top of the interface. Afterwards, click on the option "Mass and composition" from the menu located at the left side of the search page.
3. Locate the "composition" block at the bottom of the page. Glycan composition can be assembled by selecting the option "yes" from the respective drop-down list if the desired residue must be included in the structure. "No" is chosen if the residue must not be part of the structure. The default value "possible" indicates that the residue could be but does not have to be present in the displayed results. The text boxes following the lists are used to specify the number of residues (exact or interval) that are expected. For example, use the values "yes"

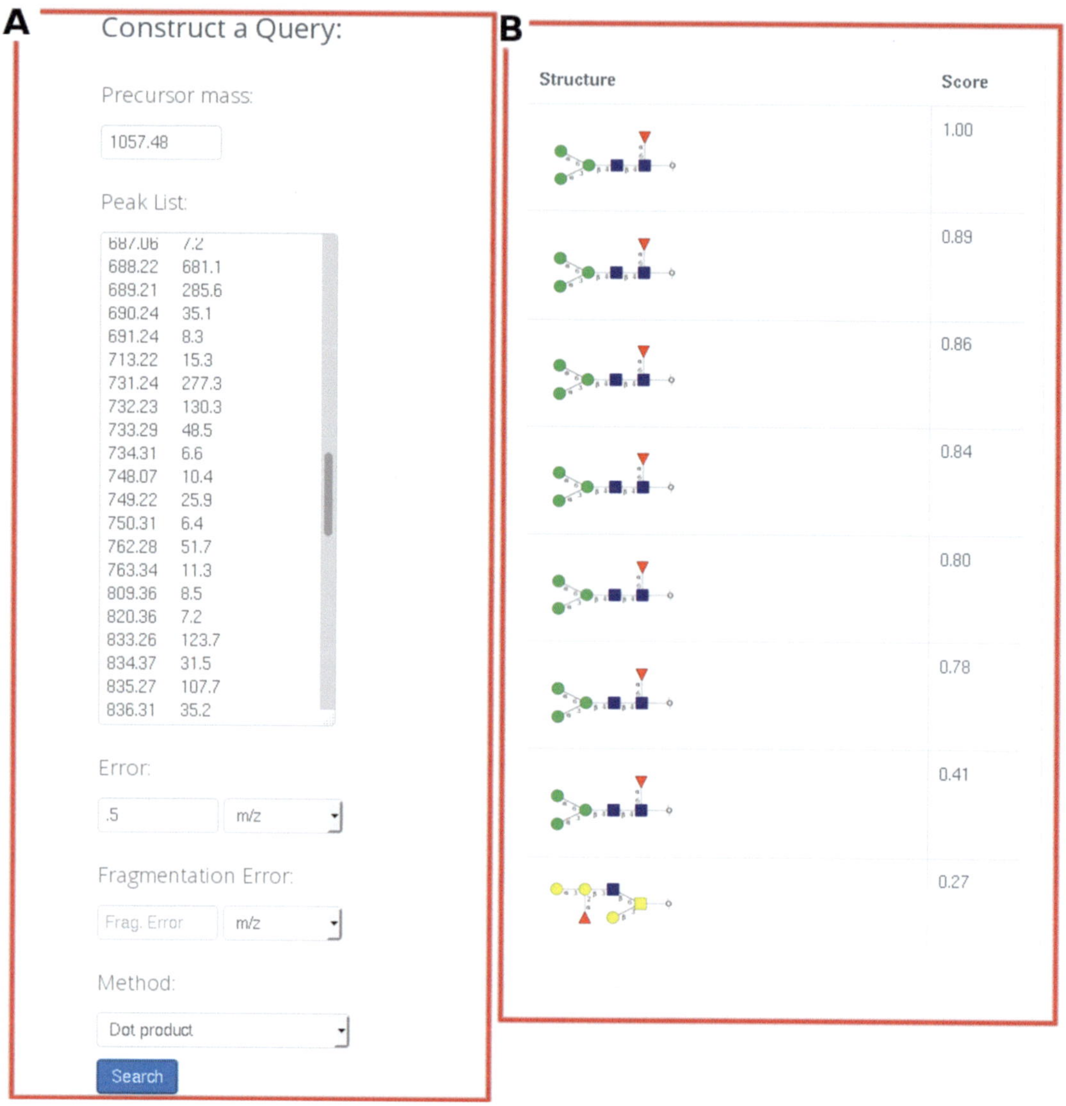

Fig. 4 UniCarb-DB MS/MS query. (**a**) Input page of the MS/MS search page on UniCarb-DB and (**b**) result page of submitted MS/MS fragment data using normalized dot product

and "1" for "dHex", "yes" and "1"-"2" for Hex, and "yes" and "2" for HexNAc.

4. Click on the "search" button.
5. The results show the list of structures that satisfy the constraints. The result can be refined using the rest of the parameters on the search form.
6. Return to the "Mass and composition" search form and perform all the previous steps.

7. Input a mass value. For example, "894.3" and select the option "Calculated".
8. The error parameter should be included only when the mass value is specified. The error is calculated for both upper and lower ranges. For example, if the value "0.5" is used, the query will select all spectra whose calculated mass values lie between 894.8 and 893.8. The drop-down list next to the text box includes three measurement formats for the error value: *m/z*, ppm, and dalton. Leave the option selected for default (*m/z*).
9. Select the option "–", "underivatised", and "reduced" from the drop-down lists "charge state", "derivative", and "reducing end", respectively.
10. Click on the "search" button.

3.3.2 Peak Matching Tool (MS/MS Search)

1. Upon accessing UniCarb-DB on www.unicarb-db.org click on the "Get Started" button at the bottom of the page.
2. Select the option "Search" from the menu located at the top of the UniCarb-DB interface. Afterwards, click on the option "MS/MS" from the menu located at the left side of the search page.
3. MS/MS spectra are represented as peak lists that consist of two columns separated by a tab stop character. The first column corresponds to the *m/z* values and the second to the respective intensities. Paste the target peak list from the excel file into the proper text box in the search form (Fig. 4a). This is the only requirement to start the search; however, the query can take a long time to display the results. The use of additional parameters can speed up the process notably.
4. To continue with the test case, use the value "1057.48" for the precursor mass.
5. The error parameter should be included only when the precursor mass is also being used. The error is calculated for both upper and lower ranges. For example, with an error value of "0.5", the query will select all spectra whose precursor mass values lie between 1056.98 and 1057.98. The drop-down list next to the text box includes three measurement formats for the error value: *m/z*, ppm, and Dalton. For the test case, leave the default option (*m/z*).
6. The fragmentation error is used to adjust for possible discrepancies among representations of numeric values.
7. The "method" drop-down list includes, for the time being, only the option "dot product". This indicates the type of algorithm that is used to perform the comparison.

8. The results are displayed as a list of structures, each one with a score (Fig. 4b). The higher the score, the more similar to the peak the lists are.

3.4 From NMR Data to Structure: CASPER

In the following example, an NMR chemical shift prediction and a structural elucidation of trisaccharide [33] and tetrasaccharide [34] *O*-mannosyl glycans present on the glycoprotein α-dystroglycan are presented using components. For structural determination, unassigned NMR spectra are also used as input. CASPER will, for a structural elucidation, produce a ranked list of structural suggestions based on the best fit of predicted chemical shifts to those from NMR experiments. Furthermore, tentative ^{1}H and ^{13}C NMR chemical shift assignments are made, as in the prediction mode, for all resonances; and a schematic chemical shift map (HSQC-type) that compares experimental and predicted chemical shifts is generated. In addition, a three-dimensional structure in pdb-format can be made seamlessly using the CarbBuilder program [35] from within the "CASPER results menu". A detailed step-by-step description of how CASPER is used was recently reported [36].

3.4.1 Input to CASPER for Prediction of NMR Chemical Shifts

In the present case, using the previously mentioned *O*-mannosyl glycan trisaccharide initiated by POMGNT2 [33], the input is prepared in the following way (Fig. 5a):

1. Go to the "Calculate Chemical Shifts" option in CASPER. (www.casper.organ.su.se/casper/calculate.php)
2. Enter a title and/or literature source for the project. (This is optional but will help with keeping track of multiple projects.) It is also possible to load a previous CCPN project [37].
3. Enter the constituents of the target structure. Anomeric configuration, residue, where the residue links, and to which residue it links must be specified. Several options for non-reducing sugars, such as O-Me, are also available. In the case of *N*-glycans and *O*-glycans, amino acids are specified as residues; however, the anomeric configuration does not apply. If it is a repeating unit, the "reducing end" should be linked to the terminal residue of the repeating unit of the backbone in the polymer. Note that the residues are specified in a biosynthetic/reducing end order, i.e., the threonine would be entered first in the example above, followed by the mannose residue. This is a truncated view, and up to 12 residues can be entered.
4. If NMR chemical shifts are known, they can be added. In the resulting output, the differences between the entered chemical shifts and the calculated ones are displayed. This is true for both ^{1}H and ^{13}C NMR data.

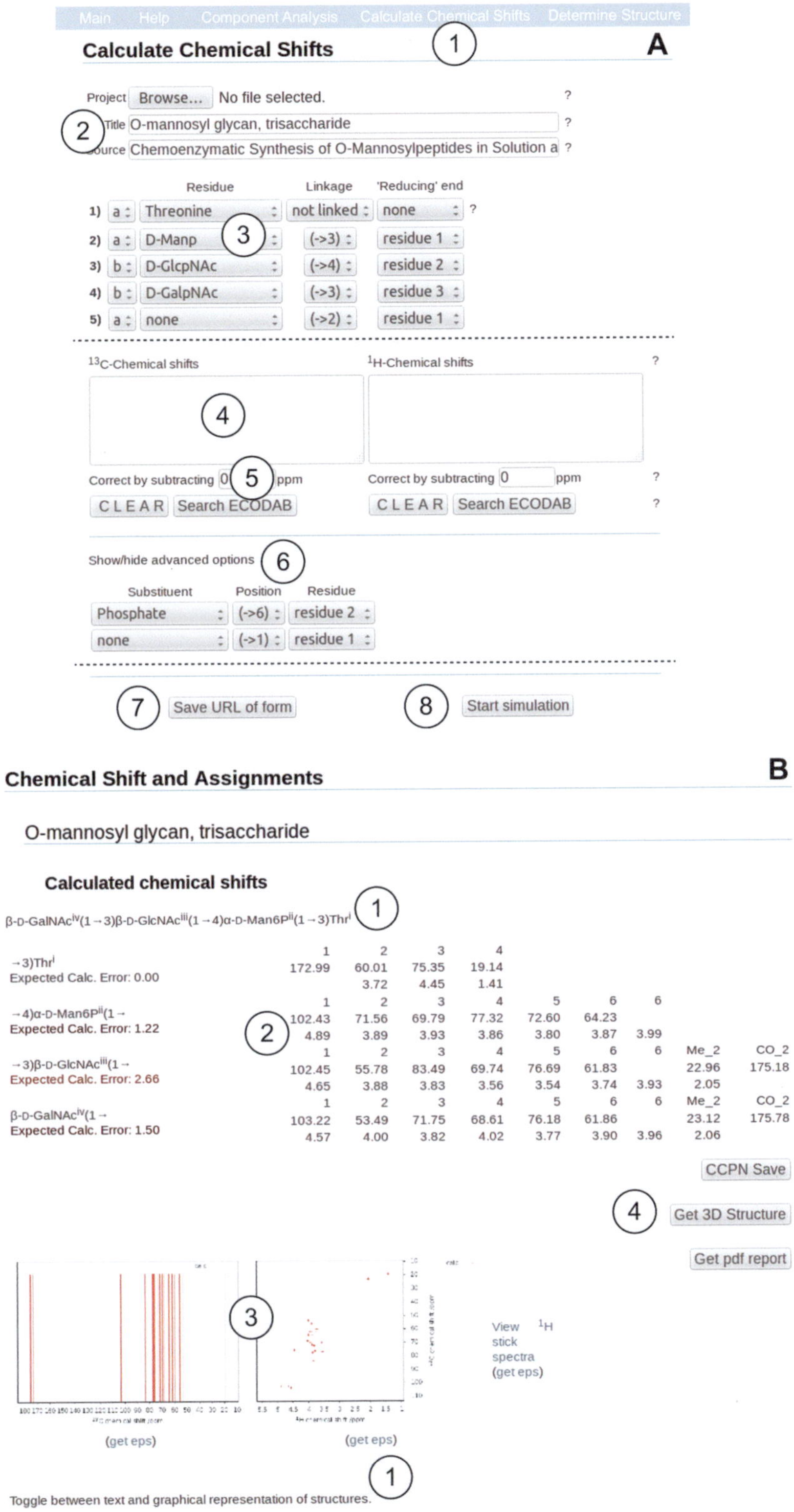

Fig. 5 CASPER input and output. (**a**) Example input for a CASPER NMR chemical shift prediction of an *O*-mannosyl glycan found on human α-dystroglycan. A *dashed line* indicates a truncated view of the part above it. (**b**) results for a CASPER NMR chemical shift prediction of an *O*-mannosyl glycan found on human α-dystroglycan. (**c**) example input for a CASPER structure determination of an *O*-mannosyl glycan found on human α-dystroglycan. The toggled window for 2D data input is not shown. A *dashed line* indicates a truncated view of the part above it

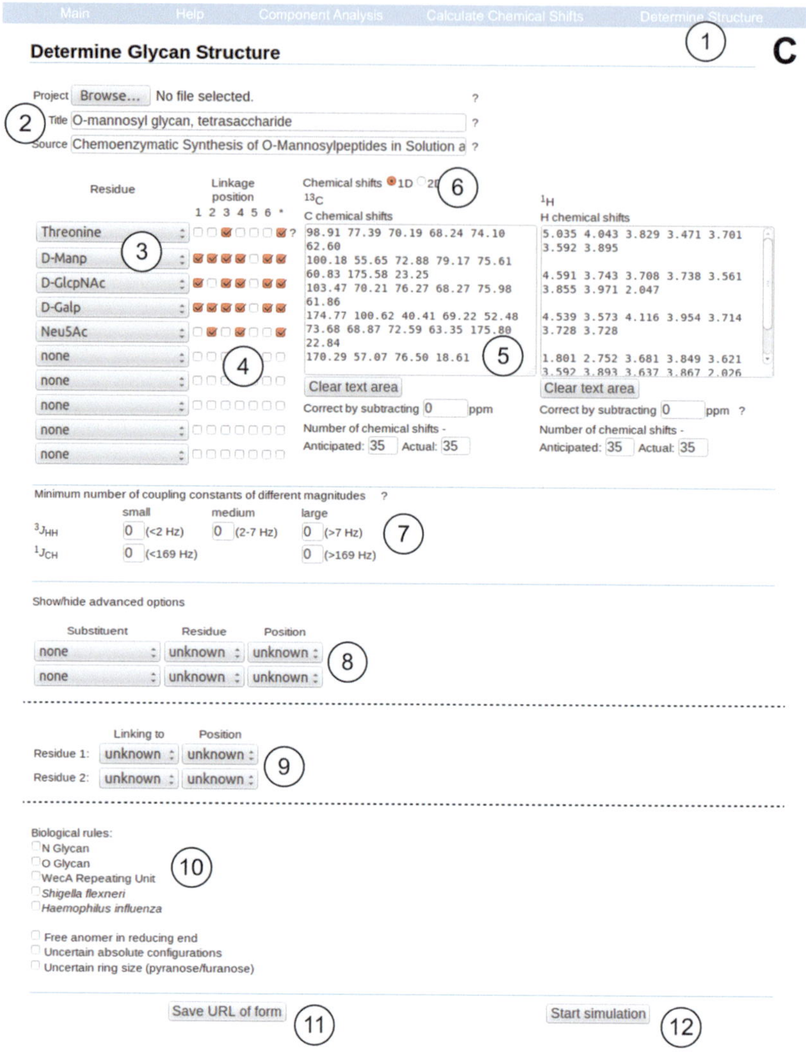

Fig. 5 (continued)

5. A systematic correction can also be added, for example, when NMR chemical shift data is obtained using different reference compounds.
6. The display for advanced options can be toggled; and, in the "Calculate Chemical Shifts" case, the option is to add substituents. This is done in a similar manner similar to specifying residues. Once again, a truncated view is shown, and up to 12 substituents can be entered.
7. The URL can be saved to quickly enter a previous input; this is done by clicking "Save URL of form".
8. Submit the job to CASPER by pressing "Start simulation". Once the calculation is complete, the page will be redirected to a results page (*see* **Note 7**).

3.4.2 Results Page from CASPER for Prediction of NMR Chemical Shifts

Once the calculation is complete, the user will automatically be redirected to a results page (Fig. 5b).

1. At the top, the structure used as input is shown in the CASPER-format [19]. The structure can also be toggled to display the CFG representation [38].
2. The NMR chemical shifts are displayed in a list for each residue along with an "expected calc. error". If chemical shift approximations to the structural entities have been made, a lower value indicates higher similarity to the predicted chemical shifts. This is a measure of the expected accuracy of the calculation.
3. Graphical representations of ^{13}C and $^{1}H/^{13}C$-correlated data are generated. If needed, the ^{1}H data can also be displayed in stick representation. It is also possible to download these in the Encapsulated PostScript format.
4. The results can be saved as a CCPN project or as a pdf-report. The option "Get 3D Structure" generates a 3D structure in the pdb format using a local version of CarbBuilder [35].

3.4.3 Using CASPER to Predict Structures from NMR Chemical Shifts

It is also possible to use CASPER to determine a structure (Fig. 5c) from NMR chemical shifts along with any known constituents, for example, from sugar analysis, methylation analysis, or putative glycosyltransferase functions retrieved from a gene sequence comparison [21, 22]. In this example, the structure of an *O*-mannosyl glycan tetrasaccharide is determined based on previously published data [34].

1. Go to the "Determine Structure" option in CASPER. (www.casper.organ.su.se/casper/determine.php)
2. As in the "Calculate Chemical Shifts" mode, a brief title and source can be entered to help keep track of projects. This is optional.
3. Enter the residues, known or unknown. Unknown residues will increase calculation times significantly. Since linkages are not specified here, the order of residues does not matter; however, the first drop-down window does have some additional options, so the residue expected to be at the reducing end must be entered here; in our case the threonine residue is given.
4. Enter the possible linkage positions. The numbering 1–6 are all AND statements, but use of the wildcard (*) will allow the linkage positions to be AND/OR statements.
5. Enter the ^{13}C and/or ^{1}H NMR chemical shifts. Just as in predicting chemical shifts, if there is a systematic error, a correction to the chemical shifts can be added.

6. By toggling to 2D experiments, data from HSQC/HETCOR, TOCSY, HMBC, and/or H2BC experiments can also be supplied (toggled window not shown).
7. If the coupling constants ($^3J_{HH}$, $^1J_{CH}$) for the anomeric protons are known, they can be entered to limit the number of structures for which NMR chemical shift calculations are carried out, thereby speeding up the procedure.
8. Advanced options can be toggled to specify any substituent(s) present; and if known, where the residues are attached and in which position. This is a truncated view; up to ten substituents can be entered.
9. There is also an option to specify linkages. This will override any ambiguity given in step C4. All ten residues can have linkage positions specified.
10. Finally, there are several options to make the search more restricted, e.g., the allowed structural elements can be limited according to biological rules, or they can be permissive. For example, when the anomeric configuration at the reducing end or the absolute configurations are unknown, all options are considered.
11. The URL can be saved to quickly enter a previous input; this is done by clicking "Save URL of form".
12. Submit the job to CASPER by pressing "Start simulation". Once the calculation is complete, the page will be redirected to a results page (*see* **Note 8**).

3.5 From HPLC/UHPLC/CE Data to N-linked Structure: GlycoBase

GlycoBase (Fig. 6) contains several glycan collections; all searches can be performed on the entire collection, a selected subsection or on glycans derived from a particular sample within a collection. One can apply different search parameters, including features such as name, antennary composition (e.g., A2), GU value, particular glycan feature (e.g., sialic acid), stoichiometry (e.g., number of hexoses), mass, or biological context (e.g., tissue name or disease.)

3.5.1 Preliminary Annotations Based on Undigested Profile

1. Launch GlycoBase
2. As a first approach, a search of the entire collection is recommended to ensure complete coverage of the glycan spectrum.
3. You can arrange all glycan structures based on increasing or decreasing GU value by simply clicking "HILIC-HPLC".
4. Locate the "Simple Search" field and search for a GU value that you have previously obtained, or one from the example data set.
5. The first glycan peak in the example data (Fig. 7, Table in Figure) has a GU value of 7.15. By searching the database, you will discover that one potential glycan of the same GU value is the A2(4)G2 structure.

Fig. 6 An overview of GlycoBase components: (**a**) GlycoBase's homepage with a simple and intuitive interface; (**b**) Search tool that allows you to search for glycan structures in a number of ways; (**c**) a table obtained from an example search; (**d**) the description and summary for A2G2 structure. Each glycan entry contains information about properties, isomers, average GU values from different modes of separation, an option to display a glycan cartoon in different notation, and a link to view an entry in UniCarb-DB; (**e**) exoglycosidase digestion data including published digestions, and records of internal digestions

6. Continue searching the database for the next peak with a GU value of 8.06. The glycan structure with the most similar GU value is FA2G(4)2S(6)1; however, it is advisable to make a note of other potential glycans with similar GU values.
7. Continue searching the database for remaining glycan peaks, and write down all possibilities.
8. For a single HPLC peak, there could be several options with potential structures of similar GU value. Preliminary annotations obtained from the GlycoBase search can be verified with exoglycosidase digestions.

3.5.2 Confirmation of Annotations with Exoglycosidase Digestions

1. Glycan sequencing by digestion with exoglycosidases is a very powerful tool for elucidating glycan structures. The enzymes may be used in a single or panel digestion array. They cleave specific glycosidic bonds between individual monosaccharides

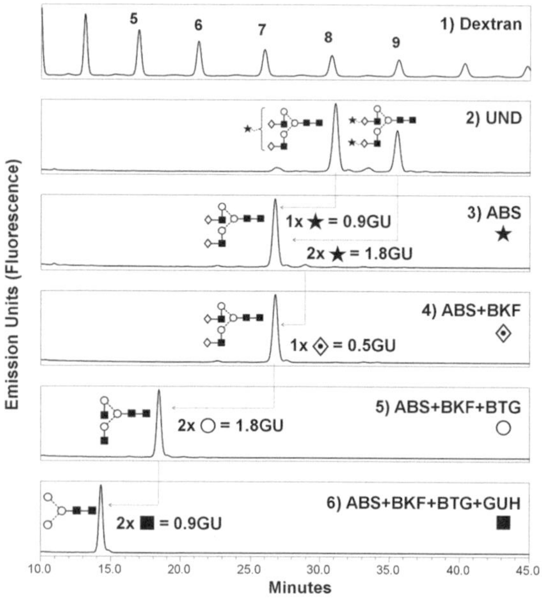

1_Fibrinogen_UND	% Area	GU	Structure
Peak 1	4.42	7.15	A2G2
Peak 2	55.06	8.06	A2G2S1
Peak 3	1.44	8.26	A2BG2S1
Peak 4	4.18	8.56	FA2G2S1
Peak 5	33.65	9.00	A2G2S2
Peak 6	0.73	9.19	A2BG2S2
Peak 7	0.51	9.41	FA2G2S2
2_Fibrinogen_ABS	**% Area**	**GU**	
Peak 1	94.09	7.15	A2G2
Peak 2	2.43	7.32	A2BG2
Peak 3	3.48	7.61	FA2G2
3_Fibrinogen_ABS+BKF	**% Area**	**GU**	
Peak 1	97.09	7.15	A2G2
Peak 2	2.91	7.32	A2BG2
4_Fibrinogen_ABS+BKF+BTG	**% Area**	**GU**	
Peak 1	97.86	5.36	A2
Peak 2	2.14	5.51	A2B
5_Fibrinogen_ABS+BKF+BTG+GUH	**% Area**	**GU**	
Peak 1	97.74	4.32	M3
Peak 2	2.26	4.45	M3B

Fig. 7 Example data acquired from 2AB-labeled glycan separation from fibrinogen containing (1) dextran ladder with indicated GU values, (2) undigested profile of fibrinogen with major peaks shown, (3) sialidase digested, (4) sialidase and fucosidase digested, (5) sialidase, fucosidase, and galactosidase digested, (6) sialidase, fucosidase, galactosidase, *N*-acetylhexosaminidase digested. *Symbols* denoted on the *right* of each chromatogram indicate the individual monosaccharides removed through exoglycosidase digestion. The *table* on the *right* contains detected peaks with assigned GU values and relative % areas for all peaks, as well as their annotated structures

units from the non-reducing end, and are highly specific for both the terminal residue and its linkage type.

2. An example of an exoglycosidase digestion array from fibrinogen is shown in Fig. 7.
3. Confident assignments of structures can be made by comparing chromatographic profiles obtained before and after digestion with exoglycosidases of known specificity. Apply the strategy of top-down digestion and bottom-up identification.
4. Start with the first digestion: ABS (*Arthrobacter ureafaciens* sialidase) releases α2-3, α2-6, and α2-8 linked non-reducing terminal sialic acid residues. Observe the peaks that are eliminated with sialidase treatment. Peaks that are removed are susceptible to sialidase treatment, which implies that they contain a sialic acid residue at the terminal end. To further validate linkage type, additional digestions with an NANI sialidase may release α2-3 linked non-reducing terminal sialic acid.
5. Continue evaluating the subsequent digestions and observe shifts in peak positions that correspond to particular glycan residues: BKF (bovine kidney α-fucosidase) releases α1-2,3,4 and 6-linked non-reducing terminal fucose residues.

BTG (bovine testes β-galactosidase) hydrolyses non-reducing terminal β1-3- and β1-4-linked galactose, and GUH (hexosaminidase cloned from *Streptococcus pneumoniae* expressed in *E. coli*) releases GlcNAc residues linked to Man (Fig. 7).

6. Confirm potentially identified structures by checking their GU value with the database.
7. Always double check the exported table containing GU values and the corresponding relative percentage areas. Area values of species before and after a digest should match.
8. In the case of complex samples it is possible to collect individual peaks. The volatile buffers can be removed and the glycans can be analyzed by mass spectrometry to validate assignments. The combination of different orthogonal techniques increases the accuracy of structure identification.

3.6 Validation of Monosaccharide Sequence and Linkages with the Aid of Exoglycosidase Digestion: GlycoDigest

A description of the tool including example data and details of exoglycosidases can be found at www.glycodigest.org. From the homepage, links guide users to the "Getting Started Guide" hosted on GitHub at bit.ly/1lZlcFk, and to "Exoglycosidase Information" options and example data. The web application version of GlycoDigest has been integrated into UniCarbKB.

3.6.1 Step One: Build Glycan Structure

1. Launch UniCarbKB-GlycoDigest (unicarbkb.org/builderDigest) (*see* **Note 9**).
2. The latest version of GlycanBuilder offers the same functionality as that developed by the EUROCarbDB project. To start building a structure, choose one of the following three options: (1) select monosaccharides from the canvas panel or "Structure" menu; (2) extend a structure from a predefined list; or (3) build a substructure/epitope.
3. Structures can be built by adding new monosaccharides or predefined groups to the highlighted residue. New groups can be selected either from the medium toolbar or from the structures menu. The list of structural constituents comprises an exhaustive collection of saccharides, substituents, reducing-end markers, and saccharide modifications. All stereochemical information about a saccharide, such as anomeric conformation, chirality, ring configuration, and linkage position, can be specified.
4. After building the appropriate glycan structure, select the "Digest" button.

3.6.2 Step Two: Selecting Exoglycosidases

The next step is to theoretically digest the structure with a defined array of exoglycosidases. Careful design and formulation of exoglycosidase arrays is critical to the success of correct glycan structural assignments.

1. To create an array of exoglycosidases, use the "Build Exoglycosidase Array" box to select supported exoglycosidases. (*see* **Note 10**)
2. After selecting a single endoglycosidase or panel of enzymes, select the "Digest" button.

3.6.3 Step Three: Output

GlycoDigest will simulate the combinations of exoglycosidase selected in Subheading 3.6.2. When complete, the "Results" section will be updated with theoretically computed end-product(s).

3.7 Association of Structure with Biology: UniCarbKB and Association of Glycan Structures with Proteins

UniCarbKB is an initiative that provides an online information storage and search platform for glycomics and glycobiology research. The knowledgebase offers open access to a rich resource supported by querying interfaces, annotation technologies, and the adoption of common standards to integrate structural, experimental, and functional data.

To centralize and organize the features of UniCarbKB, a confluence knowledge base is available (confluence.unicarbkb.org). The knowledgebase is a repository of how-to and troubleshooting guides that provides users with access to solutions and relevant search results.

The confluence knowledge base is split into four main sections: (1) Content, (2) Browsing, (3) Nomenclature, and (4) How-to articles. Each section is further divided into specific layers that focus on a particular feature, which includes screenshots and step-by-step notes on discovering data sets and navigating the database.

1. Launch UniCarbKB Confluence (confluence.unicarbkb.org)
2. Select a section from the navigation panel on the left-hand side. For example, to find information about glycoproteins click the "How-to articles" link followed by the "Glycoprotein Page".
3. Each page summarizes the content available with documented screenshots and how users can access relevant information. For the "Glycoprotein Page" a screenshot highlights the major content sections and briefly describes the level of information available.
4. The "Related articles" section provides links to related guides.
5. The "Search" box offers a quick navigation aid as well as a full site search. When you perform a full search, Confluence will search all content and the results screen shows a list of matching items.

The "How-to articles" provide new and returning users with access to up-to-date documented guides. For example, the "Reference" guide provides an overview for selecting and searching references and a screenshot of an individual reference page.

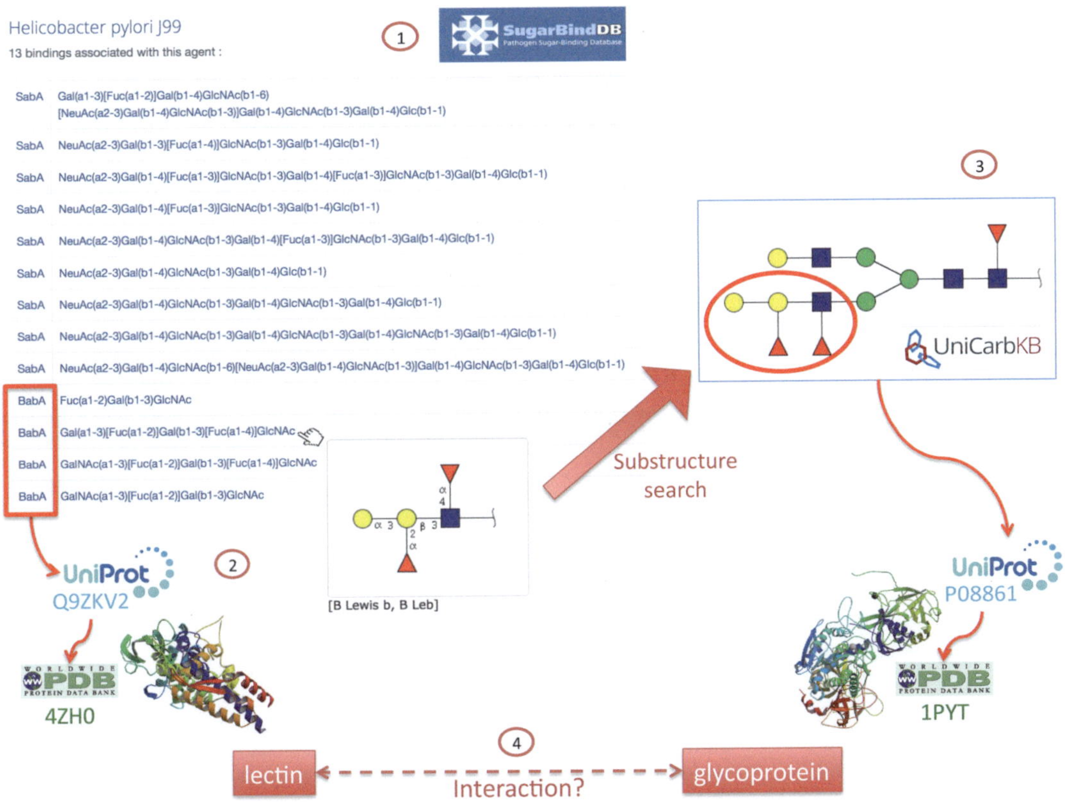

Fig. 8 Investigating glycan-mediated protein–protein interactions using SugarBind. (1) shows the output list of protein–glycan bindings when SugarBindDB is queried with "Helicobacter pylori J99". In this list, the BabA-B Lewis b (i.e., Galα1-3(Fucα1-2)Galβ1-3(Fucα1-4)GlcNAc) is selected. Mousing over the ligand written in IUPAC code prompts an image of the B Lewis b written in CFG symbols, as shown in (2). The BabA lectin is cross-referenced to UniProt entry Q9ZKV2. (3) shows the result of searching UniCarbKB with B Lewis b as a potential substructure of full glycan structures. Only one match is found. This UniCarbKB entry is linked to a single UniProt entry: P08861. Finally the lectin and glycoprotein information can be correlated via glycans and (4) highlights the potential for interaction

3.8 Association of Structure with Microbial Interactions: SugarBind

A glycan-lectin binding pair within SugarBindDB can lead to the identification of a glycan-mediated protein–protein interaction, i.e., a lectin–glycoprotein interaction. The following example describes how this information can be revealed.

1. Search SugarBindDB for a specific pathogen, for instance *Helicobacter pylori* J99. A list of glycan-binding proteins (BabA and SabA lectins) is displayed along with all stored glycan ligands.
2. In the list of output glycan-binding pairs, consider the case of BabA-B Lewis b (i.e., Galα1-3(Fucα1-2)Galβ1-3(Fucα1-4)GlcNAc) (Fig. 8). Click on the glycan link, which prompts a new page that displays the structural details of B Lewis b.

3. From the B Lewis b page, follow the cross-link to UniCarbKB via the substructure search, which reveals the structure of a corresponding full glycan entry (UniCarbKB 7563) that contains B Lewis b. According to UniCarbKB information, this glycan is reported to be attached to Asn-114 of human elastase 1 (P08861).
4. Hypothesize the potential interaction between *H. pylori* BabA lectin (Q9ZKV2) with the human glycoprotein elastase 1 (P08861).

Needless to say, this result is hypothetical and requires confirmation. However, it provides a lead for further investigation of correlations between Lewis B epitope expression and the severity of diseases caused by *Helicobacter pylori* strains expressing BabA adhesins [39].

4 Notes

1. The GlycoMod tool is specifically designed for *N*-linked and *O*-linked glycoproteins. The possible allowed *N*-linked compositions are more restricted by the presence of the core pentasaccharide (GlcNAc2, Man3) than the *O*-linked; so, for generic searches of other types of glycoconjugates, the *O*-linked search is recommended.
2. Since the database currently consists mainly of negative ion mode MS/MS fragment spectra, other types of data are welcome, and should be submitted as glycoworkbench files.
3. Charge deconvolution of MS data is required if the charge state of pseudomolecular ions is higher than ±1.
4. For mammalian sugars, monosaccharides in the left hand column (7) should be left as selection "possible", while the right hand column should be selected only if one has additional knowledge or indication that these sugars are present in the sample. (for example, humans usually do not express NeuGc.)
5. The term "glycoform mass" refers to an uncharged mass of a glycan without its reducing end modification, i.e., it should be considered to be the mass of the reducing glycan after subtracting the mass of a water molecule.
6. For computer operating systems that use "comma (,)" as decimal point, all occurrences of comma should be changed to "dot (.)" before pasting a MS/MS peak list to GlycoWorkbench. For Xcalibur, one can export the list directly using *Export> Clipboard* (*Exact Mass*) to GlycoWorkbench. When pasting a peak list, Crtl + V does not work.
7. An example input can be accessed from the following link: http://tinyurl.com/CalcShiftTri.

8. An example input for the tetrasaccharide above can be accessed from the following link: http://tinyurl.com/DetermineTetra.
9. To successfully model the action of an exoglycosidase, the monosaccharide anomeric form and glycosidic linkages must be defined. Exoglycosidases are highly specific, and the rules embedded in GlycoDigest will only work if the submitted structure is fully defined. To specify linkage and anomeric form, select a residue and configure the residue using the main GlycanBuilder canvas.
10. Abbreviated names of exoglycosidases are used, but a pop-up table is available that summarizes their specificity. To access the table, select the "Information on Exoglycosidases" button.

Acknowledgements

The authors acknowledge support from the European Union FP7 GastricGlycoExplorer ITN under grant agreement no.316929, Petrus and Augusta Hedlunds foundation, The Swedish Foundation for International Cooperation in Research and Higher Education, and the Swedish Research Council (621-2013-5895 and 621-2013-4859).

References

1. Doubet S, Bock K, Smith D, Darvill A, Albersheim P (1989) The complex carbohydrate structure database. Trends Biochem Sciences 14(12):475–477
2. von der Lieth CW, Freire AA, Blank D, Campbell MP, Ceroni A, Damerell DR, Dell A, Dwek RA, Ernst B, Fogh R, Frank M, Geyer H, Geyer R, Harrison MJ, Henrick K, Herget S, Hull WE, Ionides J, Joshi HJ, Kamerling JP, Leeflang BR, Lütteke T, Lundborg M, Maass K, Merry A, Ranzinger R, Rosen J, Royle L, Rudd PM, Schloissnig S, Stenutz R, Vranken WF, Widmalm G, Haslam SM (2011) EUROCarbDB: an open-access platform for glycoinformatics. Glycobiology 21(4):493–502. doi:10.1093/glycob/cwq188
3. Cooper CA, Harrison MJ, Wilkins MR, Packer NH (2001) GlycoSuiteDB: a new curated relational database of glycoprotein glycan structures and their biological sources. Nucleic Acids Res 29(1):332–335
4. Hayes CA, Karlsson NG, Struwe WB, Lisacek F, Rudd PM, Packer NH, Campbell MP (2011) UniCarb-DB: a database resource for glycomic discovery. Bioinformatics 27(9): 1343–1344. doi:10.1093/bioinformatics/btr137, btr137 [pii]
5. Campbell MP, Hayes CA, Struwe WB, Wilkins MR, Aoki-Kinoshita KF, Harvey DJ, Rudd PM, Kolarich D, Lisacek F, Karlsson NG, Packer NH (2011) UniCarbKB: putting the pieces together for glycomics research. Proteomics 11(21):4117–4121. doi:10.1002/pmic.201100302
6. Ranzinger R, Herget S, Wetter T, von der Lieth CW (2008) GlycomeDB - integration of open-access carbohydrate structure databases. BMC Bioinformatics 9:384. doi:10.1186/1471-2105-9-384
7. Royle L, Campbell MP, Radcliffe CM, White DM, Harvey DJ, Abrahams JL, Kim YG, Henry GW, Shadick NA, Weinblatt ME, Lee DM, Rudd PM, Dwek RA (2008) HPLC-based analysis of serum N-glycans on a 96-well plate platform with dedicated database software. Anal Biochem 376(1):1–12. doi:10.1016/j.ab.2007.12.012, S0003-2697(07)00794-4 [pii]
8. Hayes CA, Carta G, Karlsson NG, Duffy F, Rudd PM (2015) Informatics and Analytical Tools for Glycan Analysis and the Development of Biot. In: Cipolla L (ed) Carbohydrate chemistry: state of the art and challenges for drug development: an overview on structure,

biological roles. Synthetic methods and application as therapeutics. Imperial College Press, London, pp 173–192

9. Campbell MP, Nguyen-Khuong T, Hayes CA, Flowers SA, Alagesan K, Kolarich D, Packer NH, Karlsson NG (2014) Validation of the curation pipeline of UniCarb-DB: building a global glycan reference MS/MS repository. Biochim Biophys Acta 184(1 Pt A):108–116. doi:10.1016/j.bbapap.2013.04.018
10. Cooper CA, Gasteiger E, Packer NH (2001) GlycoMod—a software tool for determining glycosylation compositions from mass spectrometric data. Proteomics 1(2):340–349. doi:10.1002/1615-9861(200102)1:2<340::AID-PROT340>3.0.CO;2-B
11. Ceroni A, Maass K, Geyer H, Geyer R, Dell A, Haslam SM (2008) GlycoWorkbench: a tool for the computer-assisted annotation of mass spectra of glycans. J Proteome Res 7(4):1650–1659. doi:10.1021/pr7008252
12. Damerell D, Ceroni A, Maass K, Ranzinger R, Dell A, Haslam SM (2012) The GlycanBuilder and GlycoWorkbench glycoinformatics tools: updates and new developments. Biol Chem 393(11):1357–1362. doi:10.1515/hsz-2012-0135
13. Jansson P-E, Kenne L, Widmalm G (1987) CASPER—a computerised approach to structure determination of polysaccharides using information from n.m.r. spectroscopy and simple chemical analyses. Carbohydr Res 168 (1): 67-77.doi:http://dx.doi.org/10.1016/0008-6215(87)80007-1
14. Jansson PE, Kenne L, Widmalm G (1989) Computer-assisted structural analysis of polysaccharides with an extended version of CASPER using ^{1}H- and ^{13}C-n.m.r. data. Carbohydr Res 188:169–191
15. Jansson PE, Kenne L, Widmalm G (1991) CASPER: a computer program used for structural analysis of carbohydrates. J Chem Inf Comput Sci 31(4):508–516
16. Stenutz R, Jansson PE, Widmalm G (1998) Computer-assisted structural analysis of oligo- and polysaccharides: an extension of CASPER to multibranched structures. Carbohydr Res 306(1-2):11–17
17. Jansson PE, Kenne L, Widmalm G (1991) Computer-assisted structural analysis of oligosaccharides using CASPER. Anal Biochem 199(1):11–17
18. Jansson PE, Stenutz R, Widmalm G (2006) Sequence determination of oligosaccharides and regular polysaccharides using NMR spectroscopy and a novel Web-based version of the computer program CASPER. Carbohydr Res 341(8):1003–1010. doi:10.1016/j.carres.2006.02.034
19. Lundborg M, Widmalm G (2011) Structural analysis of glycans by NMR chemical shift prediction. Anal Chem 83(5):1514–1517. doi:10.1021/ac1032534
20. Lundborg M, Fontana C, Widmalm G (2011) Automatic structure determination of regular polysaccharides based solely on NMR spectroscopy. Biomacromolecules 12(11):3851–3855. doi:10.1021/bm201169y
21. Lundborg M, Modhukur V, Widmalm G (2010) Glycosyltransferase functions of E. coli O-antigens. Glycobiology 20(3):366–368. doi:10.1093/glycob/cwp185
22. Rojas-Macias MA, Ståhle J, Lütteke T, Widmalm G (2015) Development of the ECODAB into a relational database for Escherichia coli O-antigens and other bacterial polysaccharides. Glycobiology 25(3):341–347. doi:10.1093/glycob/cwu116
23. Roslund MU, Säwén E, Landström J, Rönnols J, Jonsson KH, Lundborg M, Svensson MV, Widmalm G (2011) Complete ^{1}H and ^{13}C NMR chemical shift assignments of mono-, di-, and trisaccharides as basis for NMR chemical shift predictions of polysaccharides using the computer program CASPER. Carbohydr Res 346(11):1311–1319. doi:10.1016/j.carres.2011.04.033
24. Ronnols J, Pendrill R, Fontana C, Hamark C, Angles d'Ortoli T, Engstrom O, Ståhle J, Zaccheus MV, Säwén E, Hahn LE, Iqbal S, Widmalm G (2013) Complete ^{1}H and ^{13}C NMR chemical shift assignments of mono- to tetrasaccharides as basis for NMR chemical shift predictions of oligosaccharides using the computer program CASPER. Carbohydr Res 380: 156–166. doi:10.1016/j.carres.2013.06.026
25. Royle L, Radcliffe CM, Dwek RA, Rudd PM (2006) Detailed structural analysis of N-glycans released from glycoproteins in SDS-PAGE gel bands using HPLC combined with exoglycosidase array digestions. In: Brockhausen I (ed) Meth Mol Biol, vol 347 Glycobiology Protocols. Humana Press Inc., Totowa, NJ, pp 125–143
26. Adamczyk B, Struwe WB, Ercan A, Nigrovic PA, Rudd PM (2012) Characterization of fibrinogen glycosylation and its importance for serum/plasma N-glycome analysis. J Proteome Res 12(1):444–454
27. Campbell MP, Royle L, Rudd PM (2015) GlycoBase and autoGU: resources for interpreting HPLC-glycan data. Methods Mol Biol 1273:17–28. doi:10.1007/978-1-4939-2343-4_2

28. Gotz L, Abrahams JL, Mariethoz J, Rudd PM, Karlsson NG, Packer NH, Campbell MP, Lisacek F (2014) GlycoDigest: a tool for the targeted use of exoglycosidase digestions in glycan structure determination. Bioinformatics 30(21):3131–3133. doi:10.1093/bioinformatics/btu425
29. Campbell MP, Peterson R, Mariethoz J, Gasteiger E, Akune Y, Aoki-Kinoshita KF, Lisacek F, Packer NH (2014) UniCarbKB: building a knowledge platform for glycoproteomics. Nucleic Acids Res 42(Database issue): D215–D221. doi:10.1093/nar/gkt1128
30. UniProt C (2015) UniProt: a hub for protein information. Nucleic Acids Res 43(Database issue):D204–D212. doi:10.1093/nar/gku989
31. Gaudet P, Michel PA, Zahn-Zabal M, Cusin I, Duek PD, Evalet O, Gateau A, Gleizes A, Pereira M, Teixeira D, Zhang Y, Lane L, Bairoch A (2015) The neXtProt knowledgebase on human proteins: current status. Nucleic Acids Res 43(Database issue):D764–D770. doi:10.1093/nar/gku1178
32. Mariethoz J, Khatib K, Alocci D, Campbell MP, Karlsson NG, Packer NH, Mullen EH, Lisacek F (2016) SugarBindDB, a resource of glycan-mediated host-pathogen interactions. Nucleic Acids Res 44(D1):D1243–D1250. doi:10.1093/nar/gkv1247
33. Live D, Wells L, Boons GJ (2013) Dissecting the molecular basis of the role of the O-mannosylation pathway in disease: α-dystroglycan and forms of muscular dystrophy. Chembiochem 14(18):2392–2402. doi:10.1002/cbic.201300417
34. Sardzik R, Green AP, Laurent N, Both P, Fontana C, Voglmeir J, Weissenborn MJ, Haddoub R, Grassi P, Haslam SM, Widmalm G, Flitsch SL (2012) Chemoenzymatic synthesis of O-mannosylpeptides in solution and on solid phase. J Am Chem Soc 134(10):4521–4524. doi:10.1021/ja211861m
35. Kuttel M, Yue M, Widmalm G, Lundborg M (2011) CarbBuilder: an adjustable tool for building 3D molecular structures of carbohydrates for molecular simulation. In: E-Science (e-Science), 2011 IEEE 7th international conference on, 5–8 Dec 2011. pp 395–402. doi:10.1109/eScience.2011.61
36. Lundborg M, Widmalm G (2015) NMR chemical shift prediction of glycans: application of the computer program CASPER in structural analysis. Methods Mol Biol 1273:29–40. doi:10.1007/978-1-4939-2343-4_3
37. Vranken WF, Boucher W, Stevens TJ, Fogh RH, Pajon A, Llinas M, Ulrich EL, Markley JL, Ionides J, Laue ED (2005) The CCPN data model for NMR spectroscopy: development of a software pipeline. Proteins 59(4):687–696. doi:10.1002/prot.20449
38. Varki A, Cummings RD, Esko JD, Freeze HH, Stanley P, Marth JD, Bertozzi CR, Hart GW, Etzler ME (2009) Symbol nomenclature for glycan representation. Proteomics 9(24):5398–5399. doi:10.1002/pmic.200900708
39. Gerhard M, Lehn N, Neumayer N, Boren T, Rad R, Schepp W, Miehlke S, Classen M, Prinz C (1999) Clinical relevance of the Helicobacter pylori gene for blood-group antigen-binding adhesin. Proc Natl Acad Sci U S A 96(22): 12778–12783
40. Farriol-Mathis N, Garavelli JS, Boeckmann B, Duvaud S, Gasteiger E, Gateau A, Veuthey AL, Bairoch A (2004) Annotation of post-translational modifications in the Swiss-Prot knowledge base. Proteomics 4(6):1537–1550. doi:10.1002/pmic.200300764
41. Berman HM, Westbrook J, Feng Z, Gilliland G, Bhat TN, Weissig H, Shindyalov IN, Bourne PE (2000) The Protein Data Bank. Nucleic Acids Res 28(1):235–242
42. Barrett T, Wilhite SE, Ledoux P, Evangelista C, Kim IF, Tomashevsky M, Marshall KA, Phillippy KH, Sherman PM, Holko M, Yefanov A, Lee H, Zhang N, Robertson CL, Serova N, Davis S, Soboleva A (2013) NCBI GEO: archive for functional genomics data sets–update. Nucleic Acids Res 41(Database issue):D991–D995. doi:10.1093/nar/gks1193
43. Uhlen M, Fagerberg L, Hallstrom BM, Lindskog C, Oksvold P, Mardinoglu A, Sivertsson A, Kampf C, Sjostedt E, Asplund A, Olsson I, Edlund K, Lundberg E, Navani S, Szigyarto CA, Odeberg J, Djureinovic D, Takanen JO, Hober S, Alm T, Edqvist PH, Berling H, Tegel H, Mulder J, Rockberg J, Nilsson P, Schwenk JM, Hamsten M, von Feilitzen K, Forsberg M, Persson L, Johansson F, Zwahlen M, von Heijne G, Nielsen J, Ponten F (2015) Proteomics. Tissue-based map of the human proteome. Science 347 (6220): 1260419. doi:10.1126/science.1260419
44. Watanabe K, Yasugi E, Oshima M (2000) How to search the glycolipid data in LIPIDBANK for Web; the newly developed lipid database in Japan. Trends Glycosci Glyc 12(65):175–184. doi:10.4052/tigg.12.175
45. Sud M, Fahy E, Cotter D, Brown A, Dennis EA, Glass CK, Merrill AH Jr, Murphy RC, Raetz CR, Russell DW, Subramaniam S (2007) LMSD: LIPID MAPS structure database. Nucleic Acids Res 35(Database issue):D527–D532. doi:10.1093/nar/gkl838
46. Agravat SB, Saltz JH, Cummings RD, Smith DF (2014) GlycoPattern: a web platform for glycan array mining. Bioinformatics

30(23):3417–3418. doi:10.1093/bioinformatics/btu559

47. Aoki-Kinoshita K, Agravat S, Aoki NP, Arpinar S, Cummings RD, Fujita A, Fujita N, Hart GM, Haslam SM, Kawasaki T, Matsubara M, Moreman KW, Okuda S, Pierce M, Ranzinger R, Shikanai T, Shinmachi D, Solovieva E, Suzuki Y, Tsuchiya S, Yamada I, York WS, Zaia J, Narimatsu H (2016) GlyTouCan 1.0—the international glycan structure repository. Nucleic Acids Res 44(D1):D1237–D1242. doi:10.1093/nar/gkv1041
48. Kanehisa M, Sato Y, Kawashima M, Furumichi M, Tanabe M (2016) KEGG as a reference resource for gene and protein annotation. Nucleic Acids Res 44(D1):D457–D462. doi:10.1093/nar/gkv1070
49. Hashimoto K, Goto S, Kawano S, Aoki-Kinoshita KF, Ueda N, Hamajima M, Kawasaki T, Kanehisa M (2006) KEGG as a glycome informatics resource. Glycobiology 16(5):63R–70R. doi:10.1093/glycob/cwj010
50. Maeda M, Fujita N, Suzuki Y, Sawaki H, Shikanai T, Narimatsu H (2015) JCGGDB: Japan consortium for glycobiology and glycotechnology database. Methods Mol Biol 1273:161–179. doi:10.1007/978-1-4939-2343-4_12
51. Tomiya N, Awaya J, Kurono M, Endo S, Arata Y, Takahashi N (1988) Analyses of N-linked oligosaccharides using a two-dimensional mapping technique. Anal Biochem 171(1):73–90
52. Steentoft C, Vakhrushev SY, Joshi HJ, Kong Y, Vester-Christensen MB, Schjoldager KT, Lavrsen K, Dabelsteen S, Pedersen NB, Marcos-Silva L, Gupta R, Bennett EP, Mandel U, Brunak S, Wandall HH, Levery SB, Clausen H (2013) Precision mapping of the human O-GalNAc glycoproteome through SimpleCell technology. EMBO J 32(10):1478–1488. doi:10.1038/emboj.2013.79
53. Lütteke T, Bohne-Lang A, Loss A, Goetz T, Frank M, von der Lieth CW (2006) GLYCOSCIENCES.de: an Internet portal to support glycomics and glycobiology research. Glycobiology 16(5):71R–81R. doi:10.1093/glycob/cwj049
54. Perez S, Sarkar A, Rivet A, Breton C, Imberty A (2015) Glyco3D: a portal for structural glycosciences. Methods Mol Biol 1273:241–258. doi:10.1007/978-1-4939-2343-4_18
55. Lombard V, Golaconda Ramulu H, Drula E, Coutinho PM, Henrissat B (2014) The carbohydrate-active enzymes database (CAZy) in 2013. Nucleic Acids Res 42(Database issue):D490–D495. doi:10.1093/nar/gkt1178
56. Zhang H, Loriaux P, Eng J, Campbell D, Keller A, Moss P, Bonneau R, Zhang N, Zhou Y, Wollscheid B, Cooke K, Yi EC, Lee H, Peskind ER, Zhang J, Smith RD, Aebersold R (2006) UniPep—a database for human N-linked glycosites: a resource for biomarker discovery. Genome Biol 7(8):R73. doi:10.1186/gb-2006-7-8-R73
57. Kirschner KN, Yongye AB, Tschampel SM, Gonzalez-Outeirino J, Daniels CR, Foley BL, Woods RJ (2008) GLYCAM06: a generalizable biomolecular force field. Carbohydrates. J Comput Chem 29(4):622–655. doi:10.1002/jcc.20820
58. Kunduru BR, Nair SA, Rathinavelan T (2016) EK3D: an E. coli K antigen 3-dimensional structure database. Nucleic Acids Res 44(D1):D675–D681. doi:10.1093/nar/gkv1313

Chapter 19

High-Throughput *N*-Glycan Analysis with Rapid Magnetic Bead-Based Sample Preparation

Marton Szigeti and Andras Guttman

Abstract

N-glycan profiling of therapeutic glycoproteins is essential to ensure the activity and efficacy of these promising new-generation drugs. The N-linked glycan moieties of these entities highly affect circulation half-life, immunogenicity and receptor-binding activity as well as physicochemical and thermal stability properties. In addition, more than half of the biopharmaceuticals are glycoproteins representing multibillion dollar worldwide business, further emphasizing the importance of their analysis. In the biomedical field, on the other hand, revealing disease-related glycan structure alterations holds the promise of the discovery of new biomarkers for early diagnostics. Therefore, there is a great demand for widely applicable, high-throughput sample preparation and analysis methods for N-glycan profiling of glycoproteins. One of the newest exciting developments of the field is the magnetic bead based glycoprotein sample preparation technique. A detailed protocol of this method is given in this chapter in conjunction with rapid capillary electrophoresis analysis of the prepared samples by laser induced fluorescence detection (CE-LIF). N-glycans are digested by the endoglycosidase PNGase F and the released carbohydrates are labeled with the charged fluorophore dye of aminopyrenetrisulfonate (APTS). Effective glycan capture by magnetic microparticles enabled fast, easily automated sample preparation both in individual (single vial) and 96-well plate formats, including excess dye removal. Rapid separation of APTS labeled IgG glycans is also shown utilizing an optimized CE-LIF protocol.

Key words Magnetic bead, N-glycan analysis, Fluorophore labeling, Capillary electrophoresis

Abbreviations

PNGase F	Peptide-N4-(N-acetyl-beta-glucosaminyl) asparagine-amidase
APTS	8-aminopyrene-1,3,6-trisulfonic acid
CE-LIF	Capillary electrophoresis with laser induced fluorescence detection
IgG	Immunoglobulin G

1 Introduction

N-glycosylation analysis is one of the fastest growing areas in the biopharmaceutical and biomedical fields [1, 2]. Glycosylation of the expressed polypeptide chains, e.g., in humans it is mediated by

Gordan Lauc and Manfred Wuhrer (eds.), *High-Throughput Glycomics and Glycoproteomics: Methods and Protocols*, Methods in Molecular Biology, vol. 1503, DOI 10.1007/978-1-4939-6493-2_19, © Springer Science+Business Media New York 2017

more than 600 enzymes during complex co-translational and posttranslational modification processes [3] resulting extensive micro- and macro-heterogeneity [4]. Actually, the true variability of glycan structures comes from the fact that posttranslational modifications are highly dependent on the surrounding environment, state of health, presence of inhibitors and suppressors, etc. Therefore, changes may occur frequently and rapidly. Thus, the analysis of disease-associated structural changes such as core and antennary fucosylation is key and may lead to glycan based biomarker development and personalized treatments [5]. As a matter of fact, precisely defining the position of this particular sugar residue is not easy, e.g., MS based methods are prone to in-source fucosylation rearrangement [6]. The biopharmaceutical industry represents another important segment that needs large scale N-glycosylation analysis methods. In 2015, glycosylated monoclonal antibodies represented almost half of the biotherapeutic drugs on the market with an estimated $70 billion dollar business worldwide [7]. It has been recognized that structural changes in the sugar moiety of therapeutic glycoproteins may significantly modify their activity and efficacy [8], highlighting the importance of their characterization during the entire manufacturing process. Large scale automated glycosylation analysis is of particular interest in early stages of protein drug development, e.g., during clone selection where transfection can result in numerous expressed variants.

One of the most frequently used high resolution glycan analysis techniques is capillary electrophoresis (CE) with laser induced fluorescence detection (LIF). CE separation is capable to differentiate glycans based on their molecular shape, even if their mass to charge ratio are exactly the same [9]. Due to the high sensitivity of the selective LIF detection method (only fluorophore labeled species are detected), injection of picomolar sample amounts are usually sufficient to acquire reliable data, and microliter sample volumes are enough for dozens or even hundreds of analyses [10].

In this protocol a rapid magnetic bead based sample preparation workflow is introduced for high-throughput N-glycan analysis applications in conjunction with fast CE-LIF separation. The N-linked carbohydrates were released from the glycoproteins by PNGase F (peptide-N4-(N-acetyl-beta-glucosaminyl) asparagine-amidase), a reliable and specific cleavage enzyme that works under simple and mild conditions. PNGase F cleaves the bond between the carbohydrate holding asparagine residue at the conserved consensus sequence of Asn-X-Ser/Thr (X is any amino acid except proline, serine, and threonine) and the innermost N-acetyl glucosamine of the trimannosyl chitobiose core [11]. The released glycans are captured by magnetic microparticles and tagged in situ *on bead* via reductive amination with 8-aminopyrene-1,3,6-trisulfonic acid (APTS). The labeling reaction is specific to the reductive end of the sugars, therefore, only one APTS molecule binds to one sugar molecule, allowing easy quantification. More importantly, since the sugar ring at the reducing end opens up during the derivatization step [12], no anomeric isomers are generated, such as might be the case in the so called instant labeling approaches. The excess fluorophore dye is removed by the same magnetic beads that were used in the earlier glycan capture step. Optimal separation parameters were also developed for rapid CE-LIF analysis of the released, fluorophore labeled and purified sugars.

2 Materials

HPLC grade water should be used (18 MΩ cm at 25 °C) in all buffer and reagent preparations. Some parts of sample preparation should be done under fume hood wearing powder-free nitrile gloves, lab coat, and eyeglasses. Only use new disposable vials that are free from possible environmental carbohydrate contaminations.

2.1 N-Glycan Release, APTS Labeling and Cleanup

1. 0.2 mL safety lock flat cap PCR tubes (actual total volume is 350 μL).
2. Thermostat block to provide stable temperatures at 37 °C and 65 °C.
3. Vortex mixer.
4. Automata pipettors and disposable pipette tips (2, 20, 200 and 1000 μL sizes).
5. Miscellaneous labware for buffers and reagents.
6. Neodymium magnet or magnetic stand or magnetic plate.
7. Agencourt CleanSEQ magnetic microparticles (Beckman Coulter, Part #A29151).
8. 50 mM dithiothreitol (DTT) in water. Prepare the solution fresh before use.
9. 250 mM iodoacetamide (IAA) in water. Prepare the solution fresh before use.
10. 5% sodium dodecyl sulfate (SDS) in water.
11. 0.5% Nonidet P-40 (NP-40) in water.
12. Peptide-N-glycosidase F enzyme (PNGase F) (ProZyme, Part #GKE5006B).
13. 20% acetic acid.
14. 8-aminopyrene-1,3,6-trisulfonic acid (APTS) (Beckman Coulter, Part #M501706).
15. 40 mM 8-aminopyrene-1,3,6-trisulfonic acid (APTS) in 20% acetic acid.
16. 1 M sodium cyanoborohydride ($NaBH_3CN$) in tetrahydrofuran (THF) (Sigma, Part #296813).
17. Acetonitrile (100%, HPLC grade).

2.2 CE-LIF Analysis of APTS Labeled N-Glycans

1. PA 800 plus Pharmaceutical Analysis System (SCIEX, Brea, CA) with Laser Induced Fluorescent (LIF) detector (488 nm excitation wavelength/520 nm emission filter) and 32 Karat software (version 8.0 or higher).
2. N-linked carbohydrate separation gel buffer (SCIEX, Part #477623).
3. N-CHO Capillary, 30 cm length (20 cm effective length), 50 μm ID (SCIEX, Part #M402370).
4. Capillary cartridge (SCIEX, Part #144738) with LIF aperture plug assembly (SCIEX, Part #721125) and probe guide (SCIEX, Part #721126).
5. Universal buffer vials (2 mL), vial caps, sample vials.

3 Methods

3.1 In Solution N-Glycan Release

1. Dissolve the dry glycoprotein sample in HPLC grade water. Preferred stock solution concentration is between 1 and 10 mg/mL.
2. Pipette 200 μL of Agencourt CleanSEQ magnetic bead suspension into a safety lock flat cap 0.2 mL PCR tube.
3. Place the tube with the beads on a magnetic stand and after the beads are pulled to the side of the vial (60 s), remove the storage solution by careful pipetting (*see* **Note 1**).
4. Add 10 μL of glycoprotein sample into the sample tube (1–10 mg/mL concentration).
5. Mix the sample with the beads using a vortex (10 s at maximum speed) (*see* **Note 2**).
6. Prepare the denaturation solution by mixing 2 μL each of 50 mM of DTT, 250 mM of IAA, 5% of SDS and 0.5% of NP-40 solutions.
7. Pipette 4 μL of denaturation solution into the sample containing tube and incubate at 65 °C for 10 min (denaturation step).
8. Pipette 16 μL of 0.5% of NP-40 and 1 μL of PNGase F enzyme (2.5 mU) to the sample and incubate at 50 °C for 1 h (deglycosylation step) (*see* **Note 3**).
9. After the denaturation step, mix the sample using a vortex (10 s at maximum speed).
10. Pipette 210 μL of acetonitrile into the reaction vial (this step will result in 87.5% final acetonitrile concentration in the vial) for glycan capture and vortex the sample (10 s at maximum speed).
11. Wait for 1 min at room temperature.

12. Place the sample tube on a magnetic stand and after the beads are pulled to the side of the vial (30 s), remove the supernatant by pipetting carefully (*see* **Notes 4** and **5**).

3.2 APTS Labeling

1. Add 9 μL of 40 mM APTS (fluorophore) in 20% acetic acid (catalyst) to the reaction vial containing the magnetic beads with the captured glycans from **step 12** in Subheading 3.1.
2. Pipette 1 μL of 1 M $NaBH_3CN$ in THF (reductive agent) and 1 μL of 0.5% NP-40 to the vial. Please note that the use of more $NaBH_3CN$ will not increase reaction speed.
3. Mix the sample using a vortex (10 s at maximum speed).
4. Incubate the reaction mixture at 37 °C for 2 h (labeling step).

3.3 Removal of the Excess APTS Dye

1. After the labeling reaction, pipette 9 μL of 0.5% NP-40 solution to the reaction tube and mix using a vortex (10 s at maximum speed) (*see* **Note 6**).
2. Add 150 μL of acetonitrile and vortex the sample at maximum speed for 10 s.
3. Wait 1 min at room temperature.
4. Place the sample tube on a magnetic stand and after the beads are pulled to the side of the vial (30 s), remove the supernatant by pipetting carefully.
5. Repeat **steps 1–4** three more times using 20 μL HPLC grade water and 150 μL acetonitrile in each step.
6. After the final supernatant removal, elute the labeled glycans from the beads by the addition of 50 μL of HPLC grade water and vortexing at maximum speed for 10 s.
7. Place the sample tube on a magnetic stand and after the beads are pulled to the side of the vial (30 s), transfer the supernatant (this contains the APTS labeled and purified sugars) into a CE sample vial by pipetting carefully.

3.4 CE-LIF Analysis of the APTS Labeled N-Glycans

3.4.1 Preparation of the Glucose Ladder and Setting Up the CE-LIF Equipment

1. Weight out 5 mg of maltooligosaccharide ladder standard (SCIEX, part #M405039).
2. Label the ladder standard similar as was described under Subheading 3.2. There is no need for excess APTS dye removal in this case due to the high sugar to dye ratio.
3. Stop the reaction by the addition of 90 μL of HPLC grade water.
4. Aliquot the APTS labeled maltooligosaccharide ladder sample into 10 μL portions and store at −20 °C.
5. Dilute the maltooligosaccharide ladder standard sample with HPLC grade water by 1000× before use in CE-LIF analysis.

6. Transfer the APTS labeled samples and APTS labeled maltooligosaccharide ladder standard into CE sample vials. Place the sample vials in universal buffer vials and cap them.
7. Fill universal buffer vials with the following reagents:
   (a) Three vials by 1.5 mL of HPLC grade water. Place two vials at home positions and one for capillary fill position.
   (b) 1.5 mL N-linked carbohydrate separation gel buffer into one vial (capillary refill) and 1.3 mL N-linked carbohydrate separation gel buffer into two vials (separation).
   (c) 0.5 mL of HPLC grade water into a waste vial (one universal buffer vial).
8. Separation/Detection parameters:
   (a) Detection: laser-induced fluorescence (488 nm excitation, 520 nm emission filter).
   (b) Data rate: 16 Hz.
   (c) Dynamic range: 100 RFU (relative fluorescence units).
   (d) Filter setting: normal
   (e) Peak width: 16–25.

3.4.2 Performing Rapid CE-LIF Analysis

1. Set the temperature at 25 °C.
2. Rinse the capillary with the N-linked carbohydrate separation gel buffer for 1 min at 100 psi.
3. Sample injection (pressure or electrokinetic):
   (a) Pressure injection: 1 psi for 5 s.
   (b) Electrokinetic injection (field enhanced):
      - Inject water at 1 psi for 5 s (forward).
      - Inject sample at 2 kV for 2 s using the separation gel buffer at the outlet end of the capillary in reversed polarity mode (cathode at the injection side and the anode at the detection side).
      - Inject separation gel buffer at 1 psi for 5 s (forward).
4. Wash the tip of the capillary by moving the trays into home position (universal buffer vials filled with HPLC grade water).
5. Separation: Apply 30 kV for 3 min in reversed polarity mode (cathode at the injection side and the anode at the detection side) with 5 psi pressure applied at both sides of the capillary.
6. Set autozero at 0.5 min.
7. End method by moving the trays to home position.
8. A rapid separation example of APTS labeled glycans from human IgG (Sigma Aldrich, Part #I4506) is shown in Fig. 1 (*see* **Note 7**).

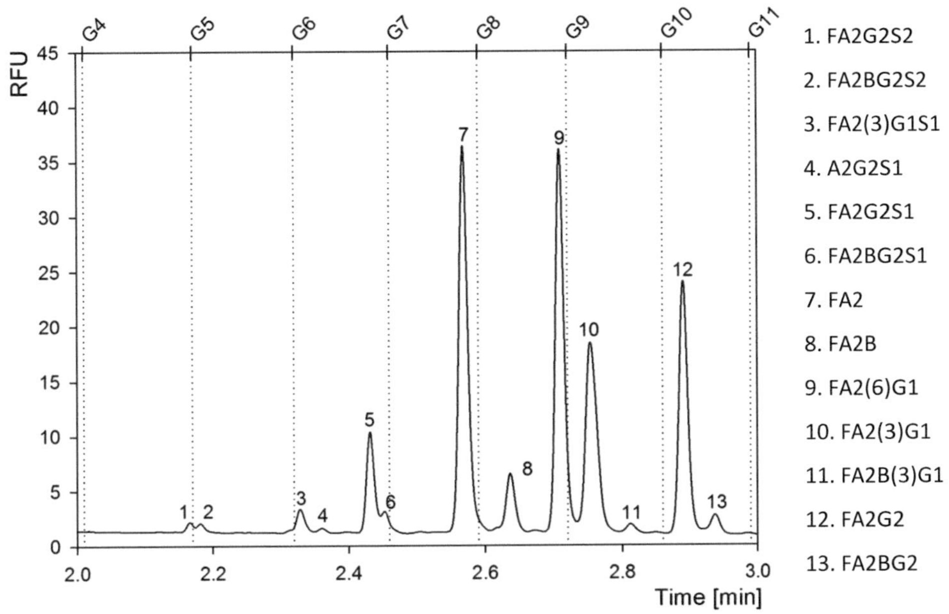

Fig. 1 CE-LIF separation of PNGase F released and APTS labeled human IgG N-glycans. Separation conditions: N-CHO capillary (total length 30 cm, 20 cm effective, 50 μm i.d.), N-linked carbohydrate separation gel buffer, $E = 30$ kV (reversed polarity), 25 °C temperature, Injection: enhanced electrokinetic by 1 psi/5 s water injection (reversed polarity) and, 2 kV/2 s sample injection, 1 psi/5 s buffer injection

4 Notes

1. Avoid touching the beads with the pipette tips to prevent bead/sample loss.
2. Beads are easy to resuspend in aqueous solutions but difficult in organics because of possible aggregation, e.g., during the acetonitrile mediated glycan capture step. Therefore, always vigorously resuspend the beads before acetonitrile addition.
3. No mixing is necessary during the deglycosylation and labeling steps.
4. Always pipette the supernatant from the bottom of the vials as in this case non-captured beads, which might be floating in the upper segment of the vials, will not be accidentally removed.
5. Use the magnets (or stand) only for the minimum time required during buffer exchange to avoid possible bead aggregation.
6. Max speed vortexing in manual sample preparation mode may push some of the sample up to the microvial cap. Applying a quick (few seconds) spin-down in a small microcentrifuge can solve the issue and prevents any potential sample loss.
7. GU unit calculation software is freely available at: www.gucal.hu.

Acknowledgement

The authors acknowledge the support of the MTA-PE Translation Glycomics (#97101) and the NKFIH (K 116263) grants of the Hungarian Government. The kind support of SCIEX and Beckman Coulter is also greatly appreciated.

References

1. Pinho SS, Reis CA (2015) Glycosylation in cancer: mechanisms and clinical implications. Nat Rev Cancer 15(9):540–555. doi:10.1038/nrc3982
2. Szekrenyes A, Partyka J, Varadi C, Krenkova J, Foret F, Guttman A (2015) Sample preparation for N-glycosylation analysis of therapeutic monoclonal antibodies by electrophoresis. Methods Mol Biol 1274:183–195. doi:10.1007/978-1-4939-2353-3_16
3. Varki A, Cummings RD, Esko JD et al (eds) (2009) Essentials of glycobiology. Cold Spring Harbor Laboratory Press, Cold Spring Harbor, NY
4. Marino K, Bones J, Kattla JJ, Rudd PM (2010) A systematic approach to protein glycosylation analysis: a path through the maze. Nat Chem Biol 6(10):713–723. doi:10.1038/nchembio.437
5. Varadi C, Mittermayr S, Szekrenyes A, Kadas J, Takacs L, Kurucz I, Guttman A (2013) Analysis of haptoglobin N-glycome alterations in inflammatory and malignant lung diseases by capillary electrophoresis. Electrophoresis 34(16):2287–2294. doi:10.1002/elps.201300041
6. Grass J, Pabst M, Kolarich D, Pöltl G, Léonard R, Brecker L, Altmann F (2011) Discovery and structural characterization of fucosylated oligomannosidic N-glycans in mushrooms. J Biol Chem 286(8):5977–5984. doi:10.1074/jbc.M110.191304
7. PhRMA The pharmaceutical research and manufacturers of America. Medicines in development for biologics. Accessed 20 Aug 2014
8. Brinks V, Hawe A, Basmeleh AH, Joachin-Rodriguez L, Haselberg R, Somsen G, Jiskoot W, Schellekens H (2011) Quality of original and biosimilar epoetin products. Pharm Res 28(2):386–393. doi:10.1007/s11095-010-0288-2
9. Guttman A, Pritchett T (1995) Capillary gel electrophoresis separation of high-mannose type oligosaccharides derivatized by 1-aminopyrene-3,6,8-trisulfonic acid. Electrophoresis 16(10):1906–1911
10. Guttman A (1996) High-resolution carbohydrate profiling by capillary gel electrophoresis. Nature 380(6573):461–462. doi:10.1038/380461a0
11. Maley F, Trimble RB, Tarentino AL, Plummer TH Jr (1989) Characterization of glycoproteins and their associated oligosaccharides through the use of endoglycosidases. Anal Biochem 180(2):195–204
12. Guttman A, Chen FT, Evangelista RA, Cooke N (1996) High-resolution capillary gel electrophoresis of reducing oligosaccharides labeled with 1-aminopyrene-3,6,8-trisulfonate. Anal Biochem 233(2):234–242. doi:10.1006/abio.1996.0034

Index

Gordan Lauc and Manfred Wuhrer (eds.), *High-Throughput Glycomics and Glycoproteomics: Methods and Protocols*, Methods in Molecular Biology, vol. 1503, DOI 10.1007/978-1-4939-6493-2, © Springer Science+Business Media New York 2017

M

N

O

P

Q

R

S

T

U

MIX
Papier aus verantwortungsvollen Quellen
Paper from responsible sources
FSC® C105338

If you have any concerns about our products,
you can contact us on
ProductSafety@springernature.com

In case Publisher is established outside the EU,
the EU authorized representative is:
Springer Nature Customer Service Center GmbH
Europaplatz 3, 69115 Heidelberg, Germany

Printed by Libri Plureos GmbH
in Hamburg, Germany